CONTAMINATED MARINE SEDIMENTS--ASSESSMENT AND REMEDIATION

Committee on Contaminated Marine Sediments

Marine Board

Commission on Engineering and Technical Systems

National Research Council

1989

National Academy Press
Washington, D.C.

NOTICE: The project that is the subject of this report was approved by the Governing Board of the National Research Council, whose members are drawn from the councils of the National Academy of Sciences, the National Academy of Engineering, and the Institute of Medicine. The members of the panel responsible for the report were chosen for their special competences and with regard for appropriate balance.

This report has been reviewed by a group other than the authors according to procedures approved by a Report Review Committee consisting of members of the National Academy of Sciences, the National Academy of Engineering, and the Institute of Medicine.

The National Academy of Sciences is a private, nonprofit, self-perpetuating society of distinguished scholars engaged in scientific and engineering research, dedicated to the furtherance of science and technology and to their use for the general welfare. Upon the authority of the charter granted to it by the Congress in 1863, the Academy has a mandate that requires it to advise the federal government on scientific and technical matters. Dr. Frank Press is president of the National Academy of Sciences.

The National Academy of Engineering was established in 1964, under the charter of the National Academy of Sciences, as a parallel organization of outstanding engineers. It is autonomous in its administration and in the selection of its members, sharing with the National Academy of Sciences the responsibility for advising the federal government. The National Academy of Engineering also sponsors engineering programs aimed at meeting national needs, encourages education and research, and recognises the superior achievements of engineers. Dr. Robert M. White is president of the National Academy of Engineering.

The Institute of Medicine was established in 1970 by the National Academy of Sciences to secure the services of eminent members of appropriate professions in the examination of policy matters pertaining to the health of the public. The Institute acts under the responsibility given to the National Academy of Sciences by its congressional charter to be an adviser to the federal government and, upon its own initiative, to identify issues of medical care, research, and education. Dr. Samuel O. Thier is president of the Institute of Medicine.

The National Research Council was organized by the National Academy of Sciences in 1916 to associate the broad community of science and technology with the Academy's purposes of furthering knowledge and advising the federal government. Functioning in accordance with general policies determined by the Academy, the Council has become the principal operating agency of both the National Academy of Sciences and the National Academy of Engineering in providing services to the government, the public, and the scientific and engineering communities. The Council is administered jointly by both Academies and the Institute of Medicine. Dr. Frank Press and Dr. Robert M. White are chairman and vice-chairman, respectively, of the National Research Council.

The program described in this report is supported by Cooperative Agreement No. 14-12-0001-30416 between the Minerals Management Service of the U.S. Department of the Interior and the National Academy of Sciences.

Library of Congress Catalog Card Number 89-62967
International Standard Book Number 0-309-04095-7

Additional copies of this report are available from:

National Academy Press
2101 Constitution Avenue, NW
Washington, DC 20418

S029
Printed in the United States of America

COMMITTEE ON CONTAMINATED MARINE SEDIMENTS

KENNETH S. KAMLET, *Chairman*, Senior Program Manager, A.T. Kearney, Inc., Alexandria, Virginia
WILLIAM J. ADAMS, Associate Fellow, Monsanto Company, St. Louis, Missouri
A. KARIM AHMED, Environ Corp., Princeton, New Jersey
HENRY J. BOKUNIEWICZ, Marine Sciences Research Center, State University of New York, Stony Brook
THOMAS A. GRIGALUNAS, Department of Resource Economics, University of Rhode Island, Kingston
JOHN B. HERBICH, Civil Engineering Department, Texas A&M University, College Station
ROBERT J. HUGGETT, Virginia Institute of Marine Science, College of William and Mary, Gloucester Point
HOWARD L. SANDERS, Scientist Emeritus, Woods Hole Oceanographic Institution, Woods Hole, Massachusetts
JAMES M. THORNTON, Department of Ecology, State of Washington, Olympia

Staff

CELIA Y. CHEN, Staff Officer
ANDREA CORELL, Editor
DELPHINE D. GLAZE, Administrative Secretary

MARINE BOARD

SIDNEY WALLACE, *Chairman*, Hill, Betts and Nash, Washington, D.C.
BRIAN J. WATT, *Vice-Chairman*, TECHSAVANT, Inc., Kingston, Texas
ROGER D. ANDERSON, Cox's Wholesale Seafood, Inc., Tampa, Florida
ROBERT G. BEA, NAE, University of California, Berkeley
JAMES M. BROADUS III, Woods Hole Oceanographic Institution
F. PAT DUNN, Shell Oil Company, Houston, Texas
LARRY L. GENTRY, Lockheed Advanced Marine Systems, Sunnyvale, California
DANA R. KESTER, University of Rhode Island, Kingston, Rhode Island
JUDITH T. KILDOW, Massachusetts Institute of Technology, Cambridge, Massachusetts
WARREN LEBACK, Puerto Rico Marine Management, Inc., Elizabeth, New Jersey
BERNARD LE MEHAUTE, University of Miami, Florida
WILLIAM R. MURDEN, NAE, Murden Marine, Ltd., Alexandria, Virginia
EUGENE K. PENTIMONTI, American President Lines, Ltd., Oakland, California
JOSEPH D. PORRICELLI, ECO, Inc., Annapolis, Maryland
JERRY R. SCHUBEL, State University of New York, Stony Brook
RICHARD J. SEYMOUR, Scripps Institution of Oceanography, La Jolla, California
ROBERT N. STEINER, Atlantic Container Line, New York, New York
EDWARD WENK, JR., NAE, University of Washington, Seattle, Washington

Staff
CHARLES A. BOOKMAN, Director
DONALD W. PERKINS, Associate Director
CELIA Y. CHEN, Program Officer (through July 1988)
SUSAN GARBINI, Program Officer
ALEXANDER B. STAVOVY, Program Officer
DORIS C. HOLMES, Staff Associate
AURORE BLECK, Administrative Secretary
DELPHINE D. GLAZE, Administrative Secretary
GLORIA B. GREEN, Senior Secretary
CARLA D. MOORE, Senior Secretary

PREFACE

The problem of contaminated marine sediments has emerged as an environmental issue of national importance. The pervasive and widespread nature of the problem has resulted from decades of using coastal waters intentionally or unintentionally for waste disposal. Harbor areas in particular have been found to contain high levels of contaminants in bottom sediments due to wastes from urban, industrial, and riverine sources, as well as navigation.

Legislative authority for the management of contaminated marine sediments falls largely under three statutes: the Comprehensive Environmental Response, Compensation, and Liability Act of 1980 (CERCLA), the Marine Protection, Research, and Sanctuaries Act (MPRSA), and the Clean Water Act (CWA). The Comprehensive Environmental Response Compensation and Liability Act of 1980, as amended by the Superfund Amendments and Reauthorization Act (SARA) of 1986, is aimed at the clean-up and remediation of inactive or abandoned hazardous waste sites, regardless of location. Superfund sites are currently ranked by the Environmental Protection Agency (EPA) based on the hazard they may pose to human health and the environment via releases to groundwater, surface water, and air. Underwater accumulations of hazardous wastes in marine environments are unlikely to threaten human health except by way of food chain exposure, which is not currently addressed in EPA's hazard-ranking process. Under the 1986 Superfund amendments, however, EPA was required to modify its Hazard Ranking System to address "the damage to natural resources which may affect the human food chain and which is associated with any release (of a hazardous substance)" (Section 105(a)(2)). It is likely, therefore, that once this amendment is there will be a significant increase in the number of "underwater Superfund sites" in both coastal and inland areas.

Meanwhile, the Clean Water Act of 1970, as amended by the Water Quality Act of 1987, gives EPA lead responsibility for safeguarding the quality of U.S. coastal and inland waters. This includes regulating the disposal of dredged and fill materials (shared with the U.S. Army Corps of Engineers, under Section 404), and removing in-place toxic pollutants in harbors and navigable waterways (under Section 115). The 1987 amendments added new authorities requiring EPA to study and conduct projects relating to the removal of toxic pollutants from Great

Lakes bottom sediments (Section 118(c)(3)); and to identify and implement individual control strategies to reduce toxic pollutant inputs into contaminated waterway segments (Section 304(1)).

In response to Title II of the Marine Protection, Research, and Sanctuaries Act of 1972 (PL 92-532) and the National Ocean Pollution Planning Act, the National Oceanic and Atmospheric Administration (NOAA) Office of Marine Pollution Assessment conducts comprehensive interdisciplinary assessments of the effects of human activities on estuarine and coastal environments. Among these assessment activities is the National Status and Trends Program (NST), which attempts to create, maintain, and assess a long-term record of contaminant concentrations and biological responses to contamination in the coastal and estuarine waters of the United States. This assessment provides some insight into the extent of contamination nationally.

As a result of legislative responsibility and programmatic interests, a wide variety of federal agencies have shown active interest in this subject. EPA's responsibilities under Superfund and the CWA are the source of its interests in water quality concerns and remediation of uncontrolled hazardous waste sites. The U.S. Army Corps of Engineers (COE) is involved because of its responsibility to dredge and maintain navigable rivers and harbors. The COE also assists in the design and implementation of remedial clean-up actions under Superfund. NOAA has responsibility for assessing the potential threat of Superfund sites to coastal marine resources as a natural resource trustee as well as under its NS&T program. The U.S. Fish and Wildlife Service has legal authority for various endangered coastal species, food chain relationships, and habitat considerations, all of which are potentially impacted by contaminated sediments. The Navy has had experience in assessing contaminated sediments and now must grapple with such problems in locating and maintaining homeports for Navy vessels.

In response to this emerging problem, the National Research Council convened the Committee on Contaminated Marine Sediments. The members of the committee were selected for their expertise and to ensure a spectrum of viewpoints. Their expertise spanned the fields of aquatic toxicology, dredging technology, resource economics, sediment dynamics and transport, benthic ecology, environmental law, and public policy. Biographies of committee members appear in Appendix A. Consistent with the policy of the National Research Council, the composition of the committee reflected the competing biases that might accompany expertise vital to the study in an effort to seek balance and fair treatment of the subject.

The committee convened a symposium and workshop with invited papers in order to determine the extent and significance of contaminated sediments, review the state of practice of technology for clean-up and remediationm identify and assess alternative management strategies, and identify research and development needs and issues for subsequent technical assessment.

The committee agreed that contaminated sediments should not be defined simply on a generic basis or as those sediments containing some level of synthetic chemicals or background substances above normal concentrations due to human activities. If so, all sediments would be defined as contaminated. However, for the sake of this report, the committee believed the following could be used as a working definition:

> Contaminated sediments are those that contain chemical substances at concentrations which pose a known or suspected environmental or human health threat.

The committee also recognized the importance of issues related to ultimate sources of contamination and competing uses of the affected areas; however, these subjects were beyond the scope of this study.

The invited papers of the symposium and workshop focused on the extent of contamination nationwide, methods for classification of sediment contamination, risks to human health and the ecosystem, sediment resuspension and contaminant mobilization, remedial strategies and technologies for handling contaminated sediments, and lastly, five case studies of the different ways in which a variety of sediment contamination problems are being handled. They included the PCB problem in New Bedford Harbor, Massachusetts; PCBs in the upper Hudson River, New York; kepone contamination of the James River, Virginia; the variety of chemicals contaminating Commencement Bay, Washington; and the Navy Homeport Project in Everett Bay, Washington.

The committee met three times, including once after the two and one-half day symposium and workshop in Tampa, Florida. The symposium lasted one and a half days, during which presentations of invited papers were made. The subsequent one-day workshop was composed of two consecutive work groups. The first one discussed the extent, classification, and significance of contaminated sediments, and the resuspension of sediments. The second work group discussed the selection of management strategies and remedial technologies for handling contaminated sediments and the case studies exemplifying marine sediment contamination problems and their remediation.

This report is in part a proceedings of the symposium and workshop. It contains all of the invited papers and a summary of the deliberations of the work groups. It also contains a discussion of the major findings and recommendations of the committee with regard to the issues covered in the meeting.

The entire report has been reviewed by a group other than the authors, but only the summary and workshop reports have been subjected to the report review criteria established by the National Research Council's Report Review Committee. The papers have been reviewed for factual correctness.

ACKNOWLEDGMENTS

The committee would like to express its gratitude to a number of individuals whose assistance has been of great benefit in the development of this report. The committee thanks each of the invited speakers whose informative papers were the focus of the symposium and workshop. A special thanks is also extended to Walter Kovalick, Jr., Deputy Director of EPA's Office of Emergency and Remedial Response for his delivery of a stimulating and informative luncheon address.

Appreciation is also conveyed to the rapporteurs of each work group, Dr. Michael Palermo of the U.S. Army Engineer Waterways Experiment Station and Dr. Jack Anderson of the Southern California Coastal Water Research Project, for their assistance in preparing the workshop summaries. The committee also thanks all of the workshop participants (listed in Appendix B) for their participation and valuable remarks during the workshop discussions.

Finally, the committee extends its thanks to the government liaison representatives whose participation in the committee meetings, the symposium and workshop, and contributions to this report were invaluable: Robert R. Bersson, Naval Facilities Engineering Command; John Cunningham, U.S. Environmental Protection Agency; Kim Devonald, U.S. Environmental Protection Agency; Norman R. Francingues, U.S. Army Engineer Waterways Experiment Station; David B. Mathis, U.S. Army Corps of Engineers; Andrew Robertson, National Oceanic and Atmospheric Administration; John Rogers, U.S. Fish and Wildlife Service; and, Christopher H. Zarba, U.S. Environmental Protection Agency.

CONTENTS

Executive Summary. 1

Findings and Recommendations 4
    Extent of Contamination, 4
    Classification Methodologies, 6
    Risks to Human Health and the Ecosystem, 10
    Mobilization and Resuspension of Contaminants, 12
    Contaminated Sediment Management Strategies, 14
    Remedial Technologies, 15
    Remediation and Source Control: Economic Considerations, 18

Workshop Summaries 20
    Work Group I--Extent, Classification and Significance of
        Contamination, 20
    Work Group II--Assessment and Selection of Remedial
        Technologies, 28
    Case Studies, 34

Presented Papers

Extent of Contamination. 37
    National Perspective on Sediment Quality, 38
        Chistopher Zarba
    National Status and Trends Program for Marine Environmental
    Quality, 47
        Andrew Robertson and Thomas P. O'Connor

Classification of Contaminated Sediments 63
    Use of Apparent Effects Threshold Approach (AET) in Classifying
    Contaminated Sediments, 64
        Robert Barrick, Harry Beller, Scott Becker, and Thomas Ginn
    The Use of the Sediment Quality Triad in Classification of
    Sediment Contamination, 78
        Edward R. Long
    A Review of the Data Supporting the Equilibrium Partitioning
    Approach to Establishing Sediment Quality Criteria, 100
        Dominic M. Di Toro

Marine Sediment Toxicity Tests, 115
   Richard C. Swartz

Significance of Contamination. 131
   Effects of Contaminated Sediments on Benthic Biota and
   Communities, 132
      K. John Scott
   Sediment Contamination and Marine Ecosystems: Potential
   Risks to Human Health, 155
      Donald C. Malins

Mobilization and Resuspension. 165
   Predicting the Dispersion and Fate of Contaminated Marine
   Sediments, 166
      Y. Peter Sheng
   Computer Simulation of DDT Distribution in Palos Verdes
   Shelf Sediments, 178
      Bruce E. Logan, Robert G. Arnold, and Alex Steele

Assessment and Selection of Remedial Technologies. 199
   Management Strategies for Disposal of Contaminated
   Sediments, 200
      M. R. Palermo, C. R. Lee, and N. R. Francingues
   Alternatives for Control/Treatment of Contaminated Dredged
   Material, 221
      M. John Cullinane, Jr., Daniel E. Averett, Richard A. Shafer,
      James W. Male, Clifford L. Truitt, and Mark R. Bradbury
   Developments in Equipment Designed for Handling of Contaminated
   Sediments, 239
      John B. Herbich
   Monitoring the Effectiveness of Capping for Isolating
   Contaminated Sediments, 262
      Robert W. Morton
   Remedial Technologies Used at International Joint Commission
   Areas of Concern, 280
      Ian Orchard
   Economic Considerations of Managing Contaminated Marine
   Sediments, 291
      Thomas A. Grigalunas and James J. Opaluch

Case Studies 311
   New Bedford Harbor Superfund Project, 312
      Allen J. Ikalainen and Douglas C. Allen
   Physical Transport Investigations at New Bedford,
   Massachusetts, 351
      Allen M. Teeter
   PCB Pollution in the Upper Hudson River, 365
      John E. Sanders

Contamination of the Hudson River--The Sediment Record, 401
    Richard F. Bopp and H. James Simpson
Kepone and the James River, 417
    Robert J. Huggett
Assessment of Contaminated Sediments in Commencement Bay
(Puget Sound, Washington), 425
    Thomas C. Ginn
St. Paul Waterway Remedial Action and Habitat Restoration
Project, 440
    Jerry K. Ficklin, Don E. Weitkamp, and Ken S. Weiner
Dredging and Disposal of Contaminated Marine Sediment for the
U.S. Navy Carrier Battlegroup Homeport Project,
Everett, Washington, 462
    Edward Lukjanowicz, J. Richard Faris, Paul F. Fuglevand,
    and Gregory L. Hartman

Appendix A   Biographies of Committee Members, 483
Appendix B   Coastal States Survey, 486
Appendix C   Workshop Participants, 488
Appendix D   Agenda, 491

EXECUTIVE SUMMARY

Contamination of marine sediments poses a potential threat to marine resources and human health (through consumption of seafood) in numerous sites throughout the country--particularly near metropolitan areas. Improving the nation's capability to assess, manage, and remediate these contaminated sediments is critical to the health of the marine environment as well as to its use for navigation, commerce, fishing, and recreation. As widespread as the problem of sediment contamination appears to be, understanding of the geographical extent and ecological significance of the problem is not well developed. In addition, management and remediation of contaminated marine sediments requires grappling with dynamic aquatic environments in which contaminant mobilization can occur in response to remediation itself, or as a result of natural resuspension, transport, and deposition of the bottom sediments.

This report, prepared by the Committee on Contaminated Marine Sediments of the Marine Board of the National Research Council, examines the extent and significance of marine sediment contamination in the United States; reviews the state of the art of contaminated sediment clean-up and remediation technology; identifies and appraises alternative sediment management strategies; and identifies research and development needs and issues for subsequent technical assessment. The report contains the results of a symposium and workshop, with supplementary discussion and recommendations by the convenors.

The committee members concluded that sediment contamination is widespread throughout U.S. coastal waters and potentially far reaching in its environmental and public health significance. A report sponsored by the U.S. Environmental Protection Agency (EPA), although limited in its data sources, estimated that there are "hundreds of sites in the United States with in-place pollutants at concentration levels that are of concern to environmental scientists and managers. More than one-third involve marine or estuarine waterways." The National Oceanic and Atmospheric Administration's (NOAA) National Status and Trends Program, which selectively excluded "hot spots" from its sampling, found high levels of contamination in samples from sites in major urban areas, including Boston, New York, San Diego, Los Angeles, San Francisco, and Seattle. However, adequate data do not currently exist for comprehensively pinpointing or prioritizing candidates for remedial action. Even so, the means and methods for making such determinations are available (or close at hand), albeit needing much improvement. They include several that were evaluated by the committee.

At present, no single technique is widely accepted and each has its advantages and disadvantages. A number of approaches may be needed to evaluate the significance and extent of contamination at any given site. Ultimately the methods used should be able to be conducted routinely and cost-effectively.

In terms of risk to human health, transfer of contaminants from marine sediments to humans is poorly documented and underassessed. However, it appears that there may be cause for concern with regard to persistent bioaccumulative chemicals contaminating seafood. The impact of this type of contamination needs further investigation.

Despite the widespread extent of the contaminated sediment problem, remedial actions directed at excavating, treating, or otherwise manipulating contaminated marine sediments have been extremely rare. Under the Superfund law, only sites designated on the National Priorities List can be funded for remediation. The Hazard Ranking System score, which determines placement on this list, gives heavy weight to potential contamination of drinking water sources, but little or no weight to sediment-mediated contamination of edible fish and shellfish. Furthermore, little effort has been made to identify contaminated sites in coastal environments under Superfund.[1]

In its examination of state-of-the-art clean-up and remediation technology, the committee determined that existing technology is adequate in most situations. However, the committee noted that some specialized dredging equipment--e.g., to allow excavation of contaminated sediments with a minimum of turbidity--is difficult to obtain in the United States (due to cabotage laws). To alleviate this problem, government support is encouraged for efforts to acquire or develop dredging equipment with features that make it well-suited to the excavation of contaminated sediments.

The committee also found that the time required for EPA or its contractors to make a clean-up decision was more often a limiting factor in accomplishing effective clean-up than any constraints imposed by limitations in clean-up science or technology. The time required for a decision was sometimes speeded up, however, where the need for navigational dredging was a driving force.

Remediating underwater sediment contamination can be a complex problem. Failure to make a decision may cause the problem to spread. Although in many instances the problem may correct itself given enough time, it is usually desirable to isolate and contain the contaminated area to the extent possible. Allowing the affected area to expand will generally only serve to increase the cost and complexity of the eventual clean-up. More attention needs to be focused on the design of

---

[1] Although as many as 141 of 1,100 (13 percent) present and proposed Superfund National Priorities List sites may be located adjacent to coastal areas and may or may not involve coastal sediments, no remedial action has been selected for the great majority of these sites.

rapid short-term actions to limit the spread of contamination at the same time that more elaborate long-term remedies are assessed and developed.

In some cases, no action can be the alternative of choice, assuming measures have been adopted to control contamination sources. This may be particularly true when natural sedimentation or dispersal may mitigate the problem or when natural detoxification of contaminants is occurring. During an evaluation process, the effects due to remediation should be compared with those associated with the no-action alternative.

The committee recommended that future research and development be focused on

- establishing better better biological and chemical techniques for rapidly and reliably assessing the presence and severity of bottom sediment contamination,
- delineating the practical limits of capping as an efficacious remediation technology,
- identifying interim measures to limit the spread of contaminated sediments while long-term remedies are assessed, and
- formulating procedures and guidelines that adequately evaluate and prioritize health and environmental risks associated with sediment contamination, and against which effectiveness and clean-up needs can be measured.

The committee also believed that in view of the high cost of most remedial actions, greater use should be made of benefit-cost comparisons. This would place investment in this area on the same economic footing as investments in other public projects. Cost-effectiveness analysis of alternative remedial actions, including "no action," should consider both short- and long-term costs, comparisons at and among sites, and incremental costs of additional levels of clean-up of contaminated sediments.

Finally, increased emphasis on sediment assessment and clean-up practices has caused rapid changes and developments in state-of-the-art technologies. Developments and experience in methods for applying these technologies are also occurring at a rapid rate. Therefore, it is an important and appropriate role for the federal government (either through individual concerned agencies or, preferably, through a coordinated interagency committee) to frequently review and evaluate the effectiveness and scientific basis for newly developed sediment assessment and clean-up technologies and procedures.

FINDINGS AND RECOMMENDATIONS

EXTENT OF CONTAMINATION

Findings

Many marine sites are known to contain sediments with high levels of anthropogenic chemicals or to have altered biological characteristics. However, there are no generally accepted definitions of contamination that trigger consideration of remedial action. The working definition of contaminated sediments used in this report is those which contain chemical substances at concentrations that pose a known or suspected environmental or human health threat. The sites that require the most urgent attention are those reservoirs of contamination that affect regions or that have the most severe impacts on health and the environment. Pending revisions of the Superfund Hazard Ranking System will facilitate the assessment and prioritization of human health and ecological risks associated with contaminated sediments.

Many contaminated marine sediments are located along all coasts of the contiguous United States, both in local "hot spots" and distributed over large areas. Some of these sites, but not many, have been well characterized. Existing data on individual sites and their contamination vary widely in content and organization. Assessments using available data have been conducted on the national extent of contamination and have identified a partial picture of the total contaminated sediment problem. These studies have shown that a wide variety of contaminants are found in sediments, including heavy metals, polychlorinated biphenols (PCBs), DDT, and polynuclear aromatic hydrocarbons (PAHs). However, no federal agency has assumed the full responsibility of establishing a national inventory of sites with contaminated sediments or a comprehensive assessment of the extent of contamination on a national basis.

A number of state and federal agencies collect data for different purposes and use different approaches. However, sediment contamination data collected for one purpose may be of little relevance or applicability for another because of parameters measured, methods used, or temporal and spatial scales designated. For example, sediment data assembled for setting regulatory criteria or for following national or regional trends may be of little value in detecting site-specific problems or in defining site-specific remediation requirements.

This can be illustrated by NOAA's National Status and Trends Program. As part of this program, NOAA has acquired sediment data from approximately 200 sites around the coasts of the United States (see Robertson and O'Connor, pages 47-62). This information is used to determine broad national- and regional-scale status and trends in sediment contamination levels. However, the network of stations is not sufficiently dense to allow the data to be used to set clean-up priorities or to make site-specific judgments. Indeed, the NOAA program intentionally excluded from its database, sampling stations deemed to be reflective of localized hot spots rather than of broad regional contamination trends. In short, care should be exercised to ensure that data generated by monitoring programs are not inappropriately used beyond the limits or intent of the original monitoring program.

At present, there are no generally accepted and validated sampling techniques, testing protocols, or classification methodologies for determining sediment contamination. A certain uniformity in parameters measured and data reported is desirable to facilitate intercomparisons. This must be accomplished by setting some national standards, criteria, or guidelines.

In general, efforts by states to address potential marine sediment contamination are diffuse and not well focused. For example, most state water quality agencies focus on discharges and impacts to the water column. Thus, little effort is being expended by state agencies on identifying and remediating contaminated marine sites. State hazardous waste agencies are, in most cases, directing their efforts to upland areas so their involvement in marine sediments problems is limited.

Recommendations

Search for Contaminated Sites

The location and extent of contaminated marine sediments have not been comprehensively assessed on a national basis to identify site-specific remediation targets. The federal government should initiate such a program to delineate areas with contaminated sediment. The objective should be neither detailed mapping nor duplication of NOAA's regional National Status and Trends Program. In regions of concern, or in areas of known hot spots, special attention should be directed to identifying and characterizing specific contaminated sites. The search for new sites or the reclassification of known sites should proceed concurrently with remedial action.

Utilization of Federal, Regional, and Local Expertise

Due to the variability in environmental conditions among sites, well-informed local specialists provide a critical complement to our national expertise. Neither federal, regional, nor local managers can operate effectively in a vacuum. Managers at all levels of government should interact and cooperate and remain receptive to the expertise and concerns of other specialists in assessing or remediating contamination at a particular site.

Coordination of Efforts

An interagency technical committee, including nongovernmental as well as state and federal experts, should be established to evaluate existing and emerging data on sediment contamination. This committee would assemble data, prepare reports, and make recommendations as to the need for and direction of sediment research and monitoring activities, including sediment and sampling assessment methodologies. The objective of the committee would be to focus the limited resources on the most needed research and monitoring, reduce redundancy, and help eliminate improper uses of data.

CLASSIFICATION METHODOLOGIES

Findings

A variety of biological and chemical sediment classification methods are available. Individually or in combination, they attempt to systematically characterize marine sediments with elevated levels of contaminants, and correlate such concentration increases with adverse biological effects. With one possible exception (the acute amphipod bioassay), none of these techniques are routinely used and each has its limitations. Indeed the cost and complexity of a number of these tests virtually ensures that they will be used routinely only at large sites.

Several contaminated sediment classification techniques were examined by the committee: sediment bioassays, sediment quality triad approach, apparent effects threshold technique, and equilibrium partitioning. Each technique is discussed in detail in a presented symposium paper (in this volume) and some of the advantages and disadvanges of each (for remedial action screening and sediment quality criteria development) are set forth in Table 1.

From a remedial clean-up standpoint, the most useful sediment testing and classification procedures would be those that are simple and inexpensive, with rapidly available test results. If sediment quality criteria methodologies are adopted by EPA, a routine basis for establishing the presence of unacceptably high levels of sediment contaminants may be available. The design and implementation of remedial action for contaminated sediments are likely to be delayed and frustrated unless one can readily determine "how clean is clean." Development of an interim working methodology to establish such a criterion would alleviate the delay.

TABLE 1 Assessment of Sediment Classification Methodologies

Classification method	Advantages	Disadvantages
Bioassay	• follows toxicological methods developed for water quality criteria • a direct measure of sediment toxicity • does not require identification of individual contaminants • does not assume a specific route of uptake • acute results available quickly • established test procedures in use for dredged material characterization	• requires development of standard chronic bioassay methodologies • may be more costly than some chemical analyses • difficult to translate laboratory results to natural conditions • difficult to determine chemical effects • does not address human health impacts • results of chronic tests may not be timely • may not identify causative contaminants
Sediment Quality Triad	• based on a combination of laboratory and field data indicating effects of actual contaminated sediments • based on observed biological effects • does not assume a specific route of chemical uptake • applicable to complex mixtures	• limited by the availability of existing data or by the ability to collect large amounts of new data • available data may be of highly variable quality • difficult to translate laboratory results to natural conditions • does not address human health impacts • may not identify causative contaminants

TABLE 1 (Continued)

Classification methods	Advantages	Disadvantages
Sediment Quality Triad (cont.)		• indicators are not independent; covary with grain size and organic carbon content • potentially not comparable between geographic locations • does not consider chemical bioavailability from site to site
Apparent Effects Threshold	• uses existing data (from field and laboratory; e.g., Sediment Quality Triad) • applicable to all chemicals and all biological effects • most useful for prioritizing contaminated areas within a large site • based on observed biological effects • does not assume a specific route of chemical uptake • applicable to complex mixtures	• limited by the availability and quality of existing data • varies with choice of biological effects indicator • relies on correlations/ may not identify causative contaminants • potentially not comparable between geographic locations • may be both over- and under-protective • difficult to translate laboratory results to natural conditions • does not address human health impacts • multicompound interactions not accounted for • Indicators are not independent; covary with grain size and organic content

TABLE 1 (Continued)

Classification method	Advantages	Disadvantages
Equilibrium Partitioning	• provides a chemical specific criterion • utilizes large toxicological data base incorporated in water quality criteria and other toxicological endpoints • relies on well-developed partitioning theory • accounts for the bioavailability of the chemical interest • provides a standard basis for comparison within and among sites • where data are available allows quick and inexpensive characterization • incorporates a built-in "how clean is clean" standard • is a direct measurement of sediment characteristics • can be readily incorporated into existing regulatory frameworks	• does not consider complex mixtures and chemical interactions • currently limited to hydrophobic neutral organic compounds • does not address human health impacts • limited to contaminants for which both water quality criteria (or other suitable toxicological endpoints) and sediment-water partitioning coefficients are available • relies on K_{oc}[a] measurements which are often variable • does not account for contaminant uptake by ingestion of particles or direct absorption/adsorption from sediments • sediment and water may not be at equilibrium with respect to contaminant concentration • does not use toxicological data derived from the sediment of interest • assumption of constant bioaccumulation factor for various contaminants and organisms is questionable

[a] K_{oc}--carbon normalized sediment-water partition coefficient.

Although a variety of methods for assessing contamination are available, there is no single method that is widely accepted and some may be more suited to a particular situation than others. Approaches that develop single numeric criteria often do not provide sufficient data for assessing the overall significance of contamination at a site. A number of approaches may be needed to evaluate the significance and extent of contamination at any given site.

Recommendations

Improved Methodologies

In order to ensure that decision making is informed and scientifically based, continued research and use of assessment methodologies should provide information to determine

- a range of concentrations of chemicals in sediments that will result in biological effects, and
- whether in-place sediments are causing biological impacts.

Additionally, increased efforts should be made to refine methods for sediment classification to be used by regulatory agencies.

Tiered Testing

A tiered approach to the assessment of contaminated sediments should be used. The approach would progress from relatively easy and less expensive (but perhaps less definitive) tests to more sensitive methods as needed.

RISKS TO HUMAN HEALTH AND THE ECOSYSTEM

Findings

The most significant human health risk associated with marine sediment contamination may be ingestion of contaminated fish and shellfish. Many compounds, such as some polyaromatic hydrocarbons (PAHs), may be readily metabolized by enzymatic systems in higher aquatic organisms such as fish, although there is uncertainty about whether they are detoxified. Some invertebrates, such as bivalve mollusks, have only a limited ability to metabolize PAHs and tend to accumulate them to higher concentrations and retain them more. Therefore, consumption of these animals may be a source of human exposure. Trace metals are not degraded and may be bioaccumulated by aquatic organisms and then transferred to humans via consumption of seafood. Reports of "fin rot" and tumors in finfish, particularly bottom-feeding fish in Puget Sound and the New York Bight in recent years, provide further evidence that there may be substantial risk to the ecosystem and potentially to human health due to the contamination in marine sediments. Although there is general consensus that seafoods present a route of transfer of contaminants to humans from contaminated sediments, the extent of risk that is posed is unknown.

In addition to the carcinogenic nature of many of these contaminants, reproductive impairments and other sublethal effects in humans are concerns that require increased attention. Risk assessments of these latter endpoints have not been conducted. Furthermore, inadequate attention has been given to mammalian studies of the long-term chronic effects of ingesting contaminated fish and shellfish. Epidemiological studies of human populations living near contaminated sediment sites also have been under-emphasized.

Assessment of the ecological effects resulting from sediment contamination is an area that needs additional study. This is especially true for soft-bottom communities in trying to correlate ecological impacts with chemical-specific factors. Accumulation of contaminants in marine sediments can cause death, reproductive failure, growth impairment, or other detrimental changes in the organisms exposed to these contaminants. Such changes can impact not only individuals but also entire benthic populations and communities.

Both localized and widespread contamination has in the past resulted in significant population and community changes. Typically this involves the elimination of less tolerant species and an increase in more tolerant species. Such changes can have far reaching, long-term effects on a given ecosystem. Generally, those species that are eliminated have not received the attention they deserve in the assessment of ecological effects. Furthermore, the technical capability has not evolved for interpreting population and community responses in relation to specific chemicals.

Sublethal and chronic effects of contaminants on the marine ecosystem are a significant environmental concern. However, at the present time there are no widely accepted sublethal and/or chronic effects tests available. Much research is being conducted on tests for growth, reproduction, or biological abnormalities. Interpretation of such tests is often difficult and there are few established criteria available to judge the sublethal and chronic effects of contaminants on the marine ecosystem.

Recommendations

Assessment of Risk Due to Contamination

Although the assessment of human health risk is important, a more balanced approach requires greater emphasis on ecosystem impacts. This will require regulatory agencies to utilize new assays being developed to detect and gauge the effect of contamination on physiology (assays such as immune suppression, enzyme induction, and DNA adduct formation), life stage impacts (using parameters such as reproductive success, growth, and recruitment), pathological effects, and changes in community structure.

In terms of risks to human health, consideration should be given to conducting available retrospective human epidemiology studies of exposed populations in the development of an overall assessment and remedial plan.

MOBILIZATION AND RESUSPENSION OF CONTAMINANTS

Findings

The decision to manage contaminated marine sediments in place or to remove and relocate them on land involves consideration of the potential for contaminant mobilization and release to the environment. There is a tendency for heavy metals in marine sediments placed in on-land disposal sites to desorb under changing geochemical conditions (such as decreased pH due to acid formation) and potentially allow chemicals to leach into groundwater. Organic chemicals found in marine sediments tend to maintain relatively constant solubility and mobility potential when disposed of on land. When contaminated sediments are excavated and placed in contact with the air, relatively low concentrations of volatile organics can contaminate the air. The most obvious difference in risks associated with on-land and aquatic disposal of contaminated marine sediments is the greater significance of food chain contamination as an exposure pathway in aquatic disposal.

Estimates of both deposition rates and erosion rates are needed in order to decide whether to remove contaminated sediments. If natural sedimentation causes the rapid burial of contaminated sediments in place, then other remediation may not be needed. However, if the contaminated sediment is subject to resuspension and dispersion, in-place capping or removal may be necessary, even if the contamination is distributed over large areas or long distances.

Where the environmental impact potential is severe (e.g., downstream shellfish beds or drinking water intakes) a significant erosion or resuspension potential may suggest the need for quick remedial or removal action while sediment contaminants are still relatively localized and concentrated.

Our understanding of the transport of coarse-grained, noncohesive sediments is relatively well developed. Unfortunately, contaminants are most often associated with fine-grained cohesive sediments and the ability to forecast their behavior with confidence is very poor. Significant research is under way by the Army Corps of Engineers and the Environmental Protection Agency to try to define the sediment-water boundary layer conditions that limit the use of predictive models. With information concerning the strength of the currents, some general statements can be made concerning whether a site is likely to be one of scour or of deposition. However, the rates of either erosion or deposition cannot now be estimated from measured parameters. General statements are usually not an adequate basis for management decisions. A more complete understanding of the sediment transport processes for fine-grained cohesive sediments is needed.

Present practice, based on state-of-the-art knowledge, is to employ empirical models. For example, several major studies have been conducted by the Corps of Engineers for Mississippi Sound in the Gulf of Mexico, Los Angeles and Long Beach harbors, and Chesapeake Bay. These investigations have attempted to modify and adapt three-dimensional models to site-specific conditions. Resuspension rate, settling velocity, deposition rate, critical erosion velocity, rate of consolidation, rate of biological mixing, and other variables must be empirically determined for each site. The relevant processes are described by direct measurements in the field to determine a set of empirical parameters that are then applied to the site. Measured site-specific data then provide the quantitative examples that are assumed to be typical of that site at all times. Although the models rely on highly empirical approaches, they are the best tools presently available for making predictions of sediment resuspension and transport.

Empirical models for predicting the resuspension and mixing of contaminated sediments have serious limitations which include the following:

1. Relying on measurements made at a specific time and place under a particular set of conditions. There is no guarantee that the measured rates will be accurate if any of the conditions change. Small changes in the environment can lead to very large discrepancies between the empirical forecast and the actual phenomenon. As a result, the empirical models are accompanied by potentially large, and usually, unspecified uncertainties. In many cases, the magnitude of the uncertainties may be acceptable in the management decision if it is known with confidence.

2. Development of empirical models can be extremely costly. There are many types of data needed and the measurements have to be made at many locations over long time periods to improve confidence in the results. Additionally, measurements have to be made for every site of interest. This would not be a serious disadvantage if there were only a few contaminated sites. Unfortunately, there are many sites that need attention.

Recommendations

Contaminant Transport and Partitioning

Continued and expanded support should be given to understanding the partitioning of contaminants among sediments, soils, water, organisms, and the atmosphere, as well as the transport of substances in the various phases.

Research in Sediment Transport

To keep costs of modeling fine-grained sediment transport reasonable, models built on basic processes need to be developed. While empirical models continue to be used to reach management decisions, effort should be simultaneously directed to understanding the basic processes to be modeled and the validation of models in the field. Specifically, support should be expanded for research to determine the fundamental processes responsible for sediment cohesion and the factors controlling their resuspension. There is also a need to improve the reliability of estimates of both deposition and resuspension. Research programs in this area should be expanded and diversified.

Tiered Response Strategy

A tiered strategy is needed to address contaminated sediment problems in situations in which high erosion rates or resuspension potential may rapidly alter the distribution of contaminants and there is no time to carry out more detailed assessments. Problems in high-energy environments should be assessed promptly.

CONTAMINATED SEDIMENT MANAGEMENT STRATEGIES

Findings

Although the dredged material management strategy developed by the Corps of Engineers may be relevant to severely contaminated sediments, it is important from a management standpoint to differentiate them from less contaminanted sediments. In particular, most highly sophisticated remedial technologies (i.e., those involving treatment or destruction of associated contaminants) are likely to be cost-effective only in small areas and for sediments with relatively high contamination levels. Sediment contamination problems often involve large volumes of sediment with relatively low contamination levels. As a result, some highly sophisticated technologies may be inapplicable or inefficient for remediating contaminated sediments.

"No action" may be the preferred alternative in cases in which the remedy may be worse than the disease--e.g., where dredging or stabilizing contaminated sediments results in more biological damage than leaving the material in place. Contaminants generally accumulate in depositional zones, and, if the source is controlled, new sediments will deposit and cap the contaminated material over time. In effect, no action alternatives in such cases may result in natural capping.

Extensive preremediation studies, as practiced at very large sites (e.g., Commencement Bay, New Bedford Harbor, upper Hudson River) may not be practical at much smaller sites. Routine screening procedures and validated sediment assessment methods may be especially valuable in such cases. Large-scale remedial technologies are often not applicable to small sites for a variety of reasons. In such cases, regional sites or facilities may provide a means for handling sediments from several smaller sites.

There are existing management alternatives that have been effectively used for dealing with contaminated sites.

1. No action may be an acceptable option if the contamination degrades or is buried by natural deposition of clean sediment in a short period of time.
2. In-place capping may be a useful option if the sediments are not in a navigation channel or if groundwater is not flowing through the site.
3. Removal and subaqueous burial off-site may be a viable option, although the experience with this technique is limited to relatively shallow water (< 100 ft).
4. Incineration seems to be viable only for sites with relatively small amounts of sediments containing high concentrations of combustible contaminants.
5. Other techniques to assist in remediation of contaminated sediment may be appropriate in special cases. Examples include a variety of sediment stabilization or solidification techniques, and biological and/or chemical treatment.

Recommendations

Dredged Material Management Strategy

Additional evaluation should be conducted to determine the applicability of the Corps of Engineers' dredged material management strategy to more severely contaminated sediments.

No Action

No action should always be considered as an alternative strategy for minimizing biological damage. In using the no-action strategy as a form of natural capping of contaminated material, consideration should be given to the length of time it takes for contaminants to be isolated from the food chain.

REMEDIAL TECHNOLOGIES

Findings

From a remediation standpoint, the most important factors are likely to be defining of the clean-up target, technical and cost feasibility, natural recovery estimates, and ability to distinguish and/or control continuing sources of contaminants.

Dredging technology exists that is capable of greatly reducing turbidity and resuspension in connection with dredging of bottom sediments in most applications. However, because of legal (i.e., Jones Act) and practical restrictions that limit access to foreign-built vessels domestically, it may be difficult to secure access to this technology in the United States--except as equipment fitted onto U.S.-built vessels or supplied through U.S. subsidiaries of foreign dredging companies. U.S. government policies have not provided adequate encouragement to domestic firms to construct innovative dredges.

Although silt curtains can prevent movement of sediment in the top two or three feet of water column, they allow movement of sediment under the silt curtain. Silt curtains cannot operate with currents faster than one knot and are ineffective in waves. Thus, the use of the silt curtains is contined to low-energy areas.

Capping of contaminated sediments--whether in place, as mounds, or in subaqueous pits--in many cases offers a promising means of effectively isolating and containing associated contaminants. A potentially significant legal and policy issue is whether capping with clean sediments is to be deemed a preferred treatment approach under SARA, Section 121(b). On the one hand, capping can be done on site (which is favored over offsite transport) and it can "significantly reduce the . . . mobility of the hazardous substances, pollutants, and contaminants" present. On the other hand, it is not treatment in the usual chemical, biological, or physical sense, but rather containment or permanent storage. If capping materials are modified with the addition of carbon or other materials they may sorb contaminants and thus could more reasonably be defined as a treatment alternative.

While widely applicable, there are practical limits to the feasibility of capping. Among the factors that may preclude or constrain the use of capping are water depth; low sediment density; high sediment water content; active erosional area; active navigational channel requiring periodic maintenance dredging; and the use of trawls, draglines, or oyster dredges, which would destroy the integrity of the cap. Although the sediment properties needed for an effective cap are not well-defined, both clay and sand have been used successfully. Attention must also be paid to any subsequent disturbance of the cap either by natural processes (e.g., storm erosion or bioturbation) or human activity (e.g., fishing).

There are several examples of capping of dredged sediment mounds on subaqueous disposal sites. These provide very useful experience for guiding future decisions. There are, however, few general standard criteria for evaluating the likely success of a planned capping operation. Where capping is clearly feasible, prudence (and/or SARA) may dictate well-directed monitoring. Such monitoring can constitute a significant proportion of the total remedial action cost.

Recommendations

Source Control

Source control measures must be considered in all cases, including no action. Federal and state regulatory agencies requiring remedial action should implement source control measures as a component of remedial action when applicable and appropriate. Use of financial incentives through strict liability for assessment costs, remedial actions, and damages also may play an important role in source control, provided that trustees make aggressive efforts to hold responsible parties liable for releases into the environment.

Technology and Information Transfer

Aggressive technology and information transfer mechanisms are needed to ensure that knowledge gained and lessons learned from all remedial actions are available and accessible to managers confronting new remediation problems at federal, regional, and local levels. Knowledge gained should be systematically compiled in guidance documents. Lessons learned regarding the feasibility of sophisticated remedial technologies under varying conditions of contamination severity and extent should be documented and made widely available to facilitate future decision making. Lastly, experience gained through the use of screening procedures at large sites should be distilled and generalized into routine methodologies for economically assessing smaller sites.

Remediation and Navigational Dredging

When possible, remediation projects should be designed to take advantage of existing navigational dredging activities that may already be authorized in conjunction with the Clean Water Act, Section 115 or Section 10/404.

Remedial Technologies

Research and development should be encouraged by the federal government to develop technology and equipment for efficiently removing contaminated sediments and to make it available in the United States. Foreign technologies should continue to be examined relative to their appropriateness in this country. Efforts to conduct and fund research and development as a partnership between government and industry should be encouraged.

Use of Capping

Although capping might not, in the strictest terms, be considered a remedial technology, it should not be ignored because it can play a valuable role in remediating contaminated sites.

Well-focused Monitoring

Monitoring programs should be well-focused on testing forecasts made during design of the remediation plan. To the extent possible, monitoring should be extended to remove uncertainties in the basic understanding of contaminated sediment behavior. For example, monitoring of capped areas might focus on changes of cap thickness, erosion around boundaries, and leakage of contaminants through the cap.

REMEDIATION AND SOURCE CONTROL: ECONOMIC CONSIDERATIONS

Findings

Remedial actions are costly and become more expensive as additional levels of clean-up or treatment are pursued. The role of tradeoffs between possible technologies at and among sites must be considered, given the scarcity of funds to clean up contaminated sites and the potentially great number of sites.

The use of benefit-cost analysis as part of the remedial action decision process would provide perspective on the issues involved. It would place investments in this area on the same footing as other public investments. However, difficulty in quantifying benefits from remedial actions in monetary terms makes reliance on benefit-cost analysis infeasible in a number of cases. Nonetheless, in light of the high cost of remedial actions, it is important that implicit (if not explicit) consideration be given to potential benefits before remedial actions are undertaken.

Cost-effectiveness analysis is also a valuable technique for helping to guide clean-up efforts at and among sites when a decision to remediate has been made. However, to be applied correctly, both short- and long-term costs must be included, and costs must be estimated consistently for alternative actions at and among sites.

The process of assessing the need for remediation and evaluating alternative remedial actions for a site appears to be excessively long and costly. In many cases, millions of dollars and several years are expended before a decision is made. If remedial action is excessively delayed, benefits may diminish over time.

Removal of contaminated sediments can be very expensive, varying widely from several hundred thousand dollars to tens of millions of dollars. Data on 15 clean-up sites indicate that total clean-up costs can reach $500,000 to $1,000,000 per acre.[1,2] This compares with

---

[1] For purposes of comparison, assume that a one-acre cleanup involved removing overburden to a depth of one yard, or a total of 43,560 yds^3 of contaminated material. In that event, total cleanup costs would range from $11.50 to $23.00 per yd^3.

[2] U.S. Congress Office of Technology Assessment. 1988. Are we cleaning up? 10 Superfund case studies. Special Report OTA-ITE-362. Washington, D.C.: U.S. Government Printing Office.

an average unit cost of navigation dredging of $1 to $2 per cubic yard of sediment dredged. The average unit cost of all dredging, both government and private, is estimated at $1.67 per cubic yard of material dredged.[3] Onsite incineration, one of the remedial measures proposed at various sites, is also very expensive. The estimates quoted are from $186 to $750 per cubic yard.[4]

Recommendations

Use of Benefit-Cost Comparisons

In view of the high cost of remedial actions in most cases, greater use should be made of benefit-cost comparisons over ecologically relevant time periods in order to place investments in this area on the same economic footing as investments in other public projects.

Cost-Effectiveness Analysis

Cost-effectiveness analysis of alternative remedial actions should consider both short- and long-term costs. Comparisons at and among sites should be based on costs estimated using a consistent approach.

Degree of Remediation

In evaluating the degree of remediation to be conducted at a site, it should be recognized that incremental costs typically will increase rapidly as additional levels of clean-up are sought.

Economic and Environmental Considerations

The decision as to whether or not remedial actions are undertaken should be based on a balanced comparison of the anticipated environmental and public health benefits of actions with their costs, including possible environmental and health risks.

Infeasible Remedial Options

Clearly infeasible options should be eliminated at the outset, before alternative remedial actions are considered in depth.

---

[3] Pequegnat, W.E. 1987. Relationship between dredged material and toxicity. TERRA et AQUA 34.

[4] Op. cit., no. 1.

WORKSHOP SUMMARIES

Two consecutive workshops were held subsequent to the symposium on contaminated marine sediments in order to discuss topics presented in the invited papers. Work Group I, led by William Adams, directed its discussion to current knowledge of the extent of contamination, methods for classification of contamination, effects on biological communities and human health, and mobilization and resuspension of contaminated sediments. Work Group II, led by John Herbich, focused on the assessment and selection of remedial technologies, economic considerations, and the lessons learned from the featured case studies.

The discussions consisted of brief summaries by each speaker and questions and comments by the committee and invited work group participants. The syntheses below were based on summaries compiled by the group leaders and rapporteurs, Jack Anderson and Michael Palermo, for Work Groups I and II, respectively. No attempt was made to attribute specific comments to specific individuals.

WORK GROUP I
EXTENT, CLASSIFICATION, AND SIGNIFICANCE OF CONTAMINATION

Extent of Contamination

Work Group I began its discussion by addressing the question of the extent of contaminated marine sediment and the actual number of sites of concern. Papers presented by Christopher Zarba and Andrew Robertson were the main focus of discussion. It is not known how many sites contain sediment contaminants at concentrations that cause biological damage. It is, however, the consensus that in areas with high human populations, the potential is great that anthropogenic chemicals are present at levels high enough to cause concern.

There has been only a modest effort expended to date to systematically determine the areal extent of sediment contamination in this country. Recently EPA has begun an effort to develop methodologies to assess the biological impact of in-place sediment contamination. Various EPA coastal regions as well as coastal states are lending encouragement.

EPA Storage and Retrieval System (STORET) data have been used to develop a document entitled "National Perspective On Sediment Quality" for EPA under contract by Battelle. This document attempts to provide a list of contaminated sites and chemicals of interest and a first-cut derivation of "threshold effect levels." However, no attempt has been made to check the quality of the STORET data against primary literature sources. This document does not provide a complete list of the many chemicals that have been reported in sediments. In fact, workshop participants believed that a listing of sediment chemicals and their respective concentrations would not necessarily provide useful data, because differing sample collection techniques and analytical protocols used to obtain the data prevent comparison from site to site. It is felt that most of the STORET and literature data on sediments are best interpreted in a qualitative sense. The determination of contaminated sites was also based on very limited data and the list should not be considered complete. In an effort to encourage standardization, a manual has been developed by the U.S. Army Corps of Engineers (COE) and U.S. Environmental Protection Agency (EPA) describing a standardized method of sediment collection and handling.

It was also pointed out that the frequency of sampling and reporting values is specific for a given site and is not an indication of how contaminated a site is in relation to other sites. Certain coastal areas with low contamination have been monitored frequently. Nevertheless, these data are important for providing a frame of reference. The Battelle report concluded that "there are hundreds of sites in the United States with in-place pollutants at concentration levels that are of concern to environmental scientists and managers. More than one-third (63 out of at least 184 sites) involve marine or estuarine waterways." EPA has concluded that "some of the major sites that have been identified that contain chemicals of interest at high concentrations include Puget Sound waterways, Corpus Christi Harbor, New York Harbor, Baltimore Harbor, Boston Harbor, New Bedford Harbor, Black Rock Harbor, the California sewage outfalls at Palos Verdes and parts of San Francisco Bay" (Zarba, page 45).

The NOAA National Status and Trends Program currently provides the most comprehensive and systematic national data set on sediments. Chemical concentrations in marine sediments at approximately 200 sites have been monitored since 1984. The program was set up to evaluate the quality of the marine environment over time and systematically excluded hot spots of contamination. Criteria to eliminate hot spot sites were based on historical data and personal knowledge about specific sites. The Status and Trends Program found that high levels of contaminants in sediments occurred at virtually all of the sampling sites near Boston and New York and at some of the sites near San Diego, Los Angeles, San Francisco, and Seattle, as well in Choctawhatchee and St. Andrews bays in Florida.

Workshop discussion also centered on what was an appropriate definition of contaminated sediments. Many participants believed that a generic definition was difficult to establish since it was necessary to judge contamination on the basis of both chemical concentrations and biological effects. Currently, it is not always possible to ascribe a particular biological effect to a given chemical concentration. It was the belief of many workshop participants that a determination of whether or not a given site is contaminated should be made on the basis of appropriate biological tests. It is clear that the science as a whole needs to be able to relate biological effects to chemical concentrations if the state of the art of evaluating sediments for extent and degree of contamination is going to be advanced.

EPA is in the process of establishing sediment quality criteria. At this time, there are no apparent specific guidelines regarding use of EPA's sediment quality criteria. A technical oversight committee (of the Science Advisory Board) has been formed to address this issue. Newly derived criteria for sediments have been used by the Superfund office as a guideline to help determine when additional biological testing is needed.

Contaminant hot spots will draw increased attention in the future, both as a result of the Comprehensive Environmental Response, Compensation, and Liability Act of 1980 and because of the need to dredge navigational channels. Existing regulations cover the extent of biological and chemical testing that must be conducted before dredged material can be placed back into the marine environment. However, there are also many highly contaminated areas that do not fall under COE navigational authority. Comparable procedures to determine the extent and significance of contamination in areas outside of established navigation channels are needed.

The work group noted that whether or not a contaminated site qualifies for Superfund designation may have little meaning relative to the degree of contamination. Since a direct link to human health is required for Superfund status, those sediments impacting only aquatic biota do not currently qualify.

Classification of Contamination

A discussion of available methods for evaluating and classifying contaminated sediments followed, focusing on papers presented by Robert Barrick (pages 64-77), Edward Long (pages 78-99), Dominic Di Toro (pages 100-114), and Richard Swartz (pages 115-129). The speakers discussed the Apparent Effects Threshold method, the Sediment Quality Triad, the Equilibrium Partitioning approach, and the Sediment Bioassay approach. Advantages and disadvantages of these approaches are listed in Table 1, page 7.

The Sediment Quality Triad and Apparent Effects Threshold (AET) are methods of evaluating and classifying the extent of contamination associated with the sediments in a given geographical area. These approaches incorporate data from chemical analyses, biological toxicity tests of the sediments, and in situ measurements of ecological diversity. The data can be used to

1. rank and classify the relative quality of sediments among sample sites;
2. prioritize sites for remedial action and estimate the size of the area to be remediated;

3. provide a descriptive ecological evaluation of the study site based on chemical, biological, and ecological data;
4. rank sediments based on each component and evaluate differences between each of the descriptors;
5. compare the combined or individual data for each descriptor against similar data collected from a reference site; and
6. establish numerical criteria for contaminants found in the study area.

The strength of these two approaches is that they incorporate both biological and chemical measures of contamination, the data are extensive and little follow-up work is needed, they do not assume a specific route of uptake by the organisms, and contaminant indices can be calculated. The approaches appear to be particularly suited for sites where remedial action is anticipated. Weaknesses include the following:

- need for a large data base and development of statistical evaluations of the developed criteria;
- results are strongly influenced by the presence of unknown covarying toxic contaminants; and
- a poor understanding of the bioavailability of the chemicals present.

Furthermore, the cost of developing criteria can be quite significant.

The Equilibrium Partitioning approach uses existing water quality criteria effects data, together with estimated concentration of a specific contaminant in the sediment interstitial water, to determine if the contaminant will be toxic to benthic invertebrates. This approach is designed to provide data on specific chemicals of interest and to provide numerical endpoint criteria that can be used as a guideline for assessing the safety of chemicals in sediments. It is based on the assumption that for nonpolar organics, ecological effects are most often observed as a function of the concentration of the chemical in the interstitial water. It also assumes that the bioavailability of nonpolar organics is controlled by the amount of organic carbon present in the sediment. Furthermore, it assumes that interstitial water concentrations can be estimated by knowing the organic carbon content of a specific sediment, the sediment bulk concentration of a specific chemical, and the carbon normalized sediment-water partition coefficient (K_{oc}) for the chemical. This approach is currently being evaluated by EPA for development of sediment quality criteria.

Once water quality criteria have been established, or a chronic test with one or more sensitive aquatic organisms has been performed, a sediment quality value can be calculated. The only data needed to make this calculation are the water quality criterion or chronic effect level and the carbon normalized sediment-water partition coefficient for the chemical of interest. The Equilibrium Partitioning method is chemical specific, like the Water Quality Criteria values, and addresses the issue of bioavailability. To date, laboratory data developed for approximately five chemicals provide empirical confirmation of the Equilibrium Partitioning approach. The disadvantages of this approach are as follows:

- it does not address the issue of complex mixtures and chemical interactions;
- at the present time it is available only for nonionic organics;
- it uses partition coefficients, which can vary significantly;
- it is limited to only a few chemicals for which water quality criteria values exist; and
- it does not incorporate toxicological data for the specific sediments of interest.

The Equilibrium Partitioning approach assumes that the interstitial water is the primary medium through which contaminants are taken up. However, ingestion appears to be a very important uptake mechanism of contaminants for many marine worms. Carbon normalization of the sediment contaminant concentration and the organism contaminant concentration (by using organism lipid content) has provided a useful method for assessing the uptake of nonpolar organics. However, concern was expressed about the effect of grain size and the degree of hydrophobicity needed in order for this approach to be valid.

The Sediment Bioassay approach can be used in two ways to determine sediment quality values. First, bioassays can be performed with the contaminated sediments of interest, and effect levels can be compared directly with the concentration of the chemical on the sediment. Second, sediments can be spiked in the laboratory and dose-response relationships can be developed. This approach is incorporated in the AET and Sediment Quality Triad approaches and has the same main advantage in that it actually tests the sediment and chemical of interest with benthic organisms. When bioassays are applied to field samples, they provide a measure of the cumulative effect of all the chemicals present. It is thought to be an efficient method of evaluating sediments. The method does not assume a specific route of chemical uptake, and it follows the approaches used to develop water quality criteria.

Limitations of this approach as with others that incorporate bioassays are that bioassays do not always identify problem areas. Sometimes a more sensitive species is needed to detect a problem. Furthermore, chronic test methods are not well developed at this time. Field-conducted bioassays do not lend themselves to development of specific chemical criteria, and laboratory-spiked sediments often have different sorption properties than aged field samples. Discussion of sediment bioassays centered on their sensitivity and the need for standardized tests and good storage and handling procedures for field sediments. Infaunal field assessments are thought to be useful measures of ecological effects, but may become costly if detailed analysis is needed.

In summary, approaches that develop single numeric criteria often do not provide sufficient data for assessing the overall significance of contamination at a site. A number of approaches may be needed to evaluate the significance and extent of contamination at any given site. The Equilibrium Partitioning approach may be a good screening tool to determine if the concentrations of chemicals are approaching known effect levels. If so, additional biological and chemical testing, as well as in situ evaluations, may be needed. The consensus of the group was that all the methods discussed were useful and that no one method had a clear advantage. There is a need for method development and standardization for sediment bioassays, as well as long-term sensitive tests. Also, methods should be developed that evaluate mutagenicity, histopathology, bioenergetics, and other short-term indicators of chronic toxicity. A three-step site assessment approach was suggested:

1. review criteria,
2. conduct laboratory bioassays, and
3. perform infaunal surveys.

Ultimately, the method used to determine sediment quality criteria should be one that can be conducted routinely and cost-effectively.

Significance of Contamination

A discussion of effects of sediment contamination on biological communities and human health was based on the papers presented by John Scott (pages 132-154) and Donald Malins (pages 155-164). The work group focused on the use of population and community parameters in sediment quality assessment and indicators of risks to human health.

Certain population and community parameters can be useful in assessing sediment contamination. It is clear that succession occurs in the marine environment in response to contaminant stress. Most studies to date have centered on hard-bottom communities. As a result, there is less information on soft-bottom community succession. Typical succession patterns indicate a steady progression from colonizing to steady-state communities following environmental perturbation. The addition of contaminant stress on the sequence of community succession does result in measurable effects. It is possible to detect population and community responses, but the science has not evolved to the point of being able to interpret these responses in relation to specific chemicals. When changes occur, frequently the cause is not known. More chemical-specific approaches for evaluating ecological impacts are needed. There is a need to know, for instance, if observed effects are primarily due to reduction of the food source, habitat modification, or some other altered variable. The tools for evaluating community health need to be improved and become predictive.

Detection of hot spots is usually not a problem. In order to detect areas with moderate contamination and to understand its impact, better understanding of chronic effects caused by specific chemicals or mixtures is needed. This is most readily done in the laboratory. Ecological succession as a result of contaminants might be viewed as a series of chronic effects occurring in the field. A series of sensitive chronic laboratory tests would greatly advance understanding of mechanisms of toxicity and increase predictive capabilities.

Discussion by the work group centered on the need for sensitive laboratory assays with endpoints other than the traditional endpoints of growth and survival. There is a real need for short-term indicators of chronic toxicity. Certain data suggest that measurement of effects of chemicals on the immune response system might partially fulfill that role. Development of a suite of responses that could be measured in the laboratory and related to ecological effects was encouraged. There are often significant differences between the organisms studied in the laboratory and the organisms inhabiting the area of concern. This points to the need for either the development of more test methods or a better understanding of functional roles at the species level. For example, what does the loss of a single species mean for the health of the community? The answer is not easily obtained.

The COE is required to use ecologically relevant species in each region designated to receive dredged materials. Since there are no standardized sediment bioassays with ecologically relevant species for all areas and types of contaminants present, the COE has used a variety of methods--including the Equilibrium Partitioning method--in addition to bioassay testing.

The extent of contaminant transfer from the marine environment to humans is also poorly understood and underassessed. However, limited studies suggest that "significant changes in health status may occur in humans consuming contaminated fish" (Malins, page 161). The most revealing data suggesting that contaminated sediments might present human health problems are residue levels in the tissue of organisms consumed by the public. Food chain transport is the primary concern, particularly for persistent and bioaccumulative chemicals like chlorinated organics and methyl mercury. Particularly worrisome are indications from PCB research that infants born to mothers that eat a lot of fish from PCB-contaminated areas showed delays in developmental maturation at birth, were smaller, had a reduced head circumference, reduced neuromuscular maturity, and behavioral anomalies (Malins, page 159). Risk assessment for tissue residue levels requires additional study: for example, the significance of various levels of chlorinated organics that can be measured in the human blood stream needs to be understood. Typically, FDA action limits for a particular chemical in fish or shellfish are not derived using risk assessment models in which a protection level for a risk of one in a million is derived. If this were done for PCBs, the action level would be much lower than the existing one of 2.0 ppm. In fact, some researchers believe if this approach were used for a wide variety of chemicals found in seafoods, most of the U.S. nearshore commercial fisheries would have to be closed.

The work group also raised the question of whether the public is adequately protected and whether existing risk assessment models are appropriate. Additional research is needed to determine if these risk calculations are in fact real. Most risk assessments are currently driven by the risk necessary to protect against cancer. This ignores a host of other endpoints, such as reproductive effects, which may be more important than cancer effects. There was general consensus that seafoods present a method of transfer of contaminants to humans, some of which are obtained from the sediments. The extent of the risk that is posed is not known. Emphasis should be placed on epidemiological studies of populations living near contaminated sites, particularly those with a history of consuming seafood from contaminated areas.

Resuspension of Sediments

Both the NRC Committee on Contaminated Marine Sediments and the Society of Environmental Toxicology and Chemistry workshop on Priority Research Needs on Risk Assessment (August 1987, Breckenridge, Colorado) have targeted sediment resuspension and mobilization as a key research need. Papers by Peter Sheng (pages 166-177) and Bruce Logan, Robert Arnold, and Alex Steele (pages 178-198) on modeling of sediment transport dynamics provided a focus for the work group discussion.

The work group agreed that cohesive sediment transport requires more research. Many troublesome contaminants are associated with fine-grained sediment particles. Because of the complexity of fine-grained cohesive sediment transport, there are no validated, general models available to describe it. Even practical rules of thumb are lacking in some areas, although excellent studies have been done on various elements of the transport problem, and both the COE and EPA are working to develop useful models. At the outset, a reliable sediment mass-balance should be constructed for each contaminated site, and--ideally--field-validated models should be developed to describe flocculation, biological aggregation, erosion, deposition, resuspension, bioturbation, and advective diffusive transport. Unfortunately, there are no data sets large enough to aid in this task. Current models are based almost entirely on laboratory data and have required extensive site-specific calibration, such as direct measurements of resuspension rates.

Recent developments in instrumentation now make possible many of the measurements needed to establish reliable models. A suggested approach was for EPA to sponsor a long-term research program at one of the aquatic Superfund sites to derive the kinds of field data necessary to build a useful model. A large portion of the data that would be collected is necessary to meet existing EPA requirements. However, collection of additional data also would be very useful.

Based on past experiences with deposition of PCBs in the Hudson Rivers, DDT in the Palos Verdes Shelf, and kepone in the James River, knowledge about long-term burial of persistent chemicals has been gained. In each case, there have been areas where these chemicals have become buried. However, reliable predictions of stability of deposited materials under different and changing environmental conditions cannot be made.

WORK GROUP II
ASSESSMENT AND SELECTION OF REMEDIAL TECHNOLOGIES

The second work group, led by John Herbich, conducted a discussion of remedial technologies that examined the state-of-the-art strategies and technologies for control/treatment and disposal of contaminated marine sediments, as well as economic considerations of remediation. The discussion was based on papers presented by Michael Palermo et al. (pages 200-220), M. John Cullinane et al. (pages 221-238), John Herbich (pages 239-261), Robert Morton (pages 262-279), Ian Orchard (pages 280-290), and Thomas Grigalunas and James Opaluch (pages 291-310).

Selection of Remedial Alternatives

At present, a range of control measures exists (both treatment and containment technologies) that have potential or proven application to dredging and disposal of contaminated sediments. During the symposium, M. John Cullinane presented a procedure for selection of remedial alternatives called the Dredged Material Alternative Selection Strategy (DMASS). Many of the dredging and disposal technologies have been derived from the hazardous waste field and have drawn heavily from the Superfund program. Because of the variability of site and material characteristics and the wide range of control/treatment technologies available, no single technology will be the universal solution. Furthermore, many factors in selection are not easily quantified. For example, the potential need for a liner to protect groundwater resources would be determined based on site-specific evaluation.

The nature of contaminated sediment must be considered carefully in the selection of an appropriate control technology. In selecting such a technology, sediments that contain some contaminants must be distinguished from sediments that are highly contaminated and possibly categorized as hazardous waste. While most hazardous waste disposal problems deal with relatively small volumes of materials with high concentrations of contaminants, problems with contaminated sediments often involve large volumes of sediment with relatively low concentrations of contaminants. For this reason, some technologies may be technically ineffective or inefficient.

Since many of the control/treatment technologies are unproven, extensive research and evaluation needs to be conducted for a range of technologies. The research should focus on applicability to treat large volumes of sediment, the degree of treatment or control achieved, and costs.

A comprehensive management strategy for evaluation of alternatives for disposal of dredged material was developed by the COE and presented at the meeting by Michael Palermo (pages 200-220). The COE considers the strategy to be technically appropriate for dredged material, providing the necessary level of environmental protection. The strategy utilizes testing procedures specially developed for dredged material that consider the geochemical environments of aquatic, intertidal, or upland disposal areas. A decision-making framework has also been developed that allows comparison of test results with applicable standards and criteria using a consistent approach. The procedures in the strategy are consistent with regulatory requirements under the Clean Water Act and Ocean Dumping Act (Marine Protection, Research and Sanctuaries Act).

The COE's strategy was applied in several of the case studies presented at the meeting, including the New Bedford and Commencement Bay Superfund projects and Everett Homeport project. No problems with application of the strategy and associated decision-making logic have been reported. Most potential problems with applying the strategy have involved selection of appropriate criteria or standards on which to base decision making. In this respect, it is essential to involve all concerned agencies and parties at every step of the process. The formation of public involvement coordination groups, interagency steering committees, or similar mechanisms for involvement are desirable.

The COE's management strategy has been adopted by Environment Canada in the evaluation of dredged material disposal alternatives for the St. Lawrence Seaway and other projects. West Germany and the Netherlands have also adopted the strategy. The strategy covers only contaminant testing and controls. However, the COE has a broader umbrella of evaluation procedures under its Long Term Management Strategy initiative, which includes consideration of other aspects of decision making, such as cost. No other comprehensive strategies for evaluation and selection of remedial alternatives specific to contaminated sediments were identified by the work group.

The work group agreed that no action should always be considered as a potential alternative to remediation in an evaluation process assuming that the fate of contaminants has been quantified. No action may be preferable where natural detoxification of contaminants occurs or where natural sedimentation processes help to isolate the contaminated material from the environment. During an evaluation process, the effects due to remediation should also be compared to those associated with the no action alternative and consideration should be given to the time required for natural processes to isolate the contaminants. Special consideration of "no action" should be given to cases in which remediation may cause irreparable harm to the resource.

For both clean-up or no action alternatives, removal and control of additional contamination sources is of critical importance. In the Commencement Bay evaluation, selecting no action was considered viable if natural recovery through biodegradation or natural capping by sedimentation was predicted to occur within 10 years. The extent of sediment disturbance by currents and bioturbation influences the time required for permanent burial in cases of natural sedimentation.

In selecting the remedial alternative, the goal of remediation should be carefully considered. Goals such as "nondegradation" or "fishable/swimmable," which are goals for actual improvement of conditions, can result in vastly different criteria. In some cases, multiple criteria are being used in evaluations.

Public involvement is essential to success in selection and implementation of a remedial alternative. Citizen advisory groups and public notices and meetings are accepted ways to ensure appropriate public involvement. It is not always certain that all available information will reach the public; therefore, the forum should be well-established to disseminate and put all information into proper perspective.

Developments in Equipment for Removal of Contaminated Sediments

A variety of equipment and operating procedures have been developed to dredge contaminated sediments while minimizing sediment resuspension and contaminant release. Conventional dredges, such as hopper, clamshell, and cutterhead dredges, are applicable when large volumes of material are to be removed to maintain navigation. However, these dredges are not well-suited for removal of highly contaminated sediment without modifications to the equipment or operating procedures. Equipment such as enclosed clamshell buckets, auger suction heads, matchbox heads, and other specialized dredge heads have been developed by the Japanese and the Dutch. These specialty dredges, for the most part, have been developed especially for removal of sediment with minimum resuspension.

One problem with utilization of specialty dredges is the availability of the equipment in the United States. Patent agreements must be considered. Some of the dredging companies have a U.S. representative or licensee, which would facilitate acquisition of the equipment. Jones Act requirements may limit the use of equipment with foreign-made floating plants. However, the use of a foreign-made dredge head on a U.S.-made floating plant is not restricted. Use of these heads on other plants could also eliminate present constraints for operation of such equipment in shallow water areas.

There is a need for research and development in the area of equipment for contaminated sediment remediation. There is presently no such effort going on in the United States. Incentive for U.S. companies has been lacking mainly because of a perceived limited market for such equipment. Development by federal agencies, such as the COE or EPA, also has potential drawbacks. Ownership of specialty dredges by these agencies may be objected to by the dredging industry, and may be restricted by law.

Available Disposal Alternatives

In considering the removal of contaminated sediments, the work group identified a wide range of disposal alternatives. Disposal in open water, including ocean disposal, may be a viable option if appropriate control measures, such as capping, are implemented. Except for prohibited substances, disposal in open water with appropriate controls is compatible with regulatory requirements under the Clean Water Act and Ocean Dumping Act and is accepted under the London Dumping Convention.

Containment alternatives, such as in-water confined disposal facilities or upland disposal sites, involve proven technologies. These sites can be constructed as simple containments or may incorporate a wide range of control measures, such as chemical treatment, filtration, solidification/stabilization, liners, and covers.

More intensive alternatives involving various treatment or destruction technologies may be effective for sediments with high levels of contamination. However, they are normally expensive and their application to large volumes of sediment--especially if not highly contaminated--is not generally an economically favored alternative.

The work group also considered the effectiveness of capping for isolating contaminated sediments. Capping, covering contaminated sediment with clean sediment, has been shown to be a viable disposal alternative for contaminated dredged material in Long Island Sound and New York Harbor. Monitoring has shown that contaminated material can be placed in mounds and clean material placed over it to successfully cap the mounds. Care in the placement procedure is essential for success and may involve use of precision navigation, taut-wire buoys, and rigorous inspection procedures. Most projects to date have involved capping on level bottoms. Capping has been proposed for the New Bedford pilot project and Commencement Bay and has been used in Norwalk Harbor.

Research issues related to capping that need to be addressed include capping procedures for deeper water sites and mass release predictions. These research needs have become evident with the proposed Everett Navy Homeport project and the pending designation of a deep-water site for disposal of material from New York Harbor.

There are presently no mathematical models developed to evaluate or design capping technology. However, to date there has been no evidence of displacement of capped material by the capping process. Even with material on the bottom in a mounded configuration, there has been no evidence of material being squeezed out from under the cap. In general, the geotechnical information related to capping is mainly judgmental since the parameters involved are not known with certainty. Monitoring also is of vital importance with capping projects. Advances in monitoring equipment and techniques are as important as advances in equipment for dredging and placing the material. However, monitoring of capped sites should not be planned or required unless there is a clear objective and the use of the monitoring data is clearly defined.

If designed and executed correctly, capping has low risk and is generally a less expensive remedial alternative than confined disposal facilities. However, capping is generally only feasible in low-energy environments and where a source of capping material is available. Capping within the boundaries of a channel area poses special problems due to potential need for future deepening and vessel traffic and anchoring. In general, deeper water sites offer low-energy environments. However, potential dispersal of contaminated sediments during placement in deep water is greater. Determining an acceptable level of sediment contaminant dispersal is a major question.

In New York Harbor, there are several large (4 to 20 million yd^3) borrow pits that were created during sand mining operations over the past 40 years. The COE's New York District has completed a draft environmental impact statement for use of these pits as containment disposal sites for contaminated sediments. The dredged sediment would be placed in the pits and capped with clean sand to both isolate and contain the contaminated material and to restore the seafloor to its original bathymetry and composition. The capped deposit would not have surface relief. A major issue is the value of existing borrow pits in concentrating fish populations, possibly making it desirable to leave the old pits unfilled and constructing a new pit specifically for the disposal site. This option may actually be less expensive than conventional open-water disposal in the area and would probably be reserved for questionable material or material that is unacceptable for ocean disposal.

The fact that capping is considered a containment alternative and not a treatment alternative may present some legal disadvantages from the perspective of Superfund sites. The preference under the Superfund Amendments and Reauthorization Act (SARA) for treatment-based permanent solutions must be re-evaluated for cases of sediment contamination. The relatively high volumes of material in most contaminated sediment sites--as compared to most Superfund sites--dictates that the containment option in many cases may be the best remedial alternative. However, capping materials may themselves be modified, or perhaps with the addition of carbon or other sorbent materials may be able to remove contaminants. In such cases, capping could be defined as a treatment alternative for these purposes.

Economic Considerations

A major consideration in implementing most of the desired technologies is cost. Some remedial technologies, such as removal of solids and associated contaminants through gravity settling, chemical clarification, and filtration or solidification and stabilization of sediment, are relatively inexpensive ($10 to $50 per yd^3) and have proven applicability to contaminated sediment disposal (this compares to $1.67 per yd^3 for navigational dredging of clean sediments). Other more intensive technologies such as incineration or chemical extraction are much more expensive ($200 to $750 per yd^3) and have not been proven in large-scale demonstrations. Some of the intensive technologies may result in secondary pollution or a waste stream of a differing nature that will also require treatment (e.g., air pollution problems related to incineration). In such cases, pilot studies can be useful in demonstrating applicability. Few detailed estimates of costs are available. For the Commencement Bay Superfund site, an array of PCB destruction technologies was examined with costs ranging from $200 to $500 per yd^3 of sediment treated. Cost information for the New Bedford Superfund site should be available in the future.

The question of who will pay for remediation is another major consideration. In the case of Superfund, responsible parties are liable for clean-up or remediation. Voluntary clean-up efforts by responsible parties and out-of-court settlements are made in some cases. If remedial action is pursued, EPA and the states (through the Superfund) will bear costs for listed Superfund sites in which costs cannot be recovered from responsible parties. For cases requiring remediation not listed under Superfund, the question of who bears the costs remains.

In general, remedial actions are costly and increasing levels of remediation lead to rapidly increasing costs. The role of trade-offs at and among sites must be considered, particularly given the scarcity of funds to be used to clean up Superfund sites and the increasing number of sites.

Both benefit-cost and cost-effectiveness analyses can assist in making remedial action decisions at a site and in allocating efforts among multiple sites (fund balancing). Benefit-cost analysis can put the issues in perspective and is the only approach that can place public and private investments at sites on the same economic footing as investments in other environmental projects or other public projects in general. However, to use benefit-cost analysis, an explicit value (or range of values) must be assigned to human health and environmental resources. Such valuation is difficult to do in many cases, it can be controversial (particularly placing a value on mortality), and it may be rendered more complicated by the potential liability of responsible parties for damages under CERCLA.

Cost-effectiveness analysis avoids the need to value human health and environmental goods explicitly. Instead, the general goal is to select (1) the least-cost approach(es) to achieve a given objective or (2) the actions that provide the greatest returns (e.g., number of lives saved or illnesses avoided) for a given budget. However, to be applied correctly, short- and long-term costs must be included, and costs must be estimated consistently for alternative actions at a site and among sites.

Cost-effectiveness is required under SARA; however, benefit-cost analysis is not required in remedial action decisions nor is it widely applied. Reasons for this might include the difficult nature of these calculations in some cases, the reluctance to assign explicit economic values to public health, legal considerations regarding the liability for damages by potentially responsible parties, and the legal standing of some economic analyses under the current legislation.

Also discussed was the potential role of strict liability for damages in providing financial incentives for source control to avoid creation of new sites. The natural resource damage assessment regulations established under CERCLA and the Clean Water Act were described and their scope and advantages outlined. Limitations of the liability approach were mentioned, including the fact that it can only be applied if the responsible party can be identified.

International Joint Commission Areas of Concern

A number of the approaches described above are being considered by the International Joint Commission (IJC) for remediation of contaminated sediment problem areas in the Great Lakes. Two major options generally are considered feasible. A confined disposal facility, built by diking nearshore areas of the lake, is considered a proven technique. These facilities provide effective containment when properly designed. Confined facilities built on upland sites are the second major option. Other options, such as capping, strip mine reclamation, and solidification, also have been evaluated by the IJC and hold promise for specific projects.

The Canadians dispose of large quantities of contaminated sediments each year. Technologies used in projects in the Netherlands and West Germany are being evaluated for use in Canada. These technologies involve use of hydrocyclones to separate contaminated fractions of sediment for more efficient treatment.

Two projects in Canada have progressed to the implementation stage. Hamilton Harbor, which involves clean-up of contaminated sediment containing PCB and metals from an estuary basin, was scheduled to start in summer 1988. For this project, solidification was unnecessary and would require more storage areas for disposal. The disposal at Hamilton will be in a conventional confined disposal facility. A second project at Port Hope, involving approximately 25,000 yd^3 of sediment contaminated with metals and uranium, will involve reclamation of the uranium and treatment.

In the United States, a site at Waukegan Harbor will involve dredging and upland disposal of 50,000 yd^3, including a hot spot of 5,000 to 10,000 yd^3. A site at Ashtabula will involve disposal of 20,000 yd^3 at a permitted hazardous waste site, and incineration of 5,000 yd^3 of hot spot material.

CASE STUDIES

Case studies were presented during the symposium for New Bedford Harbor, the Hudson River, the James River, and Commencement Bay. Considerable effort was made in each of these studies to acquire good data on the extent and degree of contamination. In most cases, the extent of contamination was better defined than its potential transport.

For all the case studies, three main options were considered:

1. complete removal of all contaminated sediment,
2. removal of limited volumes or hotspots, and
3. no action.

Disposal and treatment of the removed material involved consideration of a wide range of alternatives. The selection of an alternative was dependent on three main factors:

1. public acceptance (the NIMBY [not in my back yard] syndrome is of importance here),
2. cost, and
3. environmental effects.

The five case studies presented a diverse set of physical, chemical, and biological characteristics related to sediment contamination. In the case of the James River, the chemical of concern, kepone, was a chlorinated pesticide that entered the aquatic environment directly from the manufacturing process. In effect, it was a point source that impacted approximately 500 km^2 of river bottom. New Bedford Harbor exemplified a relatively confined point source of PCBs and the trace metals cadmium, copper, lead, and non-point sources of PAHs. Approximately 4 km^2 of New Bedford Harbor were contaminated where tidal current velocity and range are 25 to 122 cm/sec and 1 m, respectively. Commencement Bay sediment became contaminated from both point and non-point sources by PCBs, PAHs, hexachlorobenzene, 4-methylphenol, and the trace metals arsenic, cadmium, copper, lead, zinc, and mercury. Approximately 2.1 km^2 of Commencement Bay have been deemed contaminated enough to require clean-up, mostly in sheltered waterways. Contamination in the Hudson River is predominantly PCBs and chlorinated hydrocarbon pesticides, but also includes heavy metals. The major source of PCBs to the system was discharges from two General Electric capacitor manufacturing facilities in the upper part of the Hudson River. River flow is variable due to hydroelectric plants. Sixty percent of the contamination in the Hudson is contained in 40 hot spot areas in the river sediments. The Navy Homeport project in Everett, Washington contains approximately 775,800 m^3 of contaminated sediment with concentrations of PAHs, PCBs and heavy metals (arsenic, cadmium, copper, lead, mercury, and zinc). The contaminated area covers 0.3 km^2 and is located in an urban embayment (ranging in depth from 28 to 40 ft. with 11-ft tides and quiescent near-bottom velocities (10 to 20 cm/sec).

All the case study areas exhibit stressed biological communities or organisms that have accumulated some or all of the contaminants in their tissues. There was no evidence that biota in the James River have been harmed by kepone, but they do have tissue concentrations in excess of FDA action levels for human consumption. In New Bedford Harbor, there was an apparent chronic toxicity gradient from the Acushnet River to outer New Bedford Harbor, coincident with a gradient in PCB concentrations. Historical data on Commencement Bay indicate high sediment toxicity, accumulation of toxic substances in indigenous biota, and the presence of liver abnormalities and tumors in flatfish. In the Hudson River, high levels of PCBs were detected in fish as early as 1969, and the striped bass fishery was closed. The sediments of concern in the Everett Homeport project contain stressed benthic communities with low biomass values, low diversity values, and low Infaunal Trophic Index values.

Various remedial actions have been or are being considered for the five areas. In the James River, chemical conversion, stabilization, dredging, and sorption were considered. The cost estimates ranged from $3 x 10^9 to in excess of $10 x 10^9$. None of these were chosen and natural sedimentation has decreased the biological availability of the contaminant to the point that commercial fishing restrictions have now been lifted. For Commencement Bay, in situ capping, dredging with various confined disposal options, and treatment are being considered. The choice of remedial action(s) has not yet been made, thus costs are not available. Evaluation of remedial options for New Bedford Harbor are ongoing. In situ capping, dredging-disposal, and dredging-treatment-disposal are being considered. Cost estimates range from $20 x 10^6 to $200 x 10^6. A course of action was to be chosen by June 1989.

In the Hudson River, proposed hot spot dredging was considered, followed by a number of options, including upland disposal, incineration, basic extraction sludge treatment, ozone-ultraviolet exposure in an ultrasonic bath, microbial treatment, and steam gasification. The cost of these options ranged from $20 to $160 per m^3. Alternatives continue to be considered and weighed, and the search continues for a final solution to PCB removal or destruction in the sediments to be dredged. Contaminated sediment disposal options for the Everett Homeport project were nearshore/intertidal disposal, upland disposal in a saturated or unsaturated sediment condition, or confined aquatic disposal (capping in deep water). The last of these was chosen and its estimated cost is $17.5 million, or $5.30/$yd^3$ (@ $6.90/$m^3$). Effectiveness of the capping will be determined by extensive monitoring.

Mark Brown, representing the New York State Department of Environmental Conservation, presented the department's perspective on Hudson River PCB clean-up efforts. He reported that removal of approximately 50 percent of the total PCBs, corresponding to approximately 30 percent of the erodible PCBs, was now anticipated. The contamination has been well defined and will be removed "because it is there." The contamination has accumulated in areas of low energy, and predicting its mobilization has been difficult. Removal of all contamination is not feasible. Sediments with PCB concentrations of 25 to 50 mg/kg will be left in certain areas of the Hudson River.

John Brown, of General Electric Corporation, discussed the natural PCB degradation in the Hudson River. The nature, cause, and environmental significance of biodegradation of PCBs have been investigated. The biodegradation process has been duplicated in the laboratory and follows the same processes found naturally in many systems. The no action alternative is viewed as a preferable option where there is a naturally occurring, gradual lowering of the hazard.

Investigation of the Buffalo River was described by Gerhard Jirka of Cornell University. He stated that simple, realistic prediction tools for evaluation of the no action alternative are needed. Combinations of field data, laboratory experiments, and models should be considered. For the Buffalo River, an extension of the COE's HEC-6 model was used to assess movement of contaminated sediment under expected flow conditions. The important considerations included time horizon, sequence, and sensitivity. Extreme events were found to have great influence on the results.

EXTENT OF CONTAMINATION

NATIONAL PERSPECTIVE ON SEDIMENT QUALITY

Christopher Zarba
U.S. Environmental Protection Agency

ABSTRACT

To address the growing concerns on the effects of contaminated sediments on aquatic life, wild life, and human health, the U.S. Environmental Protection Agency (EPA) placed greater emphasis on development of a regulatory mechanism to aid in evaluating and making decisions concerning contaminated sediments. Started in 1984, one of the first activities conducted under this effort was a national assessment of the scope of contaminated sediments. At that time, need and available resources did not warrant exhaustive study of the extent of contaminated sediments; however, to focus future activities a general assessment was needed. EPA Storage and Retrieval System (STORET) data was used (and supplemented with data from the literature) to provide a partial picture of distribution of some commonly found chemicals in sediments and to identify chemical concentrations associated with contaminated sediments on a national basis. The following identifies the findings of the national assessment study.

SEDIMENT QUALITY ASSESSMENT AND REGULATION

The Environmental Protection Agency's (EPA) Criteria and Standards Division (CSD) develops and revises criteria, regulations, standards, and guidelines in support of the mandates of the Clean Water Act. The CSD has published water quality criteria for 65 priority pollutants and pollutant categories. These criteria are based on an assessment of water column pollutant concentrations, which--if not exceeded--will protect designated uses of a water body and 95 percent of the aquatic life from adverse effects. The EPA recognizes that while ambient water quality criteria are an important component in assuring a healthy aquatic and human environment, contaminated sediments may be responsible for significant adverse effects even though water quality criteria are being met.

To meet the growing need for a regulatory tool that could be used in assessing and making decisions concerning contaminated sediments, a sediment criteria development effort was undertaken by CSD. One of the first activities conducted in this effort was development of a better understanding of the scope of contaminated sediments on a national basis. A national assessment on sediment quality was conducted. While this national assessment of sediment quality was not intended to be an exhaustive study, it was intended to provide those responsible for directing and focusing sediment criteria development activities with a clearer picture on the extent of the problem.

The assessment was conducted using a total of 48 chemical contaminants representing a diverse group of naturally occurring and anthropogenic materials indicative of compounds of increasing environmental concern. This list is not intended to be exhaustive, but rather is illustrative of the types of chemical data available for sediment.

The seven chemical categories identified are

1. polynuclear aromatic hydrocarbons,
2. pesticides,
3. chlorinated hydrocarbons,
4. mononuclear aromatic hydrocarbons,
5. phthalate esters,
6. metals, and
7. miscellaneous.

Threshold concentrations were then developed to judge differences in the levels of various chemicals in sediments. The majority of these values were calculated using the methodology of the Sediment-Water Equilibrium Partitioning Approach.

To do this the assumption was made that the distribution of a chemical between the organic carbon phase of the sediment and the soluble phase in interstitial water in equilibrium with the solid phase is determined by the organic carbon-water partitioning coefficient (K_{oc}) for the chemical. If the water quality criterion value for the chemical is taken to be the maximum acceptable concentration of the chemical in solution in the interstitial water, then the threshold concentration of the chemical in the bulk sediment is calculated based on the sediment organic carbon-normalized K_{oc} for the chemical.

This allows for a numerical threshold to be established against which available monitoring data can be compared. The nonjudgmental use of this approach allows the distribution of the data set into percentiles above and below the threshold even though one might question the significance of the results relative to the observed integrity or lack of integrity of biological communities.

Threshold values derived from the sediment-water partitioning approach are based on the organic carbon content of the particular sediment (Table 1). These values were adjusted to a whole sediment basis on the assumption that an average sediment contains 4 percent organic carbon. Furthermore, the values are compared to the monitoring data on a dry weight equivalent basis. For several chemicals, for which no acute and chronic water quality criterion values are available, other toxicological endpoints were used.

TABLE 1 Contaminants Reviewed and Developed Threshold Values

Contaminants	Threshold value mg/kg	Contaminants	Threshold value mg/kg
Polynuclear aromatic hydrocarbons		Monoaromatic hydrocarbons	
benzo^{a}pyrene	1,800	toluene	10
naphthalene	42	benzene	1.36
phenanthrene	56	ethylbenzene	5.6
chrysene	460	nitrobenzene	6.6
fluorene	28	dinitrobenzene	0.88
acenaphthene	66		
anthracene	44	Phthalates	
benzo^{a}anthracene	220	butylbenzyl phthalate	220
acentphthalene	24	di-N-butylphthalate	2,000
indeno(1,2,3-CD) pyrene	24,000	diethylphthalate	1.28
benzo^d fluoranthene	5,000	dimethylphthalate	1.96
		Metals	
		arsenic	33
Pesticides		cadmium	31
lindane	0.012	copper	136
DDD	13	lead	132
DDE	28	mercury^b	0.8
aldrin	0.021	zinc	760
isophrone	9.6	chromium^e	25
DDT	0.006	nickel^e	20
chlordane^b	0.020		
toxaphene^b	0.020	Miscellaneous	
heptachlor^b	0.020	cyanide	0.1
Chlorinated hydrocarbons			
hexachloroethane	14.4		
hexachlorobutadiene	1.28		
tetrachloroethylene	1.8		
trichloroethylene	6.4		
dichlorobenzene	2.8		
methylene chloride	2		
PCBs^c	0.28		

NOTES:
^{a}Threshold concentrations are those determined by the EPA/OWRS unless otherwise stated. Criteria for organic contaminants are calculated on the basis of 4 percent organic carbon content of sediment.
^{b}U.S. Geological Survey sediment alert levels.
^{c}Based on criterion for hexachlorobiphenyl.
^{d}The value of 0.8 was not corrected for organic carbon. Correction of this value would have resulted in a mercury concentration of 0.03. which is considerably lower than the concentration of this metal in most sediments.
^{e}EPA Region V guidelines for designating contaminated verses noncontaminated sediments.

The methods used to evaluate the level of various chemicals in sediments of fresh waters, marine waters, and estuaries was consistent to the extent that the data search included the same chemical and selected threshold values.

To compare the monitoring data, the concentrations were divided for convenience into four subranges:

- level 1, less than threshold value;
- level 2, 1-3 times threshold value;
- level 3, 3-10 times threshold value; and
- level 4, greater than 10 times threshold value.

THE DATA BASE

Due to the limitations of the EPA Storage and Retrieval System (STORET) data base for this type of analysis, it is likely that there are numerous additional areas with significant contaminant levels that are not represented in this report. Figures 1, 2, and 3 depict locations where contaminants were at levels 3 and 4.

Freshwater Data

A very large data base was available for evaluation of contamination of sediments in streams, rivers, lakes, and reservoirs. The STORET system was the primary source of data for this effort. Of the group of 48 chemicals identified initially, data on 22 were obtained from the STORET system and over 255,000 data records were processed. No attempt was made to judge the quality of these data or the accuracy and precision of the analytical techniques used to obtain them.

Marine/Estuarine Data

Concentrations of various chemicals in marine and estuarine sediments were obtained from the published literature and from some literature with limited distributions. An additional set of referenced data points was derived from STORET data files. Marine/estuarine STORET data were limited to median concentrations of various chemicals. No ranges of concentrations were given and the data base was sufficient to manipulate data.

OBSERVATIONS AND CONCLUSIONS

Fresh Water

There was a clustering of a variety of chemicals at certain sites rather than a qeneral scatterinq of data. In general, coastal areas were the most noticeably affected regions. New York, New Jersey,

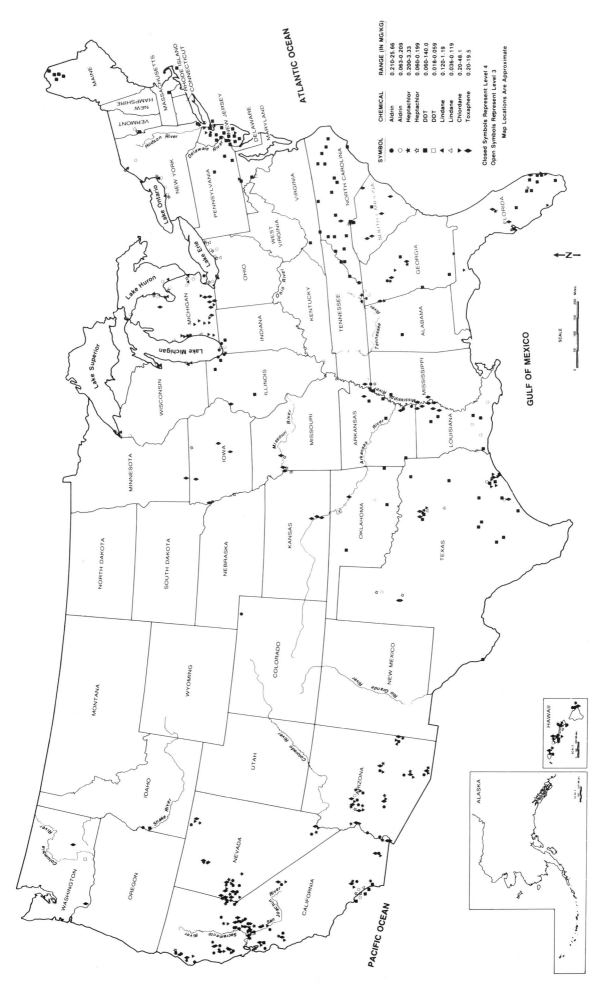

FIGURE 1 Classification levels for pesticides.

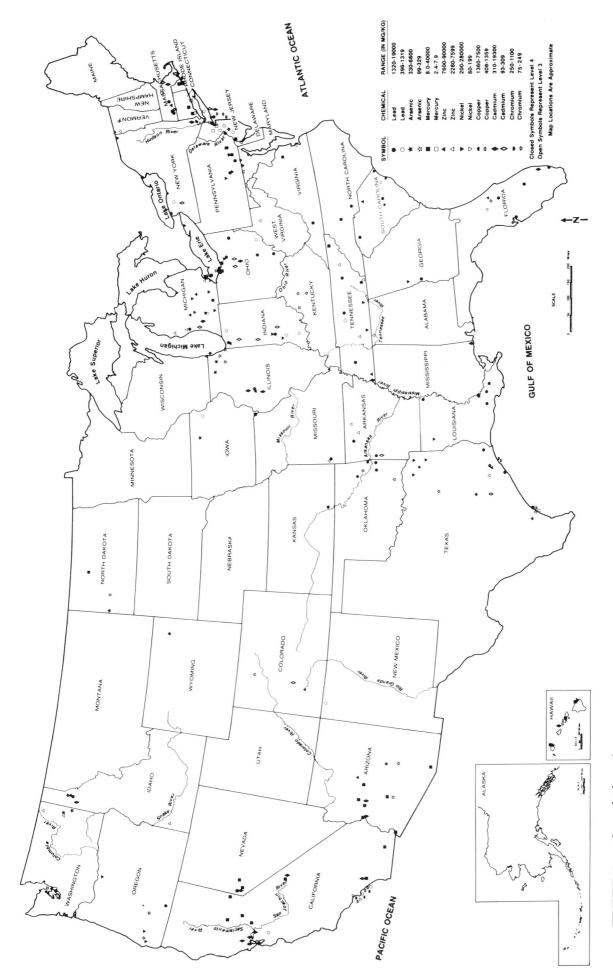

FIGURE 2 Classification levels for metals.

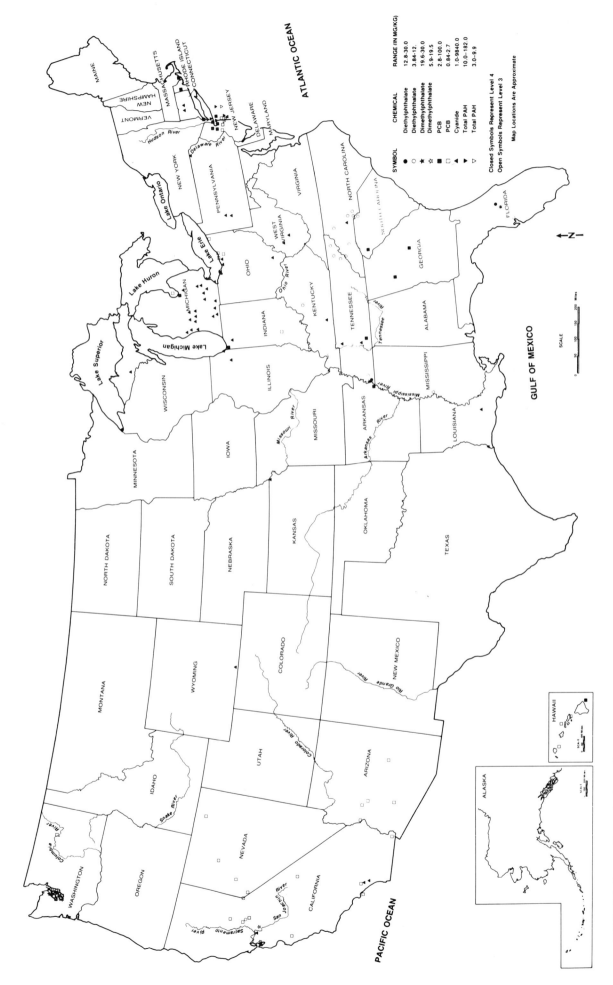

FIGURE 3. Classification levels for Phthalates, cyanide, PCBs total PAHs.

Michigan, Arizona, Nevada, Washington, and California contain areas which were repeatedly identified as contaminated. This distribution indicated a broad spectrum of contamination within certain areas.

The limited analysis of station-by-station data for the top 200 measurements indicate that (1) areas are widely scattered around the country, justifying development of sediment criteria on a national basis, and (2) the highest contamination levels for many chemicals represent potential hot spots rather than general high concentrations over a broad area.

Metal concentrations were, for the most part, classified in level 1. Only nickel, which had 42 percent of the data in the upper three concentration levels, was an exception. In situ and bioassay data suggest a need to reevaluate threshold values derived from water quality criteria.

Almost all of the PAH data were classified in level 1.

A wide span exists between the currently suggested threshold values and even the highest observed concentrations. The biological impact data suggest a need to reevaluate threshold levels.

A significant percent of phthalate data fell in the higher classification levels: 36 percent for diethylphthalate and 35 percent for dimethylphthalate. However, no biological impact data were found to indicate a need to review threshold values.

The vast majority of the pesticide data were in the level 1 range. Chlordane, DDT, and toxaphene had 16, 7, and 14 percent of the data points in the level 2 range, respectively. Biological impact data indicated a need to reevaluate threshold levels.

The PCB data were distributed with 18 percent of the data above level 1. However, only 0.04 percent would be located in level 2 at the highest proposed threshold value using alternative criteria. There is a need for further evaluation using biological impact data.

Much of the available in situ or in vivo data were inappropriate to determine sediment-related toxic effects, because parallel measurements of chemical concentrations and biological species distribution of other biological effects have not been made.

Marine/Estuarine Waters

The principal sites that contained chemicals of interest at high concentrations include Puget Sound waterways, Corpus Christi Harbor, New York Harbor, Baltimore Harbor, Boston Harbor, New Bedford Harbor, Blackrock Harbor, the California sewage outfalls at Palos Verdes, and parts of San Francisco Bay.

Chemicals of major concern were those that exceeded the provisional sediment threshold values of several coastal locations. These chemicals include toxic metals, PAH, PCBs, and DDT. Other chemicals in this inventory of coastal sites did not reach or exceed the first-cut sediment threshold values.

The marine/estuarine survey was based on a very limited data base. A more detailed literature search may reveal additional chemicals of major concern.

Threshold concentrations for chemicals in sediments based on sediment-water equilibrium partitioning are probably set too high for the majority of chemicals considered, most notably for PAH compounds and metals. This overestimation was best illustrated by the discrepancy between biological effects observed in New York Bight sediments, despite corresponding sediment contaminant concentrations of inventoried chemicals that rarely exceed threshold biological effects levels.

ACKNOWLEDGMENT

The material presented in this paper is a summary of material presented in the EPA document "National Perspective on Sediment Quality." This document was assembled in 1985 and in many cases reflects the state of knowledge on toxic sediments at that time. If this document were updated. changes in the conclusions identified above would be likely. For copies of the document or for additional information contact Christopher Zarba at (202) 475-7326.

NATIONAL STATUS AND TRENDS PROGRAM FOR
MARINE ENVIRONMENTAL QUALITY
Aspects Dealing With Contamination in Sediments

Andrew Robertson and Thomas P. O'Connor
National Oceanic and Atmospheric Administration

ABSTRACT

Since 1984, the National Oceanic and Atmospheric Administration has conducted the National Status and Trends Program, which makes systematic measurements of chemical and biological indicators of coastal marine environmental conditions to determine the current status and developing trends in environmental quality of U.S. coastal waters. As part of this program, levels of chemical contaminants in sediments have been measured at least once, and in most cases several times, at 213 locations around the U.S. coast. The program measures concentrations of DDT and its metabolites, 9 other chlorinated pesticides, 8 polychorinated biphenyl congener groupings aggregated by chlorination number, 18 polyaromatic hydrocarbons, 12 trace elements, 4 major elements, total organic carbon, coprostanol, and *Clostridium perfringens* spores, as well as grain-size distribution in sediments from each site. All sampling sites are given showing, for each site, how many and which contaminant concentrations ranked in the top 20 concentrations of the 176 sites with fine-grained sediments. Relatively few sites, all in the vicinity of major coastal cities, were found to have most of the contaminant concentrations in the upper 20.

INTRODUCTION

In order to make well-founded and balanced decisions that provide for allocation and utilization of the nation's coastal and estuarine resources, while assuring continued availability of these resources for future generations, it is necessary to have reliable information concerning the status and trends of environmental quality around our coasts. The Ocean Assessments Division of the National Oceanic and Atmospheric Administration's (NOAA) Office of Oceanography and Marine Assessment initiated the National Status and Trends Program (NS & T) for Marine Environmental Quality in 1984 to provide such information.

The purpose of this program is to make systematic observations on a suite of meaningful indicators of coastal marine environmental conditions in order to determine the current status of and detect any substantial changes occurring in environmental quality of U.S. coastal waters. Because of the level of concern with regard to anthropogenic additions of toxic substances into coastal waters, it was decided to focus this program initially on these substances and their effects.

A three-tiered approach (Figure 1) is used in the NS & T Program to evaluate the status and trends of chemical contaminants and their effects in estuarine and coastal waters. The first tier involves a monitoring program that measures the concentrations of toxic chemicals and certain associated biological effects in biota and sediments from numerous locations around the coasts of this country. This tier of the program is directed at determining existing levels of toxic chemicals and bioeffects at various sites. These data provide a baseline against which future measurements at the same sites may be compared to determine whether or not substantial changes have occurred. This tier also provides data that can be used to evaluate which of the sites have the highest levels of contamination and bioeffects. This tier is intended to provide warnings as to which locations are of greatest concern regarding potential for degradation in environmental quality.

The second tier involves a closer examination of the conditions at the locations that were identified of greatest concern in the first tier. Before initiating detailed field studies as part of tier 3, the available literature information and data relating to the substances and effects of concern in such areas are obtained and synthesized in order to make a more detailed evaluation of existing knowledge concerning the spatial and temporal extent of degradation in environmental quality in these areas. These tier 2 analyses result in hypotheses as to the levels of ecological degradation and other undesirable effects that have occurred, or are in process of occurring, in the various areas of greatest concern.

These hypotheses are then tested in the third tier for those areas where substantial levels of degradation are hypothesized. Measurements, such as detailed examinations of indigenous organisms and bioassays of sediment and water, are made to determine the extent and

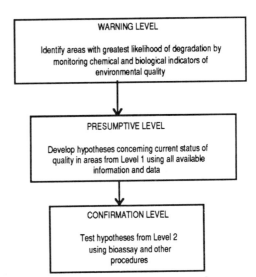

FIGURE 1 Schematic representation of the three-tiered structure of the NS & T Program.

degree of degradation caused by chemical contamination. Those areas where substantial degradation is confirmed are identified as areas where management actions may be needed if the degradation is to be reversed.

The present paper provides a preliminary description of the data being obtained by the NS & T Program, primarily through the first tier monitoring, concerning the distribution of levels of contamination in sediments in U.S. coastal waters and a summary of the procedures used to obtain these data. Examples of program results related to contaminated sediments are also included.

NATIONAL STATUS AND TRENDS MONITORING OF SEDIMENT PROPERTIES

Collection and analysis of contaminant concentrations in sediments is included in the NS & T Program to provide a long-term integrated measure of the comparative levels of the various contaminants at the individual sites. The program also measures the concentrations of contaminants in bivalves and fish for this purpose. Measurements of contaminants in both biota and sediments are included as complimentary measures of comparative levels in the environment. Although the values in biota are of more direct interest and concern, they are more difficult to compare among sites. This is because no single species that is appropriate for such measurements is found all around the coasts of the United States, so several species must be used to obtain national coverage. The two measures complement each other because concentrations in biota, in general, respond more rapidly to changes in contaminant inputs, so they provide integration of the levels of such inputs over shorter time periods than do the levels in the sediments.

Measurements on contamination in sediments are obtained primarily in the first tier of the NS & T Program in which the concentrations of toxic chemicals in biota and associated sediments are measured at numerous locations around the coasts of this country. Such measurements are included in two major components of the first tier of the program. In one, the benthic surveillance component, measurements are made of the concentrations of a wide variety of toxic substances in the livers of benthic fish and in sediments collected in association with them. In the second one, the mussel watch component, concentrations of the same substances in the tissues of sessile molluscs (i.e., mussels and oysters) and associated sediments are measured.

Samples have been collected annually since 1984 at about 50 benthic surveillance sites and since 1986 at about 150 mussel watch locations. Although most locations included in these efforts have been sampled more than once, some sites have been added and omitted each year.

SITE LOCATIONS

Benthic surveillance and mussel watch sampling sites were selected to be, as much as possible with present knowledge, representative of the general conditions within the areas in which they are located.

Efforts were made to avoid locating sampling sites near point sources of contamination, such as outfalls from industrial or sewage treatment plants, because such "hot spot" locations are not considered to be representative. Maps showing the specific sites sampled are included in a report summarizing the NS & T data concerning contamination concentrations in biota (NOAA, 1987).

Surficial sediment samples were collected at three stations at each sampling site, with the stations being spaced over an area within 500 m of the center of the site. At mussel watch sites, if only sediments composed predominantly of sand or larger grain sizes could be found in the area, an attempt was made to find stations with fine-grained sediments up to 2 km from the site center. If still no stations with fine-grained sediments could be found, sand samples from near the site center were used. In the benthic surveillance sampling, no effort was made to find fine-grained sediments if these were not detected within 500 m of the site center. An exception to the general procedure for selecting sediment stations was followed by the benthic surveillance program for the 18 sites along the northeast coast (i.e., from Maine through Virginia). For these sites the stations were generally much farther apart than at the other sites, up to about 5 km.

FIELD AND ANALYTICAL METHODS

In the Benthic Surveillance Project, sediment samples were obtained with a specially constructed box corer or a standard Smith-MacIntyre bottom grab. In the Mussel Watch Project, the samples were obtained with the box corer or with a Kynar-coated Van Veen grab sampler. Three samples were obtained at each of the three stations at a site, resulting in a total of nine samples at each site.

In the Benthic Surveillance Project, a surface skim was taken from the top 3 cm of each sediment sample for analysis for organic substances. The resulting three skim subsamples were composited in the laboratory, so only one analysis for organics was carried out for each station. A small corer was used on deck to get a subsample from the top 3 cm of each box core or grab sample for analysis for trace metals and other elements, and as with the organics, the three subsamples from each station were composited in the laboratory. Two other core subsamples were obtained from each sample, one for analysis of sediment texture and one for storage.

In the Mussel Watch Project, two samples were obtained at each station--one for analysis of organics and the other for analysis of trace metals and other elements--by taking two surface skims from the top 1 cm of each box corer or grab sample and compositing, in the field, the two sets of three subsamples from each station.

Samples for analyses of organic components were stored in Teflon jars or glass jars with lids lined with aluminum foil. Those to be used for analyses of major and trace elements were stored in Teflon jars or ziplock bags. A more detailed presentation of the sampling protocols is included in Shigenaka and Lauenstein (1988).

The sedimentary properties measured in the NS & T Program are listed in Table 1. The methods used for the analysis of organic chemicals in sediments collected in the Benthic Surveillance Project are described in a technical report prepared by NOAA's National Analytical Facility (MacLeod et al., 1985); those for major and trace metals will be described in a report currently in preparation. The methods used for chemical analyses of sediments in the Mussel Watch Project are described in reports to NOAA prepared by Battelle Ocean Sciences (1987) and the Geochemical and Environmental Research Group, Texas A&M University (1988). In addition to undergoing analyses for certain organic compounds and major and trace elements, the sediments were analyzed for two biological properties, *Clostrium perfringens* spores (only in the Benthic Surveillance Project) and coprostanol, that can serve as indicators of the level of contamination with sewage as well as for total organic carbon and grain-size distribution. The measurements for these latter two properties as well as those for the major elements (i.e., aluminum, iron, manganese, and silicon) were included primarily to be used as normalizing factors to help explain the observed distributions of the toxic organics and trace metals.

Quality assurance (QA) protocols were included as an integral part of the NS & T Program. The QA efforts were designed to produce nationally uniform analytical results of known and accepted quality, thereby ensuring comparability among data sets. Attainment of this goal involved five major activities:

1. developing and using standardized field sampling procedures and analytical protocols;
2. conducting interlaboratory comparisons of analytical methods;
3. conducting periodic QA workshops;
4. developing Standard Reference Materials and Interim Reference Materials for marine sediments and tissues; and
5. developing and using a standardized data base for QA data and information.

RESULTS

The chemical measurements on the benthic surveillance sediment samples collected in 1984 and 1985 have been completed, as have the measurements on the mussel watch sediment samples from 1986 and 1987. A complete listing of the data resulting from these measurements and an analysis of these data and their significance is being prepared for publication. The preliminary results discussed in the present report are based on these data, but with results from different years at a site combined. Year-to-year differences are not considered in this preliminary look at the results in order to focus on geographical distribution of contamination levels in sediments.

The concentrations of contaminants in sediments are influenced by a number of factors besides inputs. As most important contaminants are strongly associated with particle surfaces, the particle-size distribution at a site is an especially important factor in determining

TABLE 1 Chemicals measured in Sediments as Part of the NS & T Program

DDT and its metabolites[a]	Polyaromatic hydrocarbons[d]	Major elements	
o,p'-DDD	Acenaphthene	Al	Aluminum
p,p'-DDD	Anthracene	Fe	Iron
o,p'-DDE	Bena[a]anthracene	Mn	Manganese
p,p'-DDE	Benzo[a]pyrene	Si	Silicon
o,p'-DDT	Benzo[e]pyrene		
p,p'-DDT	Biphenyl	Trace elements	
	Chrysene		
	Dibenzanthracene		
Chlorinated pesticides other than DDT[b]	2,6-Dimethylnaphthalene	Sb	Antimony
	Fluoranthene	As	Arsenic
	Fluorene	Cd	Cadmium
Aldrin	1-Methylnaphthalene	Cr	Chromium
Alpha-chlordane	2-Methylnaphthalene	Cu	Copper
Trans-nonachlor	1-Methylnaphthalene	Pb	Lead
Dieldrin	Naphthalene	Hg	Mercury
Heptachlor	Perylene	Ni	Nickel
Heptachlor epoxide	Phenanthrene	Se	Selenium
Hexachlorobenzene	Pyrene	Ag	Silver
Lindane (gamma-BHC)		Sn	Tin
Mirex		Zn	Zinc
	Other parameters		
Polychlorinated biphenyls[c]	*Clostridium perfringens* spores		
	Coprostanol		
	Grain size		
Dichlorobiphenyls	Total organic carbon		
Trichlorobiphenyls			
Tetrachlorobiphenyls			
Pentachlorobiphenyls			
Hexachlorobiphenyls			
Heptachlorobiphenyls			
Octachlorobiphenyls			
Nonachlorobiphenyls			

NOTES:
[a] Combined and reported in this paper as total DDT (tDDT).
[b] Combined and reported in this paper as total chlorinated pesticides other than DDT (tChlP)>
[c] Combined and reported in this paper as total polychlorinated biphenyls (tPCB).
[d] Combined and reported in this paper as total polyaromatic hydrocarbons (tPAH).

sedimentary contaminant levels, with fine sediments tending to concentrate the contaminants. To compensate for this influence, the sediment data collected by the NS & T Program are normalized by dividing the raw contaminant concentrations by the fraction by weight of the sediment particles in the sample that are less than 64 μ in diameter (i.e., the fine- grained or silt and clay fraction). This method of normalizing can yield misleading results for sediments that are composed primarily of sand or larger particles, however. When such sediments contain detectable levels of contaminants, these levels will often appear very high when the values are normalized based on the small amounts of fine sediments present. To avoid such distortions, contaminant data based on analyses from sediments with less than 20 percent fine-grained sediments are omitted from further consideration.

There are 213 sites around the coasts of the United States from which sediments were collected and analyzed by the NS & T Program. For 176 of these sites, there are contaminant data for at least one sample composed of 20 percent or more fine-grained sediments. In fact most of these sites were occupied in two different years and yielded three samples with sediments with 20 percent or more of fines both times they were visited.

For each contaminant the concentrations at the individual sites have been plotted on a bar graph in descending order of concentration, and examples are presented in Figure 2 for mercury and Figure 3 for total PCBs. These plots have shown that the ranges in concentrations vary quite greatly among the various contaminants, as is summarized in Table 2. These ranges can provide an indication of the amount of influence human activities are having on the presence of the contaminant in the environment, with contaminants that exist naturally in the marine coastal environment only at very low levels or not at all tending to be most subject to and indicative of human activity. Table 2 lists, in descending order, the magnitude of the concentration ranges for the various contaminants. As expected, the substances that do not occur naturally have wide ranges (i.e., the synthetic organic compounds, total PCB, total DDT, and total chlorinated pesticides other than DDT). Wide ranges (factors greater than 50) are also indicated for total PAH, mercury, silver, and tin. While natural scales of variabilities in concentrations of these substances are not known, these are the naturally occurring substances that seem the most affected by human activities. On the other hand, the ranges for arsenic, zinc, total organic carbon (TOC), selenium, nickel, lead, and chromium vary by less than a factor of 20 over the national grid and therefore seem not so greatly affected by, nor indicative of, human perturbations.

Table 3 presents an overall summary of the levels of contamination at the NS & T national network of sites. It includes a listing of the locations at which sediments have been gathered and analyzed by the program. For each site the number of contaminants for which the concentration is in the upper 20 of the 176 concentrations found at NS & T sites is listed, as are the specific contaminants that were present at this level. It can be noted that relatively few sites have most of the contaminant concentrations that are in the upper 20, and that most sites have few if any contaminants in the upper 20. Further, the sites where the upper 20 contaminant values are concentrated tend to lie close to one another along the coast in the vicinity of major coastal cities, e.g., Boston, New York, San Diego, and Los Angeles. These observations provide verification for the generally held assumption that contaminated coastal sediments are most likely to be found near major metropolitan centers.

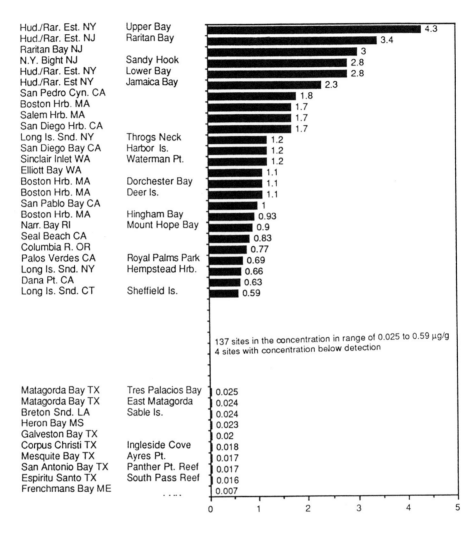

FIGURE 2 Concentrations of mercury in sediments from NS & T sites around the coast of the United States (μg/g dry weight normalized for % fines).

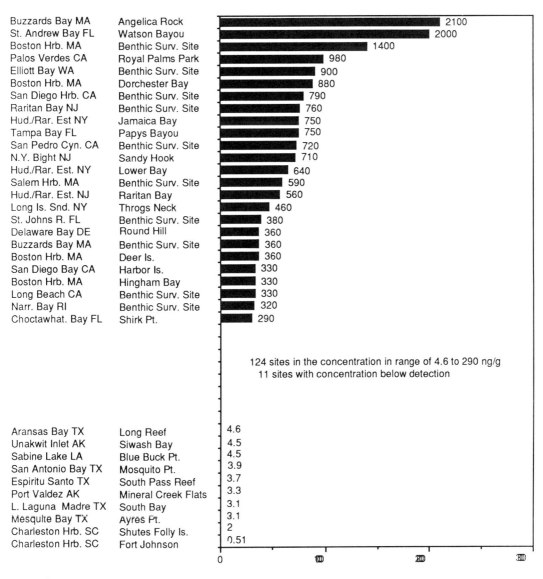

FIGURE 3 Concentrations of total PCB in sediments from NS & T sites around the coast of the United States (λ/g dry weight normalized for % normalized fines).

TABLE 2 Ranges Defined by Ratios of High to Low Concentrations of Contaminants in Sediments

Chemical	Number of sites with detectable concentrations[a]	Range (ratio of 15th highest to 5th lowest)[b]
		Range > 1000
tDDT	164	1500
		100 < Range < 1000
tPAH	162	290
tPCB	159	280
tCHlP	161	150
Mercury	171	110
		50 < Range < 100
Silver	175	93
Tin	156	60
		20 < Range < 50
Cadmium	175	32
Antimony	138	26
Copper	175	22
		10 < Range < 20
Lead	174	18
TOC	176	16
Chromium	168	16
Selenium	146	13
Nickel	175	13
Zinc	172	10
		Range < 10
Arsenic	175	7.5

NOTES:
[a]Only the 176 sites with at least one sample displaying a detectable concentration and > 20 percent fine-grained material have been considered. In some cases the fact that the number in this column is less than 176 is due to the chemical not having been measured rather than its not having been detected.
[b]Ranges calculated on the basis of fifth highest and fifth lowest concentrations to avoid possible distortions from extremely high or low concentrations.

TABLE 3 Locations of All NS & T Sites with Mean Percent Fines (% f) in Samples with > 20 Percent Fine-grained Sediments, Total Number (T) of Contaminant Concentrations (Normalized to Fines) Ranking in the Top 20, and the Specific Contaminants with These Ranks

SITE LOCATION		%f	T	CONTAMINANTS IN TOP 20 (RANK)
Machias Bay ME	Benthic Surveillance Site	68	0	
Frenchmans Bay ME	Benthic Surveillance Site	93	0	
Penobscot Bay ME	Benthic Surveillance Site	97	0	
Penobscot Bay ME	Sears Is.	90	0	
Penobscot Bay ME	Pickering Is.	54	2	tPAH(19),TOC(9)
Casco Bay ME	Benthic Surveillance Site	73	0	
Merrimack R. MA	Benthic Surveillance Site	nfgs [a]	0	
Salem Harbor MA	Benthic Surveillance Site	69	12	Ag(17),Cd(2),Cr(1),Hg(9),Pb(2), Sb(15), Se(9), Sn(8),Zn(12),tChlP(10), tPCB(14), tPAH(4)
Cape Ann MA	Straightsmouth Is.	27	3	Pb(20), Sb(9), Sn(20)
Boston Harbor MA	Deer Is.	78	11	Ag(13),Cr(12),Cu(14),Hg(16),Pb(13), SB(4),Sn(5),tChlP(19),tPCB(20) tPAH(20),TOC(12)
Boston Harbor MA	Dorchester Bay	85	13	Ag(15),Cd(19),Cr(14),Cu(12),Hg(14), Pb(12),Sb(5),Sn(6),TChlP(8),tDDT(17), tPCB(6),TPAH(11),TOC(13)
Boston Harbor MA	Hingham Bay	27	8	Ag(9),Cr(16),Hg(18),Pb(14),Sb(1), Sn(10),tChlP(16)TOC(18)
Boston Harbor MA	Brewster Is.	nfgs	0	
Boston Harbor MA	Benthic Surveillance Site	63	14	Ag(1),Cd(5),Cr(9),Cu(2),Hg(8),Pb(8), Sb(3),Se(18),Sn(1),Zn(6),TChlP(4), tDDT(6),tPCB(3),tPAH(1)
Buzzards Bay MA	Round Hill	65	1	tPCB(19)
Buzzards Bay MA	Angelica Rock	29	4	Ag(18),tChlp(14),tPCB(1),tPAH(9)
Buzzards Bay MA	Goosebury Neck	37	0	
Buzzards Bay MA	Benthic Surveillance Site	73	0	
Narragansett Bay RI	Mount Hope Bay	91	2	Hg(19),Sn(18)
Narragansett Bay RI	Conanicut Is.	60	0	
Narragansett Bay RI	Dyer Is.	38	3	Sb(16),Sn(19),tPAH(16)
Narragansett Bay RI	Benthic Surveillance Site	69	1	Sn(17)
Block Is. RI	Block Is.	71	1	TOC(15)
E. Long Is. Sound CT	Benthic Surveillance Site	nfgs	0	
Long Is. Sound CT	Connecticut R.	50	0	
Long Is. Sound CT	New Haven	nfgs	0	
Long Is. Sound CT	Housatonic R.	nfgs	0	
Long Is. Sound CT	Sheffield Is.	63	2	Cu(15),tPAH(8)
W. Long Is. Sound NY	Benthic Surveillance Site	80	2	Cu(16),SN(15)
Long Is. Sound NY	Huntington Harbor	nfgs	0	
Long Is. Sound NY	Port Jefferson	nfgs	0	
Long Is. Sound NY	Mamaroneck	74	4	Cd(16),Cu(19),Pb(19),tChlP(20)
Long Is. Sound NY	Hempstead Harbor	90	7	Ag(8),Cd(13),Cu(10),Pb(15),Zn(18), tChlP(9),TOC(6)
Long Is. Sound NY	Throgs Neck	74	11	Ag(10),Cd(18),Cu(9),Hg(11),Pb(10), Zn(20),tChlP(12),tDDT(14),tPCB(16), tPAH((7),TOC(3)
Hudson/Raritan Estuary NY	Jamaica Bay	64	16	Ag(7),As(18),Cd(14),Cr(19),Cu(11), Hg(6),Pb(7),Sb(8),Se(13),Sn(3),Zn(17), tChlP(2),tDDT(11),tPCB(9),tPAH(18), TOC(8)

NOTE
[a] nfgs = No Fine-grained Sediments

TABLE 3 (Cont.)

Location	Site	Value	N	Analytes
Hudson/Raritan Estuary NY	Upper Bay	77	8	Ag(14),As(13),Cu(18),Hg(1),Pb(11), Sb(10),Sn(7),tPAH(2),
Hudson/Raritan Estuary NY	Lower Bay	66	16	Ag(2),As(19),Cd(7),Cr(17),Cu(6),Hg(5), Pb(3),Sb(11),Se(12),Sn(13),Zn(10), tChlP(6),tDDT(12),tPCB(13),tPAH(5), TOC(7)
Hudson/Raritan Estuary NJ	Raritan Bay	70	14	Ag(3),As(12),Cd(12),Cu(5),Hg(2),Pb(1), Sb(7),Se(8),Sn(2),Zn(2),tChlP(11), tDDT(16),tPCB(15),tPAH(13)
Raritan Bay NJ	Benthic Surveillance Site	77	15	Ag(5),As(7),Cd(4),Cr(18),Cu(4),Hg(3), Pb(5),Sb(18),Se(3),Sn(4),Zn(1), tChlP(15),tDDT(18),tPCB(8),tPAH(17)
N.Y. Bight NJ	Sandy Hook	65	16	Ag(4),As(9),Cd(8),Cr(15),Cu(8),Hg(4), Pb(4),Sb(6),Se(4),Sn(14),Zn(5), tChlP(7),TDDT(15),TPCB(12),tPAH(12), TOC(5)
N.Y. Bight NJ	Long Branch	nfgs	0	
N.Y. Bight NJ	Shark R.	nfgs	0	
Moriches Bay NY	Tuthill Pt.	57	1	TOC(10)
Great Bay NJ	Benthic Surveillance Site	71	0	
Delaware Bay DE	Benthic Surveillance Site	46	2	tChlP(5),tPCB(18)
Delaware Bay DE	False Egg Is. Pt.	41	1	TOC(2)
Delaware Bay DE	Ben Davis Pt. Shoal	56	0	
Delaware Bay DE	Arnolds Pt. Shoal	75	0	
Delaware Bay DE	Kelly Is.	58	0	
Upper Ches. Bay MD	Benthic Surveillance Site	73	2	Ni(17),Zn(14)
Chesapeake Bay MD	Mountain Pt. Bar	98	2	Zn(9),TOC(17)
Chesapeake Bay MD	Hackett Pt. Bar	98	0	
Chesapeake Bay MD	Hog Pt.	nfgs	0	
Mid. Chesapeake Bay VA	Benthic Surveillance Site	48	0	
Chesapeake Bay VA	Ingram Bay	77	0	
Chesapeake Bay VA	Cape Charles	71	0	
Chesapeake Bay VA	Stony Pt.	nfgs	0	
Chesapeake Bay VA	Dandy Pt.	46	0	
Lower Ches.Bay VA	Benthic Surveillance Site	50	0	
Chincoteague Bay VA	Chincoteague Inlet	nfgs	0	
Quinby Inlet VA	Upshur Bay	38	0	
Roanoke Sound VA	John Creek	nfgs	0	
Pamlico Sound NC	Wysoching Bay	nfgs	0	
Pamlico Sound NC	Benthic Surveillance Site	78	0	
Cape Fear NC	Battery Is.	56	2	As(8),TOC(11)
Charleston Harbor SC	Fort Johnson	81	0	
Charleston Harbor SC	Shutes Folly Is.	51	0	
Charleston Harbor SC	Benthic Surveillance Site	80	0	
Savannah R. Estuary GA	Tybee Is.	52	0	
Sapelo Sound GA	Sapelo Is.	nfgs	0	
Sapelo Is. GA	Benthic Surveillance Site	49	1	Sn(11)
St. Johns R. FL	Chicopit Bay	76	1	TOC(4)
St. Johns R. FL	Benthic Surveillance Site	47	1	tPCB(17)
Matanzas R. FL	Cresent Beach	nfgs	0	
Biscayne Bay FL	Princeton Canal	85	1	TOC(1)
Everglades FL	Faka Union Bay	82	0	
Rookery Bay FL	Henderson Creek	72	0	
Naples Bay FL	Naples Bay	60	0	
Charlotte Harbor FL	Bird Is.	44	0	
Charlotte Harbor FL	Benthic Surveillance Site	26	0	
Tampa Bay FL	Benthic Surveillance Site	49	1	Se(14)
Tampa Bay FL	Mullet Key Bayou	25	2	Se(2),tChlP(3)
Tampa Bay FL	Cockroach Bay	nfgs	0	
Tampa Bay FL	Hillsborough Bay	54	2	Pb(17),Se(5)
Tampa Bay FL	Papys Bayou	53	2	Se(19),tPCB(10)
Cedar Key FL	Black Pt.	46	1	Se(6)

TABLE 3 (Cont.)

Location	Site	Value	Count	Contaminants
Apalachicola Bay FL	Cat Pt. Bar	59	0	
Apalachicola Bay FL	Dry Bar	50	0	
Apalachicola Bay FL	Benthic Surveillance Site	76	0	
St. Andrew Bay FL	Watson Bayou	46	4	tChlP(18),tDDT(10),tPCB(2),tPAH(3)
Choctawhatchee Bay FL	Shirk Pt.	52	5	Pb(6),Se(7),tChlP(1),tDDT(3),tPAH(6),
Choctawhatchee Bay FL	Off Santa Rosa	66	2	As(4),tDDT(7)
Pensacola Bay FL	Benthic Surveillance Site	81	1	Se(11)
Pensacola Bay FL	Indian Bayou	34	0	
Mobile Bay AL	Cedar Pt. Reef	74	0	
Mobile Bay AL	Benthic Surveillance Site	93	0	
Round Is. MS	Benthic Surveillance Site	57	0	
Heron Bay MS	Benthic Surveillance Site	58	0	
Miss. Sound MS	Pascagoula Bay	60	0	
Miss. Sound MS	Biloxi Bay	74	1	tPAH(15)
Miss. Sound MS	Pass Christian	76	0	
Miss. Delta LA	Benthic Surveillance Site	77	0	
Lake Borgne LA	Malheureux Pt.	77	0	
Breton Sound LA	Sable Is.	88	0	
Breton Sound LA	Bay Garderne	28	0	
Barataria Bay LA	Bayou St. Denis	83	0	
Barataria Bay LA	Middle Bank	42	1	Se(10)
Barataria Bay LA	Benthic Surveillance Site	49	0	
Terrebonne Bay LA	Lake Felicity	77	0	
Terrebonne Bay LA	Lake Barre	86	0	
Caillou Lake LA	Caillou Lake	67	0	
Atchafalaya Bay LA	Oyster Bayou	83	0	
Vermillion Bay LA	Southwest Pass	82	0	
J. Harbor Bayou LA	Joseph Harbor Bay	70	0	
Calcasieu Lake LA	St. Johns Is.	84	0	
Sabine Lake LA	Blue Buck Pt.	57	0	
E. Cote Blanche LA	South Pt.	nfgs	0	
Galveston Bay TX	Hanna Reef	80	0	
Galveston Bay TX	Yacht Club	62	1	Se(16)
Galveston Bay TX	Todd's Dump	67	0	
Galveston Bay TX	Confederate Reef	53	0	
Galveston Bay TX	Benthic Surveillance Site	56	0	
Matagorda Bay TX	East Matagorda	52	0	
Matagorda Bay TX	Tres Palacios Bay	60	0	
Matagorda Bay TX	Gallinipper Pt.	74	0	
Matagorda Bay TX	Lavaca R. Mouth	63	0	
Espiritu Santo TX	South Pass Reef	87	0	
Espiritu Santo TX	Bill Days Reef	24	0	
San Antonio Bay TX	Mosquito Pt.	48	0	
San Antonio Bay TX	Panther Pt. Reef	46	0	
San Antonio Bay TX	Benthic Surveillance Site	58	0	
Mesquite Bay TX	Ayres Pt.	91	0	
Copano Bay TX	Copano Reef	96	0	
Aransas Bay TX	Long Reef	45	0	
Corpus Christi TX	Ingleside Cove	47	0	
Corpus Christi TX	Neuces Bay	56	0	
Corpus Christi Bay TX	Benthic Surveillance Site	74	0	
Lower Laguna Madre TX	South Bay	56	0	
Lower Laguna Madre TX	Benthic Surveillance Site	33	0	
Imperial Beach CA	Imperial Beach	nfgs	0	
San Diego Bay CA	Benthic Surveillance Site	34	2	As(20),Cd(9)
San Diego Bay CA	Harbor Is.	29	5	As(1),Cu(13),Hg(12),Pb(18),Zn(8)
San Diego Harbor CA	Benthic Surveillance Site	66	7	Cu(1),Hg(10),Pb(16),Sn(12),Zn(3), tPCB(7),tPAH(14)
Pt. Loma CA	Lighthouse	31	1	As(17)
Mission Bay CA	Ventura Bridge	nfgs	0	
La Jolla CA	Pt. La Jolla	57	0	
Oceanside CA	Beach Jetty	79	1	tDDT(19)

TABLE 3 (Cont.)

Location	Site			Contaminants
Dana Pt. CA	Benthic Surveillance Site	33	1	As(14)
Newport Beach CA	Balboa Channel Jetty	51	1	tDDT(20)
Anaheim Bay CA	West Jetty	58	0	
Seal Beach CA	Benthic Surveillance Site	56	2	Ag(19),Hg(20)
Long Beach CA	Benthic Surveillance Site	63	5	Cd(15),Pb(9),Zn(16),tChlP(17),tDDT(9)
San Pedro Bay CA	Benthic Surveillance Site	91	0	
San Pedro Canyon CA	Benthic Surveillance Site	26	11	Ag(11),Cd(3),Cr(7),Cu(17),Hg(7),Ni(19),Se(1),Sn(9),Zn(4),tDDT(2),tPCB(11)
San Pedro Harbor CA	Fishing Pier	92	4	Cd(11),Cu(7),tDDT(4),TOC(20)
Palos Verdes CA	Royal Palms State Park	59	9	Ag(12),Cd(1),Cr(13),Cu(20),Sn(16),Zn(15),tChlP(13),tDDT(1),tPCB(4)
Santa Catalina Is. CA	Bird Rock	nfgs	0	
Santa Monica Bay CA	Benthic Surveillance Site	nfgs	0	
Marina Del Ray CA	South Jetty	40	3	Ag(20),As(10),tDDT(8)
Pt. Dume CA	Pt. Dume	36	2	As(16),tDDT(5)
Santa Cruz Is. CA	Fraser Pt.	nfgs	0	
Pt. Santa Barbara CA	Pt. Santa Barbara	40	3	As(2),Cd(17),tDDT(13)
Pt. Conception CA	Pt. Conception	nfgs	0	
San Luis Obispo Bay CA	Pt. San Luis	nfgs	0	
San Simeone Pt. CA	San Simeone Pt.	nfgs	0	
Pacific Grove CA	Lovers Pt.	nfgs	0	
Monterey Bay CA	Pt. Santa Cruz	28	0	
Monterey Bay CA	Benthic Surveillance Site	nfgs	0	
S. San Francisco Bay CA	Benthic Surveillance Site	nfgs	0	
SouthamptonShoal CA	Benthic Surveillance Site	nfgs	0	
Castro Bay CA	Benthic Surveillance Site	nfgs	0	
Oakland Estuary CA	Benthic Surveillance Site	91	1	Ni(12)
Hunters Pt. CA	Benthic Surveillance Site	74	2	Cr(4),Ni(6)
San Francisco Bay CA	Dunbarton Bridge	91	1	Ni(14)
San Francisco Bay CA	San Mateo Bridge	90	1	Ni(10)
San Francisco Bay CA	Emeryville	93	2	Ni(11),Se(15)
San Pablo Bay CA	Benthic Surveillance Site	35	4	Cr(3),Hg(17),Ni(1),Zn(13)
San Pablo Bay CA	Semple Pt.	68	3	As(5),Cr(10),Ni(3)
San Pablo Bay CA	Pt. St. Pedro	90	1	Ni(9)
Tomales Bay CA	Spanger's Restaurant	97	1	Ni(4)
Bodega Bay CA	Bodega Bay Entrance	nfgs	0	
Bodega Bay CA	Benthic Surveillance Site	nfgs	0	
Pt. Arena CA	Pt. Arena	nfgs	0	
Pt. Delgada CA	Shelter Cove	nfgs	0	
Humboldt Bay CA	Jetty	nfgs	0	
Humboldt Bay CA	Benthic Surveillance Site	31	4	As(3),Cr(2),Ni(2),Sb(12)
Pt. St. George OR	Pt. St. George	nfgs	0	
Coos Bay OR	Benthic Surveillance Site	46	2	As(11),Cd(20)
Coos Bay OR	Coos Head	23	4	As(6),Cr(6),Ni(7),Se(17)
Coos Bay OR	Russell Pt.	33	2	Cr(11),Ni(15)
Yaquina Bay OR	Oneata Pt.	51	1	TOC(16)
Yaquina Head OR	Yaquina Head	34	1	Cr(5)
Tillamook Bay OR	Hobsonville Pt.	30	4	As(15),Cr(8),Ni(8),TOC(14)
Columbia R. OR	Youngs Bay	31	2	Sb(13),Zn(19)
Columbia R. OR	Benthic Surveillance Site	27	4	Ag(6),Cd(6),Se(20),Zn(11)
Gray's Harbor WA	Westport Jetty	nfgs	0	
Strait Juan de Fuca WA	Neah Bay	49	3	Cr(20),Sb(20),TOC(19)
South Puget Sound WA	Budd Inlet	98	1	Sb(17)
Nisqually Reach WA	Benthic Surveillance Site	nfgs	0	
Commencement Bay WA	Benthic Surveillance Site	81	1	Ag(16)
Commencement Bay WA	Tahlequah Pt.	87	1	Sb(14)
Elliott Bay WA	Four-Mile Rock	nfgs	0	
Elliott Bay WA	Benthic Surveillance Site	46	8	Cd(10),Cu(3),Hg(15),Ni(16),Sb(19),Zn(7),tPCB(5),tPAH(10)

TABLE 3 (Cont.)

Sinclair Inlet WA	Waterman Pt.	63	3	Hg(13),Ni(20),Sb(2)
Whidbey Is. WA	Possession Pt.	95	0	
Bellingham Bay WA	Squalicum Marina Jetty	98	1	Ni(5)
Pt. Roberts WA	Pt. Roberts	79	0	
Lutak Inlet AK	Benthic Surveillance Site	89	0	
Nahku Bay AK	Benthic Surveillance Site	nfgs	0	
Unakwit Inlet AK	Siwash Bay	82	0	
Port Valdez AK	Mineral Creek Flats	100	0	
Oliktok Pt. AK	Benthic Surveillance Site	nfgs	0	
Prudhoe Bay AK	Benthic Surveillance Site	34	1	Ni(18)
Barber's Pt. HI	Barber's Pt. Boat Basin	48	1	Ni(13)
Honolulu Harbor HI	Keehi Lagoon	47	0	

REFERENCES

Battelle Ocean Sciences. 1987. Phase 2, Work/Quality Assurance Project Plan for Contract No. 50-DGNC-5-0263, Collection of Bivalve Molluscs and Surficial Sediments and Performance of Analyses for Organic Chemicals and Toxic Trace Elements. Report to National Oceanic and Atmospheric Administration. Duxbury, Mass.: Battelle Ocean Sciences. 111 pp. + Appendices A-O.

MacLeod, W. D., Jr., D. W. Brown, A. S. Friedman, D. G. Burrows, O. Maynes, R. Pearce, C. A. Wigren, and R. G. Bogar. 1985. Standard Analytical Procedures of the NOAA National Analytical Facility, 1985-1986: Extractable Toxic Organic Compounds, 2nd edition. NOAA Tech. Memo. NMFS F/NWC-92. Rockville, Md.: NOAA. 121 pp.

National Oceanic and Atmospheric Administration, Ocean Assessments Division. 1987. National Status and Trends Program for Marine Environmental Quality: Progress Report--A Summary of Selected Data on Chemical Contaminants in Tissues Collected During 1984, 1985, and 1986. NOAA Tech. Memo. NOS OMA 38. Rockville, Md.: NOAA. 23 pp. + Appendices A-E.

Shigenaka, G. and G. G. Lauenstein. 1988. National Status and Trends Program for Marine Environmental Quality: Benthic Surveillance and Mussel Watch Projects Overview. NOAA Tech. Memo. NOS OMA 40. Rockville, Md.: NOAA. 12 pp.

Texas A&M University, Geochemical and Environmental Research Group. 1988. Second Annual Report, Analyses of Bivalves and Sediments for Organic Chemicals and Trace Elements. Report to National Oceanic and Atmospheric Administration. College Station, Tx.: Texas A&M Research Foundation.

CLASSIFICATION OF CONTAMINATED SEDIMENTS

USE OF THE APPARENT EFFECTS THRESHOLD APPROACH (AET) IN CLASSIFYING CONTAMINATED SEDIMENTS

Robert Barrick, Harry Beller, Scott Becker, and Thomas Ginn
PTI Environmental Services

ABSTRACT

The Apparent Effects Threshold (AET) approach is a tool for deriving sediment quality values for a range of biological indicators used to assess contaminated sediments. The AET is the contaminant concentration in sediment above which adverse effects are always expected for a particular biological indicator. Application of this approach in 13 embayments of Puget Sound, Washington is described. Approximately 85 percent of the 198 benthic infauna stations and 283 amphipod bioassay stations are in accordance with the predictions of the proposed AET values for these indicators (i.e., they do not exhibit adverse effects when all concentrations are less than AET values, and do exhibit adverse effects at chemical concentrations above the AET values). Similarly, approximately 95 percent of 56 oyster larvae bioassay stations and 56 Microtox bioassay stations are in accordance with predictions of AET for these indicators evaluated in a single urban bay and associated reference area (Commencement Bay and Carr Inlet; additional data for other embayments are not available). The integration of AET as one tool for environmental decision making is discussed, including the need for more than a single number that defines "clean sediments." Specific management requirements may dictate that low sediment quality values (e.g., the lowest AET for a range of biological indicators) be applied to ensure sensitive identification of potential problems, while high values (e.g., the highest AET for a range of biological indicators) be applied to ensure that remedial action is efficiently focused on problem sediments for which there is a preponderance of evidence.

INTRODUCTION

Sediments are a primary reservoir of contaminants released by industrial, commercial, and residential activities in coastal urban bays. The management of contaminated sediments requires that biological or chemical criteria be developed to distinguish effects or concentration levels above which the sediments are considered to be a problem. Regulatory sediment criteria for defining "clean sediments" have not yet been adopted, but the assessment of toxic effects associated with contaminated sediments has been approached by environmental scientists in two general ways (Figure 1).

The first general approach is based on empirical relationships between laboratory sediment bioassays, in situ biological effects observed in organisms associated with sediments, and chemical concentrations measured in sediments. Examples of this "effects-based" approach include the Sediment Quality Triad (Long and Chapman, 1985), the Apparent Effects Threshold (AET) (Barrick et al., 1985), and the Screening Level Concentration (SLC) (Battelle, 1986). The second approach emphasizes theoretical models to predict the partitioning of sediment contaminants to interstitial water (a major exposure pathway for organisms associated with sediments). The predicted interstitial water concentrations are then compared to water quality criteria based on laboratory measurements of biological effects (e.g., the equilibrium partitioning approach; Pavlou, 1987). None of the available approaches is fully capable of addressing all concerns over interactive effects among chemicals; hence, field verification using diverse environmental samples is important to the evaluation of each approach.

The following presentation summarizes the concept of the AET approach and its application and verification in multiple areas of Puget Sound, Washington. The information has been abstracted from a manuscript in preparation (Barrick et al., in prep.). The AET approach assumes a dose-response relationship between increasing chemical contamination and biological effects. Specifically, for each chemical of concern, the AET is the chemical concentration in sediments above which statistically significant biological effects are always expected for one or more biological effects indicator. AET can be developed for any measured chemical (organic or inorganic) that spans a wide concentration range in the data set used to generate AET. The AET concept

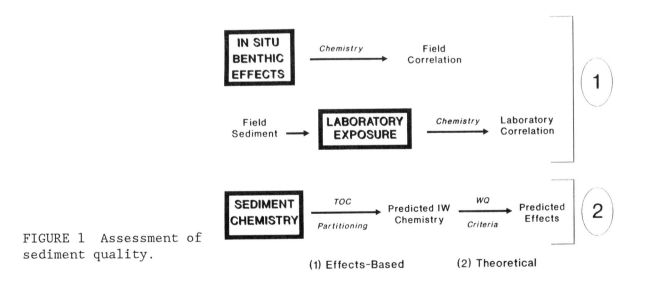

FIGURE 1 Assessment of sediment quality.

can be applied to matched field data for sediment chemistry and any observable biological effects (e.g., bioassays, benthic infaunal abundance, bioaccumulation).

By using these different indicators, application of the resulting sediment quality values addresses a wide range of biological effects in the management of contaminated sediments. A single biological test or single concentration for a chemical may not always define "acceptable contamination." Hence, a range of sediment quality values such as provided by the AET approach may be necessary to serve the needs of different programs.

DESCRIPTION OF AET APPROACH

The focus of the AET approach is to identify concentrations of contaminants that are associated exclusively with sediments exhibiting statistically significant biological effects relative to reference sediments. The calculation of AET for each chemical and biological indicator is straightforward:

1. collect "matched" chemical and biological effects data--conduct chemical and biological effects testing on subsamples of the same field sample (because of subsampling concerns, benthic infaunal analyses may require chemical analyses to be conducted on a separate sediment sample collected concurrently with the chemical sample);
2. determine "impacted" and "nonimpacted" stations--statistically test the significance of adverse biological effects relative to suitable reference conditions for each sediment sample and biological indicator;
3. determine AET using only "nonimpacted" stations--for each chemical, determine the AET for a given biological indicator as the highest *detected* concentration among sediment samples that do not exhibit statistically significant effects (if all values for the chemical are always undetected in these samples, no AET can be established for that chemical and biological indicator);
4. check for tentative AET--verify that statistically significant biological effects are observed at a chemical concentration higher than the AET, otherwise the AET is only a tentative minimum estimate (or may not exist.)

A pictorial representation of the AET approach for two chemicals is presented in Figure 2 based on amphipod bioassay results in Puget Sound. Two subpopulations of all sediments analyzed for chemistry and subjected to an amphipod bioassay are represented by bars in the figures, and include

- sediments that did not exhibit statistically significant (P > 0.05) amphipod toxicity ("nonimpacted" stations), and
- sediments that exhibited statistically significant amphipod toxicity in bioassays ("impacted" stations).

FIGURE 2 The AET approach applied to sediments tested for lead and 4-methyl phenol concentrations and amphipod mortality during bioassays.

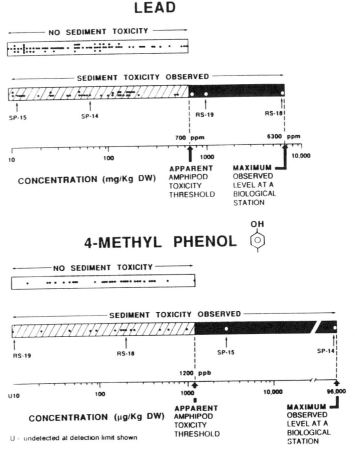

The horizontal axis in each figure represents sedimentary concentrations of contaminant of concern (i.e., lead or 4-methyl phenol) on a log scale. For the amphipod bioassay under consideration, the AET for lead is the highest lead concentration corresponding to sediments that did *not* exhibit significant toxicity. Above this lead AET, significant amphipod toxicity was *always* observed in the data set. The AET for 4-methyl phenol was determined analogously.

The Potential Effect Threshold (Figure 2) is the concentration below which no statistically significant biological effects were observed in any sample. Note that this threshold for 4-methyl phenol is equal to the detection limit for the compound. The threshold is designated as "potential" because toxicity was observed in some, but not all, of the samples from stations with higher lead or 4-methyl phenol concentrations. The toxicity effects observed at these stations could have resulted from other contaminants or physical conditions (e.g., grain size). Because the potential effect threshold for a chemical cannot be related in a meaningful way to the observed biological effects, it is not used to set sediment quality values.

INTERPRETATION OF AET

AET correspond to the sediment concentration of a chemical above which *all* samples for a particular biological indicator were observed to have adverse effects. Thus, AET are based on noncontradictory evidence of biological effects for a given data set. Data are treated in this manner to reduce the weight given to samples in which factors other than the contaminant examined (e.g., other contaminants, environmental variables) may contribute to the biological effect.

For example, sediment from Station SP-14 shown in Figure 2 exhibited severe toxicity, potentially related to a greatly elevated level of 4-methyl phenol (7,400 times reference levels). The same sediment from Station SP-14 contained a low concentration of lead that was well below the AET for lead (Figure 2). Despite the toxic effects displayed by the sample, sediments from many other stations with higher lead concentrations than Station SP-14 exhibited no statistically significant biological effects. These results were interpreted to suggest that the effects at Station SP-14 were more likely associated with 4-methyl phenol (or a substance with an environmental distribution) than with lead. A converse argument can be made for lead and 4-methyl phenol in sediments from Station RS-18. Hence, the AET approach helps to identify different contaminants that are most likely associated with observed effects at each biologically impacted site. Based on the results for these two contaminants, effects at 4 of the 28 impacted sites shown in the figures may be associated with elevated concentrations of 4-methyl phenol, and effects at 7 other sites may be associated with elevated lead concentrations (or similarly distributed contaminants).

These results illustrate that the occurrence of impacted stations at concentrations below the AET of a single chemical does not imply that AET in general are not protective against biological effects, only that single chemicals may not account for all biological effects. By developing AET for multiple chemicals, a high percentage of all stations with biological effects are accounted for with the AET approach (see "Results of Validation Tests"). Nevertheless, unmeasured toxic chemicals may occur in the environment with a different spatial distribution than any of the measured chemicals. In such cases, it is unlikely that the AET approach could regularly predict impacts at stations where only such chemicals induce toxic effects.

AET can be expected to be most predictive when developed from a large data base with wide ranges of chemical concentrations and a wide diversity of measured contaminants. Small data sets that have large concentration gaps between stations and/or that do not cover a wide range of concentrations must be scrutinized carefully (e.g., to discern whether chemical concentrations in the data set exceed reference concentrations) before generation of the AET is appropriate.

GENERATION OF PUGET SOUND AET

AET were originally generated for a combined measure of sediment toxicity (i.e., either amphipod mortality [Swartz et al., 1985] or oyster larvae abnormality [Chapman and Morgan, 1983]), and depressions in the abundance of benthic infauna (at high taxonomic levels). These AET were based on data from 50 to 60 stations sampled during the 1984-1985 Commencement Bay Remedial Investigation (Barrick et al., 1985). In a 1986 project for the Puget Sound Dredged Disposal Analysis (PSDDA) and Puget Sound Estuary Program (PSEP), AET were generated with a larger Puget Sound data base (190 samples, including Commencement Bay data) for individual measures of toxicity (i.e., amphipod mortality, oyster larvae abnormality, and Microtox bioassays [Williams et al., 1986]), and benthic infaunal depressions (at high taxonomic levels). Matched biological and chemical data for an additional 10 stations from a joint state and federal investigation of creosote contamination in Eagle Harbor in central Puget Sound have also been incorporated (Barrick et al., 1986).

The geographic distribution of samples in this data set is shown in Figure 3. Detailed descriptions of the specific chemical tests and statistical analyses for biological indicators is provided elsewhere (Beller et al., 1986). As the result of additional refinement studies sponsored by the Environmental Protection Agency's (EPA) Region 10, the AET data base has recently been expanded to include additional data on sediment chemistry, benthic infaunal abundance, and amphipod mortality for over 100 sediment samples collected in two additional embayments of Puget Sound, Elliott Bay and Everett Harbor. A report summarizing the results of this refinement study is available from EPA (PTI, 1988).

RESULTS OF VALIDATION TESTS

Selected AET generated from a 200-sample Puget Sound data base are presented for dry-weight normalized chemical data (Table 1). AET generated from chemical data normalized to total organic carbon have also been tested, but are less or no more predictive of observed biological effects than dry-weight normalized data. Such a result was not expected based on organic carbon normalization theory, which assumes that interstitial water is the primary source of nonpolar organic contaminants to biota, and that, under equilibrium conditions, the distribution of nonpolar contaminants between sedimentary organic matter and water (i.e., K_{oc}) is constant (and predictable).

For contaminated sediments in the environment, organic carbon normalization could be less predictive than dry-weight normalization if sediment/interstitial water systems are not at equilibrium (e.g., because of overriding kinetic factors), if all sediment organic matter does not have uniform affinity for hydrophobic pollutants, or if interstitial water is not the predominant route of contaminant uptake. Dry-weight normalization assumes that mass loading of a contaminant in sediment is a predominant factor influencing toxicity to benthic organisms (although organic carbon interactions may be a secondary factor). The AET concept does not favor one of these mechanistic explanations over the other, but can operate whether one, a combination of the two, or alternative mechanistic assumptions are appropriate.

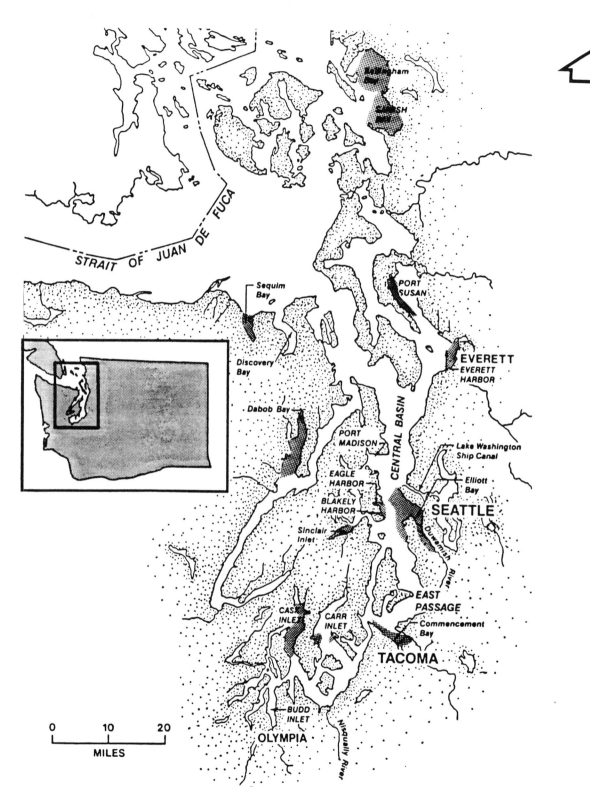

FIGURE 3 Location of sampling sites for AET data sets.

TABLE 1 Continued

NOTES:
[a] Based on data from Beller et al., 1986 and Eagle Harbor Preliminary Investigation (Barrick et al., 1986). Data for recent surveys in Elliott Bay and Everett Harbor are not incorporated in these AET because the proposed AET values were under review by the Puget Sound Sediment Criteria Workgroup during preparation of this paper (see PTI, 1988). Note: ">" indicates that a definite AET could not be established because there were no "effects" stations with chemical concentrations above the highest concentration among "no effects" stations.
[b] Based on 160 stations.
[c] Based on 56 stations (all from Commencement Bay Remedial Investigation).
[d] Based on 104 stations.
[e] Based on 50 stations (all from Commencement Bay Remedial Investigation).
[f] A higher AET (24,000 µg/kg for low molecular weight PAH and 13,000 µg/kg for anthracene) could be established based on data from an Eagle Harbor station. However, the low-molecular-weight PAH composition at this station is considered atypical of Puget Sound sediments because of the unusually high relative proportion of anthracene. Thus, the low-molecular-weight PAH and anthracene AET shown are based on the next highest station in the data set.
[g] The value shown exceeds the Puget Sound AET established in Beller et al., (1986) and results from the addition of Eagle Harbor Preliminary Investigation data (Barrick et al., 1986).
[h] The value shown exceeds AET established from Commencement Bay Remedial Investigation data (Barrick et al., 1985) and results from the addition of Puget Sound data presented in Beller et al. (1986).

Measures of Reliability

To meet the needs of ongoing sediment management programs, an ideal approach for sediment criteria would perform well on both of the following two tests of reliability based on actual field data:

1. sensitivity--the proportion of actual environmental problems that are predicted as problems (i.e., the complement of "false negatives"; are *all* sediments exhibiting biological effects identified using the predictive approach?); and
2. efficiency--the proportion of predicted problems that are actually environmental problems (i.e., the complement of "false positives"; are *only* sediments exhibiting biological effects identified using the predictive approach?).

The concepts of sensitivity and efficiency are illustrated in Figure 4. Sediment quality values that are highly sensitive may be

FIGURE 4 Measures of reliability (sensitivity and efficiency).

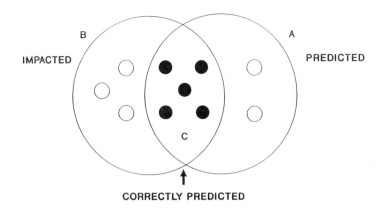

environmentally protective but are not necessarily cost-effective. Sediment quality values that are highly efficient may be cost-effective and defensible in pursuing high priority remedial action but are not necessarily protective. The concepts of sensitivity and efficiency will be used in this paper as an evaluation tool for assessing the reliability of AET.

Test of AET Predictions

As one example of a test of reliability, AET (dry weight) developed for eight Puget Sound embayments (approximately 200 samples; Beller et al., 1986) were used to independently evaluate new data from an additional five embayments (including Eagle Harbor, Elliott Bay, Everett Harbor, and two reference areas; PTI, 1988). Stations predicted to have biological effects were identified as those stations with one or more chemicals exceeding the AET. Only amphipod bioassay and benthic infaunal abundance data were collected in these new surveys. Microtox and oyster larvae bioassays have not been conducted outside of Commencement Bay.

The sensitivity of 1986 benthic infauna AET in correctly identifying impacted stations ranged from 71 to 100 percent, and totaled 81 percent (57/70 impacted stations) for the combined new surveys. This sensitivity is comparable to that reported overall for the original surveys used to calculate these AET (82 percent; Beller et al., 1986). The predictive efficiency of benthic infauna AET applied to different geographic areas in the new surveys ranged from 56 to 100 percent and totaled 74 percent (57/77 predictions) for the combined new surveys.

The sensitivity of 1986 amphipod bioassay AET in correctly identifying impacted stations generally ranged from 60 to 100 percent, and totaled 65 percent (35/54 impacted stations) for the new surveys. This sensitivity is also comparable to an overall sensitivity of 54 percent for the original surveys used to calculate these amphipod bioassay AET. The predictive efficiency generally ranged from 50 to 100 percent and totaled 52 percent for the combined new surveys (in the Eagle Harbor survey, however, significant amphipod mortality was observed at one of the five predicted stations).

Recalculation of AET to include this independent data set of five embayments in the generation of AET automatically results (by definition of AET) in 100 percent efficiency for both biological indicators. The sensitivity of the proposed amphipod AET is 58 percent, and the sensitivity of the proposed benthic infauna AET is 74 percent. Therefore, for the 13 Puget Sound embayments, approximately 85 percent of the benthic infauna stations (170 of 198) and amphipod bioassay stations (238 of 283) are in accordance with the predictions of the proposed AET values for these indicators (i.e., they do not exhibit adverse effects when all concentrations are less than AET values, and do exhibit adverse effects at chemical concentrations above the AET values).

An additional test of the reliability of AET has been conducted as they might be used in management programs. AET were developed for each chemical of concern for multiple biological indicators because different kinds of biological indicators respond in different ways to the same chemical exposure. For example, an assessment of acute lethal toxicity to contaminated sediments is expected to result in different sediment quality values than an assessment of acute or chronic sublethal toxicity to the same sediments. Acute or sublethal responses by different biological species also can differ. For example, the lowest AET (LAET) for one chemical (e.g., PCBs) may be established by the Microtox bioassay and for another chemical (e.g., lead) by benthic infaunal analyses. The LAET is expected to be protective of a range of biological effects. Used in combination, the multiple AET can also provide a preponderance of evidence for associating environmental effects and chemical contamination. Above the highest AET (HAET) for a range of biological indicators there is a high degree of confidence that sediments will fail biological testing regardless of the test. In recent (1988) evaluations with the 300-sample data base for 13 embayments, LAET were from 90 to 94 percent sensitive in correctly predicting all known biological effects in the data base (depending on the particular biological test). By definition, HAET based on this 300-sample data base were 100 percent efficient in only predicting actual problem sediments (i.e., all of the sediments actually exhibited the predicted effects).

Application of Puget Sound AET to other coastal areas of the United States would require validation through at least some site-specific chemical/biological testing. An initial evaluation based on predictions using Puget Sound AET and comparing with limited biological effects data from outside of Puget Sound has recently been reported (PTI, 1987).

APPLICATION OF AET TO SEDIMENT MANAGEMENT IN PUGET SOUND

The reliability of the AET approach (particularly AET normalized to dry weight) at predicting biological effects indicates its potential utility as a tool for sediment quality management. Uses for which the AET approach is well-suited include

- determination of the extent and relative priority of potential problem areas to be managed,
- identification of potential problem chemicals in impacted sediments,
- prioritization of laboratory studies for determining cause-effect relationships, and
- with appropriate safety factors or other modifications, for use in regulatory programs and as "trigger levels" for screening decisions on the need for further chemical or biological testing of sediments.

Sediment criteria based on definitive laboratory cause-effect studies and field verification studies will continue to be active research issues for many years. In the interim, field effects-based approaches using the AET concept provide decision tools that have the following characteristics:

1. developed empirically from field data,
2. provide chemical-specific values,
3. supported by a variety of biological indicators including acute lethal and sublethal bioassays and in situ benthic infaunal analyses reflecting acute and/or chronic effects,
4. driven by statistically significant adverse effects,
5. supported by noncontradictory evidence of adverse effects within a given data set.

Sediment quality values based primarily on AET have been integrated into several Puget Sound programs (Figure 5). These chemical values are not used in isolation; in all cases site-specific biological testing is used to supplement or verify the predictions based on sediment quality values. Problem area and problem chemical identification was the focus of the Commencement Bay Remedial Investigation, and is currently a major aspect of the urban bays program of PSEP. This problem identification establishes a basis in toxics action plans for prioritizing potential remedial actions according to the environmental significance of contamination. Problem identification requires sensitive sediment quality values to ensure that all potential problems are considered.

The purpose of subsequent sediment remedial action is to mitigate contamination in problem areas and thereby to eliminate associated adverse biological effects. In the Commencement Bay Feasibility Study, sediment quality values based on the lowest AET for a range of biological indicators were developed as potential target cleanup goals (not considering cost or technical feasibility). These goals correspond to

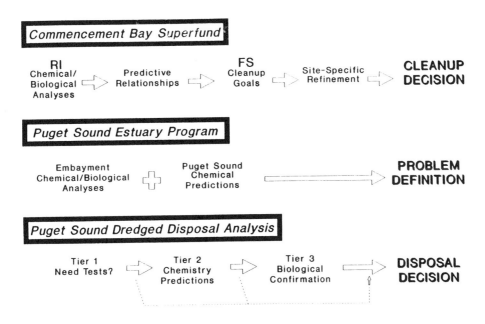

FIGURE 5 Use of sediment quality values based on AET in Puget Sound remedial action programs.

those sediment contaminant concentrations that are not predicted to result in adverse effects according to the biological effects indicators used to generate AET. Although the goals may not be totally protective of all potential environmental problems, they are sensitive to currently measurable effects, including effects originally used to identify problem areas in the remedial investigation.

A higher (i.e., less stringent) level for cleanup was identified as an alternative to the target cleanup goal. This alternative cleanup level was recommended for use should the target goal be infeasible at a particular problem area. The higher concentration alternative would be more technically or economically feasible than target goals because it would tend to require smaller volumes of sediment for remedial action. Instead of using an arbitrary multiple of the target goals, this alternative was generally based on the highest AET for the range of biological indicators (i.e., the concentration of each chemical above which *all* biological effects accounted for by AET are predicted to occur). This alternative cleanup level is expected to be efficient in addressing major contaminant problems.

Dredged material disposal guidelines developed by PSDDA incorporate sediment quality values to address both sensitivity and efficiency concerns (Phillips et al., 1988). PSDDA guidelines establish a chemical screening level (SL) above which biological testing must be performed to establish the suitability of dredged material for disposal at unconfined, open-water sites. The SL is lower or equal to the lowest AET for a range of biological indicators and is intended to be sensitive (i.e., fully protective of the environment). Contamination below the SL is assumed to be acceptable without confirming biological tests. A maximum level (ML) was also established by PSDDA as the highest AET for a range of biological indicators. The ML was intended to indicate a level of chemical contamination above which there was a preponderance of evidence for adverse effects. Biological testing above the ML is always expected to confirm the prediction of unacceptable biological effects, and is not required.

It is recognized that site-specific factors could anomalously influence predictions of biological effects based on sediment quality values. Therefore, in evaluating final requirements for sediment remedial action, selected verification of predicted effects is recommended (extensive biological testing of each sample may not be feasible). For example, in the Commencement Bay Feasibility Study an option is provided to appeal the site-specific prediction of biological effects. This biological testing program is consistent with the intent of other regional contaminated sediment management programs, including PSDDA disposal guidelines. Comparable tests and test protocols are recommended, and site-specific biological information overrides predictions of biological effects based on chemical data. Some specific differences between regional programs in the interpretation of biological test results may exist because of differing program goals (e.g., cleanup of nearshore sediments in a multiuse environment versus assessment of the suitability of potentially contaminated material for disposal at a designated deep-water site).

CONCLUSIONS AND RESEARCH RECOMMENDATIONS

Sediment remedial action is controversial because few objective criteria exist for quantitatively assessing more than the economic feasibility of remedial actions. A commonly expressed concern is that cleanup or disposal guidelines based solely on the most sensitive biological effects would likely be economically or technically infeasible. In any case, there is a strongly perceived need for a preponderance of evidence to implement remedial action.

To address these concerns, a range of sediment quality values--such as those incorporated into existing Puget Sound programs--is recommended. The low end of this range is protective of a wide range of adverse biological effects. At the high end of this range, a preponderance of evidence exists for the prediction of adverse biological effects by multiple indicators. These tradeoffs required by balancing environmental protection and remedial action feasibility are reflected in the sensitivity and efficiency of sediment quality values, which should both be evaluated as part of their validation or in any research effort.

REFERENCES

Barrick, R. C., D. S. Becker, D. P. Weston, and T. C. Ginn. 1985. Commencement Bay Nearshore/Tideflats Remedial Investigation, Final Report. Prepared for the Washington [State] Department of Ecology and U.S. Environmental Protection Agency. EPA-910/9-85-134b. Bellevue, Wash.: Tetra Tech, Inc. 2 volumes + appendices.

Barrick, R. C., H. R. Beller, and M. Meredith. 1986. Eagle Harbor Preliminary Investigation, Final Report. Prepared for Black & Veatch Engineers-Architects and the Washington Department of Ecology. Bellevue, Wash.: Tetra Tech, Inc. 247 pp.

Battelle. 1986. Sediment Quality Criteria Methodology Validation: Calculation of Screening Level Concentrations from Field Data, Final Report. Prepared for the U.S. Environmental Protection Agency, Criteria and Standards Division. Washington, D.C.: Battelle. 60 pp. + appendices.

Chapman, P. M. and J. D. Morgan. 1983. Sediment bioassays with oyster larvae. Bull. Environ. Contam. Toxicol. 31:438-444.

Beller, H. R., R. C. Barrick, D. S. Becker. 1986. Development of Sediment Quality Values for Puget Sound, Final Report. Prepared for the Puget Sound Dredged Disposal Analysis and Puget Sound Estuary Program. Bellevue, Wash.: Tetra Tech, Inc.

Long, E. R. and P. M. Chapman. 1985. A sediment quality triad: measures of sediment contamination, toxicity and infaunal community composition in Puget Sound. Mar. Poll. Bull. 16:405-415.

Pavlou, S. P. 1987. The use of the equilibrium partitioning approach in determining safe levels of contaminants in marine sediments. In Fate and Effects of Sediment-Bound Chemicals in Aquatic Systems, K. L. Dickson, A. W. Maki, and W. A. Brungs, eds. Toronto: Pergamon Press. Pp. 388-412.

Phillips, K., D. Jamison, J. Malek, B. Ross, C. Krueger, J. Thornton, and J. Krull. 1988. Evaluation procedures technical appendix. Public Review draft report prepared by the the Evaluation Procedures Work Group with assistance of Resource Planning Associates, PTI Environmental Services, Shapiro & Associates, and Tetra Tech, Inc. for the Puget Sound Dredged Disposal Analysis.

PTI Environmental Services. 1987. Policy implications of effects-based marine sediment criteria. Prepared for AMS/U.S. Environmental Protection Agency, Office of Policy Analysis. EPA Contract No. 68-01-7002. Bellevue, Wash: PTI Environmental Services. 57 pp. plus appendices.

PTI Environmental Services. 1988. Sediment quality values refinement: 1988 Update and evaluation of Puget Sound AET. Final report prepared for Tetra Tech, Inc./U.S. Environmental Protection Agency, Region 10 Office of Puget Sound. PTI Environmental Services, Bellevue, Wash.

Swartz, R. C., W. A. Deben, J. K. P. Jones, J. O. Lamberson, and F. A. Cole. 1985. Phoxocephalid amphipod bioassay for marine sediment toxicity. In Aquatic Toxicology and Hazard Assessment: 7th Symposium, R. D. Cardwell, R. Purdy, and R. C. Bahner, eds. ASTM STP 854. Philadelphia, Pa.: American Society for Testing and Materials. Pp. 284-307.

Williams, L. G., P. M. Chapman, and T. C. Ginn. 1986. A comparative evaluation of sediment toxicology using bacterial luminescence, oyster embryo, and amphipod sediment bioassays. Mar. Environ. Res. 19:225-249.

THE USE OF THE SEDIMENT QUALITY TRIAD IN CLASSIFICATION OF SEDIMENT CONTAMINATION

Edward R. Long
National Oceanic and Atmospheric Administration

ABSTRACT

A concept for use in the collection of data needed to classify sediment quality is described. This concept is based upon the observed need for information on the kinds and concentrations of potentially toxic chemicals in the sediments, the relative toxicity of the sediments as determined under controlled laboratory conditions, and the characteristics of resident benthos under in situ conditions. The concept, called the Sediment Quality Triad, has been used in numerous assessments of urban embayments and prospective dredge material. Case studies from Puget Sound and San Francisco Bay and various uses of the data are described.

DESCRIPTION OF THE METHODOLOGY

The Sediment Quality Triad (the "triad") is a concept recently developed (Long and Chapman, 1985; Chapman et al., 1987) for use in the classification and evaluation of the relative quality of surficial sediments. It consists of measures of sediment contamination quantified by chemical analyses, sediment toxicity determined with laboratory bioassays and benthos community structure described through taxonomic analyses of macrofauna. The chemical analyses provide information on the mixtures and concentrations of contaminants in the sediments that may be harmful to marine biota. The bioassays provide information on the relative bioavailability and toxicity of sediment-sorbed contaminants under laboratory conditions where the effects of many "natural" environmental factors are controlled. The benthos community data provide corroborating evidence from resident biota regarding major compositional alterations to a component of the ecosystem under in situ conditions. The data from the three measures are complimentary and provide a preponderance of empirical evidence of both contamination and effects that can be used to classify the relative quality of sediments.

Portions or aliquots of the same sediment samples are usually tested for contamination and toxicity. The macrobenthos are examined in additional portions of the same samples or, more often, are collected at the same sampling stations in separate grab samples. Chemical analyses are performed for a variety of trace metals and organic compounds. The physical/chemical characteristics of the sediments, such as sediment texture and total organic carbon, are also determined.

Sediment toxicity is determined through bioassays in which mortality, impairment of reproductive success, sublethal behavioral and/or mutagenic effects are recorded. The taxonomic analyses of the benthos provide information on the species richness, total abundance, abundance of individual species, and indices of community similarity. The relative abundance of the bioassay species in the benthos also can be recorded, providing a strong link between the bioassay results and benthos data.

The triad is a concept for measuring and classifying sediment quality; it is not an index per se. The data resulting from the three measures can be used in several ways to satisfy a variety of objectives. First, they can be used in a descriptive mode, in which the preponderance of evidence is used to evaluate and classify the relative quality of sediments among sampling sites. The relationships among the contaminant, physical/chemical, and biological data may be interpreted in descriptive ecological evaluations of the study sites. Second, site ranks can be calculated independently for each of the triad components, using a variety of techniques. Cumulative ranks, based upon the three independent ranks, have also been determined. One technique that has been used to rank sites has involved calculation of the ratios between data from the more contaminated sites and from an apparently uncontaminated reference site. By calculating these ratios, data from all the measures, which often have very different units, absolute values, and ranges can be treated with similar weight on a common, unitless scale. The independent ranks can be illustrated with triaxial plots to highlight differences among sites. Classification of sites can be performed to determine both geographic and temporal trends in sediment quality.

The triad data also can be used to determine the means and ranges in contaminant concentrations associated with modes and ranges in the biological effects data. In this type of evaluation the means and ranges in contaminant concentrations associated with the highest responses in the biological analyses can be compared with those associated with the intermediate and lowest ranges in those tests, using data from a variety of sampling sites. This type of evaluation can form the basis for predictive models in which the relationships between synoptically collected biological and chemical data are used to estimate the relative degree of contamination that is often associated with biological effects. Finally, where a sufficient amount of data exist, these types of predictive evaluations of the triad data can be used to estimate the contaminant thresholds above which biological effects are always observed. The Apparent Effects Threshold (AET) approach, one method of using triad data in a predictive mode that has been used in Puget Sound, is described by Barrick et al. in this volume.

EFFECTIVENESS AND RELEVANCE OF THE METHODOLOGY

Because the methodology provides a thorough assessment of the quality of the sediments, it is very effective at classifying sites based upon the preponderance of evidence. The chemical data provide evidence regarding whether the sites are contaminated or not and which chemicals are present in the highest concentrations. The chemical data can provide clues regarding the most likely sources of contaminants when the chemical ratios or "signals" in the sediments match those in nearby potential sources. The bioassay data provide direct evidence of whether or not the sediments are toxic to selected test organisms. If they are toxic, it can be assumed that the chemical contaminants were bioavailable to the test organisms. They also can be useful in determining the degree of toxicity and the nature (lethal, mutagenic, sublethal) of the toxicity. The benthic community data provide an in situ confirmation or denial that the sediments are toxic to biota. These data can serve as a measure of ecosystem structure and function. Evidence of severely altered benthos coupled with evidence of sediment toxicity provide a powerful argument that contaminated sediments are biologically damaging. For example, Swartz et al. (1982) showed that portions of the Commencement Bay waterways were very toxic to amphipods and that the amphipod populations in resident benthos in the same areas were severely depressed in abundance relative to other nearby areas.

Classification approaches that rely only upon chemical data provide no empirical evidence that the contaminants are (1) bioavailable and (2) biologically damaging. While predictive physical chemical models may provide theoretical estimates of single contaminant concentrations that are biologically damaging, they do not provide these estimates for the complex and variable mixtures of contaminants that usually occur in estuaries, ports, and harbors. Sediment toxicity tests are often performed under worst-case laboratory conditions with test organisms that have no chance of escape, may not be native to the sampling sites, and have no time to acclimate to the properties of the sediments. Therefore, classification approaches that rely upon bioassay data alone may overestimate the poor quality of sediments or may be received with indifference by managers. Classification approaches that rely only upon benthos data may be frustrated by the major alterations in benthic communities that can be caused entirely or partly by differences among sites in depth, sediment texture, near-bottom or interstitial salinity, predation, bottom scouring, and other biotic and abiotic factors.

While the three types of data from the triad concept provide complementary measures of sediment quality, the data from the three components may not necessarily parallel each other. Each component measures different properties of the sediments. For example, sites that are relatively contaminated may not be the most toxic, or sites with relatively altered benthic communities may not be most contaminated. If each of the components mimicked each other in the classification of sites, there would be no need to measure all three. The strength of the triad approach is the use of both chemical and biological measures that can be used in an ecological evaluation of sediment quality. The triad concept can provide the data needed by an ecologist to interpret and use in characterizing sediment quality.

Simple before and after surveys can be performed to determine an changes in sediment quality caused by a specific remedial action. Ecological evaluations of the triad data can be performed to determine if contaminant concentrations and toxicity have decreased and if measures of benthos alteration have been alleviated. Also, any cumulative indices calculated from the triad data collected before the remedial action can be compared with those calculated from data generated after the action.

FIELD VALIDATION OF THE METHODOLOGY

The utility of the triad approach has been verified in many field surveys and experiments. The triad approach, per se, was first applied in an evaluation of available data from sites in Puget Sound, Washington (Long and Chapman, 1985). This study indicated that data from the three components of the triad often showed gross parallel patterns in sediment quality among sites, but that the agreement among the three measures was not absolute. In a subsequent survey in San Francisco Bay, Chapman et al. (1986, 1987) demonstrated the differences in sediment quality among three sites, based upon a preponderance of evidence (see case study below). The "Urban Bay Approach" taken by a consortium of the U.S. Environmental Protection Agency's (EPA) Region 10 and the Washington Department of Ecology (WDOE) has used the triad as the basis for ranking contaminated sites in the urban bays and waterways of Puget Sound for remedial action. The quality of sites in Commencement Bay (see Case Study below), Elliott Bay, Everett Harbor, and Eagle Harbor has been assessed using this approach. The data from these Puget Sound studies have been used to calculate Elevations Above Reference (EAR) conditions to classify the relative quality of sites and to calculate Apparent Effects Thresholds (AET). While all these studies have shown generally good overall agreement in results among the triad components, they also indicated that, as expected, the agreement on a station-to-station basis was not perfect. Therefore, the results from any one or two of the components, if measured alone, may have not accurately predicted the results from the other component(s).

Seattle METRO assessed the quality of sediments in a baseline study of a prospective sewer discharge site in Puget Sound, using the triad of measures and other tests. Many samples from the southern portion of the central basin of the Sound were collected and analyzed (Stober and Chew, 1984). Battelle Pacific Northwest Laboratories (1986) assessed the quality of sediments in eight bays of Puget Sound for EPA Region 10 to determine the relative quality of rural and urban areas.

Off Southern California, Swartz et al. (1986) described temporal changes between 1980 and 1983 in contamination, toxicity, and benthos. Chapman (1986) summarized sediment bioassay and bottomfish histopathology data from Puget Sound and described the contaminant levels in sediments associated with high and low incidences of these measures of effects. Other uses of the triad concept are underway in studies being conducted by the Southern California Coastal Water Research Project in Southern California harbors (Karen Taberski, California State Water Quality Control Board, personal communication); in the Gulf of Mexico near oil production platforms (Peter Chapman, E.V.S. Consultants, personal communication); and in Lake Union near Seattle, Washington (Bill Yake, Washington Department of Ecology, personal communication). The biological effects of Black Rock Harbor sediments at a Long Island Sound dump site have been investigated with the triad of measures by the U.S. EPA and Army Corps of Engineers (Gentile et al., 1985; Rogerson et al., 1985).

Both the states of Washington and California are currently considering the possible development of effects-based sediment quality criteria, using AET values based upon triad data. The Washington Department of Ecology must adopt statewide sediment quality standards by June 30, 1989 in response to Element P-2 of the Puget Sound Water Quality Plan. The California State Water Quality Control Board must adopt sediment criteria by 1991 in response to provisions of Assembly Bill 3947 that would assure the protection of wildlife and humans from sediment-associated contamination.

REQUIRED EXPERTISE AND COSTS

Since the triad approach provides a comprehensive assessment, followup studies are seldom required to address unresolved questions. However, because the triad concept requires data from three scientific disciplines (analytical chemistry, toxicology, benthic ecology), a study team with broad expertise is required. It is possible that once the relationships between contaminant levels and biological effects in sediments are established for a geographic region, one or two of the triad components could be eliminated or reduced in scope. A variety of short-cut measures of chemical contamination, toxicity, or benthos alterations may help to reduce costs. For example, bacterial luminescence bioassays may prove to be very inexpensive tests of sediment toxicity (Schiewe et al., 1985). Quantification of only selected chemicals known to occur in the study area or known to be of highest toxicological concern would reduce costs. Examination of small cores for, say, presence of amphipods in the benthos in the grab samples used for chemical and bioassay analyses may reduce costs of benthic community analyses. The benthos could be examined inexpensively by a sediment profiling camera to determine selected community properties.

The availability of these types of expertise is widespread in many commercial, agency, and academic laboratories in the United States. The specific expertise and equipment needed to develop triad data, however, would vary among regions, depending upon region-specific research needs and environmental variables. Also, costs would vary among regions and among studies depending upon complexity and precision of chemical analyses, types and number of bioassays, and the complexity and density of the benthos.

In the triad case study in San Francisco Bay described below, total costs were about $100,000. For that total cost, data were collected for 66 chemicals, many physical/chemical properties, four bioassays, and complete taxonomic analyses of the benthos at nine stations. The bioassays and benthos analyses were performed with quintuplicates at each station. The costs also included thorough data analysis and report preparation steps.

Two case studies, Puget Sound and San Francisco Bay, will be briefly summarized to illustrate the use and results of the triad approach. The references cited should be studied to determine details of methods and results. In both case studies, most of the data have been presented as Ratio-to-Reference (RTR) values to facilitate comparisons of conditions in study sites with those in an apparently uncontaminated reference site and to place the three disparate types of data on the same unitless scale (Chapman et al., 1987). Since positive ratios (i.e., greater than 1) could theoretically range to infinity and negative ratios could only range from 1 to 0, differences among sites in mean RTR values may be slightly exaggerated. RTRs can be calculated and transformed to logarithms, wherein both negative and positive ratios can range from zero to negative infinity and to positive infinity, respectively. In this approach, negative and positive values are given equal weight in the calculation of means. Transformation of the RTR values determined in the case studies to logarithms slightly altered the mean RTR values, but did not change the relative ranks of stations or sites.

CASE STUDIES

San Francisco Bay

Research was conducted by Chapman et al. (1986, 1987). Sediment samples were collected at three stations at each of three sites: Islais Creek Waterway (IS), off Oakland (OA), and in San Pablo Bay (SP). The former site was in an industrial waterway that receives major discharges from combined sewer overflows and was expected to be highly contaminated. The second site was located near the Oakland Harbor maritime facilities and was expected to be moderately contaminated. The third site was located in the open waters of San Pablo Bay in the northern part of the San Francisco Bay estuary and was expected to be the least contaminated, based upon studies of sediment and bottom-fish contamination conducted at the site.

The samples were collected with a 0.1-m^2 van Veen grab sampler. The upper 2 cm were collected for the chemical and toxicity analyses. The contents of multiple grabs were composited at each station, homogenized, and aliquots taken for each of the chemical and bioassay tests. Five separate replicate grab samples were taken for the benthos evaluations at each station. The benthos samples were wet-sieved at each station and the biota retained on a 1-mm screen were kept for examination.

The chemical analyses were performed for 21 major and trace metals, 20 low- and high-molecular-weight aromatic hydrocarbons, 17 chlorinated hydrocarbons, and 8 chlorination levels of polychlorinated biphenyls. In addition, sediment texture, total organic carbon content, total volatile solids content, sulfide content, and percent solids were determined for each station. Toxicity was determined with four bioassays:

1. solid-phase bioassays of acute lethality and avoidance of sediments by the amphipod *Rhepoxynius abronius*;
2. elutriate bioassay of acute lethality and abnormal morphological development of the embryos of the mussel

Mytilus edulis;
3. solid-phase bioassay of reburial rate by the clam *Macoma balthica*; and
4. solid-phase bioassay of impairment of reproduction with the copepod *Tigriopus californicus*.

Complete taxonomic analyses of the benthos were performed and indices of total abundance, abundance of individual taxa, species richness, species diversity, proportional contribution of major taxonomic groups to total abundance, dominance, and equitability were calculated.

Mean percent silt + clay content was 69 percent at the SP site, 86 percent at the OA site, and 94 percent at the IS site. Mean total organic carbon content was 1.10 percent at the SP site, 1.22 percent at the OA site, and 2.87 percent at the IS site. Mean sulfide content was 29.7 mg/kg at the SP site, 3.1 mg/kg at the OA site, and 540.0 mg/kg at the IS site. From these data it was apparent that the IS site was highly organically enriched compared to the other two sites and possibly anoxic.

Table 1 summarizes selected chemical data as RTR values. The data have been normalized to total organic carbon content and each concentration divided by the mean values for the SP reference site. The IS site was much more contaminated than the other two sites; primarily with aromatic hydrocarbons, PCBs, DDTs, and silver. The coprostanol data provide evidence that the site was contaminated with municipal sewage. The OA site was only slightly more contaminated than the SP site. Some trace metals there were less concentrated than at the SP site.

The data from the four bioassays are summarized in Table 2, also as RTR values. The data from most of the bioassay endpoints indicate that the IS samples were significantly more toxic ($p < 0.05$, one-tailed t-test) than those from SP and OA (Chapman et al., 1987). A mean of 10.4 amphipods out of 20 died in the IS site samples, compared to means of 2.5 and 2.3 at the other sites. A mean of 55.2 percent of the mussel embryos exposed to IS samples were abnormal, compared to 12.1 percent and 19.3 percent at the other sites. Mean mortality was highest in embryos exposed to the IS samples. Mean clam reburial time in IS samples was roughly twice that in the SP samples. The number of young copepods produced did not differ substantially among the three sites, though it was lower at the OA site. Among the four types of bioassays, those with amphipods and mussel larvae appeared to be most sensitive to the sediments.

Results of the benthos analyses are summarized as RTR values in Table 3. The benthos at SP and OA were dominated by tube-dwelling amphipods (specifically *Ampelisca abdita*) and other crustaceans). The benthos at IS was dominated by *Capitella capitata*, polychaetes and molluscs. Mean total abundance was 609 organisms per 0.1 m^2 at SP, 3,502 organisms per 0.1 m^2 at OA, and 41 organisms per 0.1 m^2 at IS. Mean number of taxa was 10.3, 14.5, and 4.2 at SP, OA, and IS, respectively. Dominance was lower and species diversity higher at IS than at the other sites, reflecting the dominance by *A. abdita* at the SP and OA sites.

TABLE 1 Ratio-to-Reference Values for Nine Stations and Three Sites Sampled in San Francisco Bay, Based on TOC-normalized Sediment Chemistry Data[a]

Site	Station	\multicolumn{13}{c	}{Ratio-to-Reference (RTR) Values}
		As	Cr	Cu	Pb	Hg	Ag	Sn	Zn	LPAH	HPAH	Copros-tanol	DDTs	PCBs	Aggregate index[b]
San Pablo Bay	02	1.34	1.43	1.20	1.41	0.81	1.29	1.21	1.40	0.42	0.74	0.67	1.15	0.94	0.86
	05	0.79	0.82	0.94	0.79	1.04	0.76	1.04	0.84	1.35	1.34	1.05	1.05	0.88	1.09
	09	0.87	0.76	0.87	0.80	1.15	0.95	0.75	0.76	1.23	0.92	1.28	0.80	1.18	1.05
	Mean	1.00	1.00	1.00	1.00	1.00	1.00	1.00	1.00	1.00	1.00	1.00	1.00	1.00	1.00
	SD	0.24	0.30	0.14	0.29	0.14	0.22	0.19	0.28	0.42	0.25	0.25	0.15	0.13	0.10
Oakland	02	0.89	0.86	0.93	1.18	1.16	1.32	0.98	0.91	2.65	2.86	1.58	1.92	2.78	2.14
	05	0.96	0.92	0.93	1.24	1.03	1.33	1.19	0.91	2.91	2.75	3.20	1.58	2.39	2.32
	09	0.72	0.86	0.87	1.14	1.27	1.67	1.27	0.86	2.09	1.78	0.56	1.15	2.15	1.47
	Mean	0.86	0.88	0.91	1.18	1.15	1.44	1.14	0.89	2.55	2.46	1.78	1.55	2.44	1.97
	SD	0.10	0.03	0.03	0.04	0.09	0.16	0.12	0.02	0.34	0.49	1.09	0.31	0.26	0.37
Islais Waterway	02	0.26	0.40	0.77	2.60	0.77	1.73	1.02	0.78	6.30	5.94	19.45	1.19	4.41	6.39
	05	0.38	0.55	0.75	1.72	2.07	2.36	1.16	0.70	7.06	7.47	20.66	1.88	8.03	7.72
	09	0.91	0.91	1.13	1.60	1.39	2.40	1.34	1.06	5.14	6.17	9.40	2.56	3.93	4.76
	Mean	0.52	0.62	0.88	1.97	1.41	2.17	1.17	0.85	6.17	6.52	16.50	1.88	5.45	6.29
	SD	0.28	0.21	0.17	0.45	0.53	0.30	0.13	0.15	0.79	0.67	5.04	0.56	1.83	1.21

NOTES:
[a] Chemical concentrations for each substance divided by mean values for the San Pablo Bay reference site.
[b] Mean RTR value for the eight inorganic compounds is determined (n = 8), then combined as a single measure with the five organic compound RTR values to provide an overall mean value (n = 6).

SOURCE: Chapman et al., 1986.

TABLE 2 Sediment Bioassay Results and RTR Values for Nine Stations and Three Sites Sampled in San Francisco Bay

Site values[c]	Station	Amphipod Mean mortality No. dead	Amphipod Mean mortality RTR[a]	Amphipod Mean emergence No. emerged	Amphipod Mean emergence RTR	Mussel Larvae Mean % abnormality normal	Mussel Larvae Mean normality RTR	Mussel Larvae Mean mortality % dead	Mussel Larvae Mean mortality RTR	Clam ET 50 (min)	Clam RTR	Copepod 200 minus No. of young produced[b]	Copepod RTR	Mean RTR
San Pablo Bay	02	1.8	0.7	1.1	1.6	13.4	1.1	43.1	1.2	3.3	0.9	92.5	0.9	1.1
	05	0.8	0.3	0.5	0.7	7.7	0.6	17.3	0.5	3.9	1.1	78.8	0.8	0.7
	09	4.8	1.9	0.5	0.7	15.3	1.3	49.1	1.3	3.2	0.9	137.1	1.3	1.2
Overall (n = 3)[d]		2.5	1.0	0.7	1.0	12.1	1.0	36.5	1.0	3.5	1.0	102.8	1.0	1.0
Oakland	02	1.8	0.7	0.7	1.0	14.5	1.2	50.9	1.4	3.6	1.0	88.6	0.9	1.0
	05	2.6	1.0	0.4	0.6	24.7	2.0	76.0	2.1	3.9	1.1	86.1	0.8	1.3
	09	2.6	1.0	1.9	2.7	18.7	1.5	66.5	1.8	5.8	1.7	81.2	0.8	1.6
Overall (n = 3)		2.3	0.9	1.0	1.4	19.3	1.6	64.5	1.8	4.4	1.3	85.1	0.8	1.3
Islais Waterway	02	19.0	7.6	7.4	10.6	67.7	5.6	94.0	2.6	7.5	2.1	103.1	1.0	4.9
	05	4.8	1.9	1.7	2.4	65.9	5.4	96.8	2.7	7.0	2.1	96.2	0.9	2.6
	09	7.4	3.0	0.6	0.9	31.9	2.6	86.1	2.4	4.0	1.1	116.0	1.1	1.8
Overall (n = 3)		10.4	4.2	3.2	4.6	55.2	4.6	92.3	2.5	6.2	1.8	104.7	1.0	3.1

NOTES:
[a]RTR: Ratio-to-Reference, bioassay data divided by mean values for the San Pablo Bay site.
[b]Arbitrary calculation used to adjust data for number of young produced per adult over four weeks in order to calculate RTR values in a similar format to other bioassay responses.
[c]n = 6 separate toxicity values.
[d]Mean reference site values used to determine RTR values. Note these are based on n = 3.

SOURCE: Chapman et al., 1986.

TABLE 3 RTR Values for Indices of Benthic Communities Sampled at Nine Stations and Three Sites in San Francisco Bay

	Ratio-to-Reference[a]												
	San Pablo Bay				Oakland				Islais Waterway			
	02	05	09	Mean	02	05	09	Mean	02	05	09	Mean
1/taxa richness[b]	1.56	1.05	0.71	1.1	0.81	0.78	0.62	0.71	4.35	5.26	1.25	4.76
1/total abundance[b]	0.95	1.64	0.75	1.1	0.20	0.16	0.16	0.17	12.50	10.00	50.00	11.11
numerical dominance	0.90	0.99	1.10	1.0	1.16	1.12	1.12	1.14	0.81	0.84	0.10	0.83
1/percent amphipoda[b]	1.00	1.03	0.97	1.0	0.95	0.97	0.95	0.96	2.63	5.88	10.53	33.33
percent polychaeta	1.04	1.22	0.74	1.0	0.35	0.26	0.23	0.28	14.29	14.29	8.33	14.29
percent mollusca	0.67	2.00	0.33	1.0	0.00	1.00	0.83	0.58	0.00	0.00	254.33	84.75
Sum:	6.12	7.93	4.6	6.2	3.47	4.29	3.91	3.84	34.58	36.27	324.54	149.07
Mean[c]:	1.02	1.32	0.77	1.0	0.58	0.71	0.65	0.64	5.76	6.04	54.09	24.85

NOTES:
[a] Reference = mean San Pablo Bay site values.
[b] High values = least altered, thus these data are entered as reciprocals.
[c] Relative degree of alteration compared to mean reference values. Values greater than 1.0 indicate greater alteration, values less than 1.0 indicate less alteration.

SOURCE: Chapman et al., 1986.

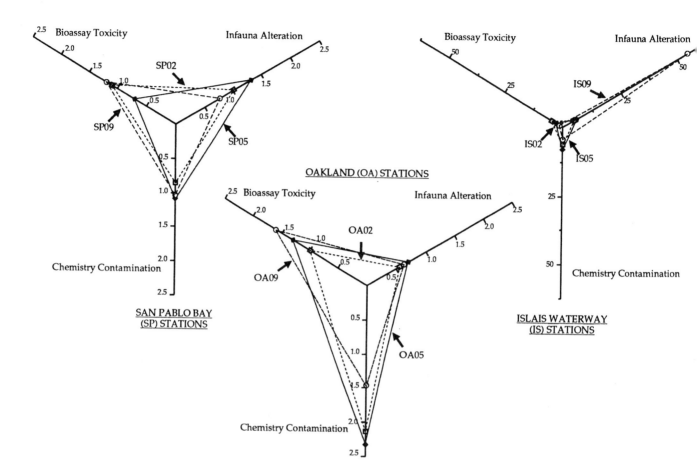

FIGURE 1 The Sediment Quality Triad determined for each station at each of the three study sites in San Francisco Bay. Chemistry RTR values are from Table 1; bioassay RTR values are from Table 2; infauna RTR values are from Table 3. The San Pablo Bay and Oakland stations are plotted on the same scale; Islais Waterway stations are plotted on a scale 1/25 the size of that for the other two sites.

The Sediment Quality Triad values for each station and for each site are illustrated in Figures 1 and 2. They are based upon mean RTR values for each triad component from the stations and sites. Note that the scales differ among stations and sites. It is apparent from these triaxial plots that the IS site had an extremely different benthos community than those observed at the other two sites, and had sediments that were more contaminated and more toxic (Figure 1). Station IS 09, in particular, had a remarkably different benthos community, whereas toxicity was highest at IS 02 and contamination was highest at IS 05 (Figure 2). At the OA site, toxicity was highest in OA 09 samples, contamination was highest in the OA 05 sample and the benthos were similar among the three stations. At the SP site contamination was similar among stations; SP 05 sediments were least toxic, but had the most altered benthos compared to the mean values for the site.

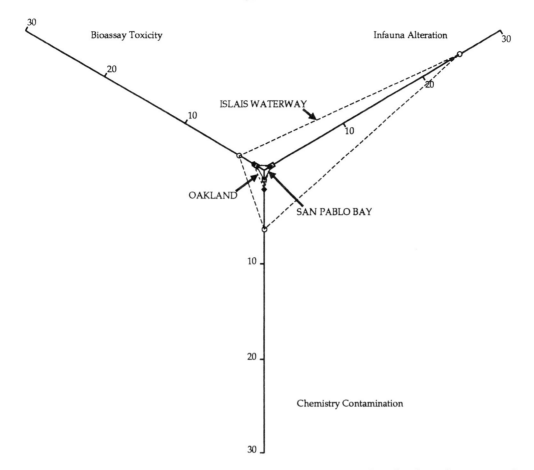

FIGURE 2 Sediment Quality Triad determined for each of the three study sites in San Francisco Bay. Chemistry RTR values are from Table 1; bioassay RTR values are from Table 2; infauna RTR values are from Table 3.

From the triad of measures taken at the three sites in San Francisco Bay, it is apparent that the IS site was more contaminated, more toxic, and had a benthic community that was very different from those at the other two sites. The preponderance of evidence clearly identified and ranked that site as having the lowest sediment quality. All three mean RTRs indicated to varying degrees that that site was most degraded. The data also indicated that the OA site was only slightly different than the reference site: toxicity and contamination were somewhat elevated, but the benthos were not apparently different.

The data indicated that the IS site benthos was highly altered compared to the benthos at the reference site. However, since the sediments at IS also had the highest organic carbon content, the modifications to the benthos, if studied alone, could have been attributed to this and other physical/chemical properties. The bioassay data confirmed that the contaminants in the sediments were bioavailable and toxic to a variety of organisms, and, therefore, posed a threat to resident organisms.

TABLE 4 Mean (and Ranges) in Contaminant Concentrations in Sediments from Nine Stations in San Francisco Bay Associated with Three Means (and Ranges) in Mortality of *Rhepoxynius abronius*

Rhepoxynius abronius mortality[a]	N	LMWPAH ppm	HMWPAH ppm	Pesticides ppm	Pb ppm	Cu ppm	Ag ppm
19 (10-19)	1	3.16 (3.16)	12.06 (12.06)	8.26 (8.26)	223 (223)	130 (130)	8.1 (8.1)
5.7 (5-9)	3	1.3 (0.22-2.76)	5.66 (0.67-11.82)	3.20 (0.92-6.24)	63 (25-115)	73 (53-98)	4.73 (1.6-8.6)
1.9 (0-4)	5	0.28 (0.03-0.43)	1.12 (.22-1.89)	1.08 (0.63-1.69)	26 (18-33)	43.6 (30-51)	1.62 (0.9-2.4)

NOTE:
[a] Number dead out of 20 animals

SOURCE: Chapman et al., 1986.

Insufficient data exist thus far to calculate AET values for San Francisco Bay. From the data collected in this case study is it apparent that results from any one of the components of the triad would have not accurately predicted the results from the other components, since the lines connecting the triaxial plots often were not parallel and crossed each other. However, patterns in co-occurrence of concentrations of selected contaminants with ranges in bioassay data have been determined (Long et al., 1988). Table 4 provides an example of these co-occurrences with the *Rhepoxynius* bioassay data from the case study. An examination of the frequency distribution of the data indicates that three modes in the results of the bioassays were observed: one in which 1 survivor in 20 was observed from one station; another in which a mean of 14.3 survivors in 20 was observed from three stations; and a third in which a mean of 18.1 survivors in 20 was observed from five stations. The low- and high-molecular-weight aromatic hydrocarbons were approximately an order of magnitude higher in concentration at the station with 1 survivor than at the stations with a mean of 18.1 survivors. All the other contaminants shown in Table 4 also increased in mean concentration between the least toxic and most toxic samples. A similar pattern was observed with the mussel larvae data (Long et al., 1988). In all cases, however, there were large ranges in contaminant concentrations within each level of bioassay response. The sample size available from this study is obviously very small, but with the addition of more data the patterns in co-occurrence of contaminant levels with biological effects measures could be established for San Francisco Bay and could, ultimately, lead to the calculation of indices such as AET for use in the bay.

Puget Sound

Data are from an assessment of the waterways of Commencement Bay near Tacoma, Washington performed by Tetra Tech, Inc. (1985). All three components of the triad were measured at 56 stations scattered among the industrialized waterways bordering Commencement Bay and at four stations in nearby Carr Inlet, an embayment selected as a reference area. Sites were selected near and away from known point sources and areas known from previous studies to be contaminated and/or toxic. The approach used in the Commencement Bay study was used in subsequent similar studies in Elliott Bay, Everett Harbor, and Eagle Harbor.

Surficial (upper 2 cm) sediments were collected with a 0.1-m^2 van Veen grab and homogenized for the chemical and toxicity analyses. Samples for benthic macroinvertebrate analyses were taken with a 0.06-m^2 van Veen grab. Four replicates were taken at each station and sieved with 0.5-mm and 1.0-mm screens.

Chemical analyses were performed for 16 elements, volatile organic compounds, polynuclear aromatic hydrocarbons, pesticides, PCBs, and a wide variety of other organic compounds. Total solids, total volatile solids, oil and grease, sulfide content, and sediment texture were also determined. Bioassays were performed with two procedures:

1. solid-phase acute toxicity with the amphipod *Rhepoxynius abronius*, and
2. suspended-phase lethality and abnormal development with the oyster *Crassostrea gigas* larvae.

Benthic infauna were identified to species when possible and enumerated in the 1.0-mm fraction only; the 0.5-mm organisms were archived.

Tables 5, 6, and 7 summarize the results of these analyses for 20 selected stations sampled in the study. All the data are presented as RTRs using the mean values from the Carr Inlet reference area as the denominator. The mean RTR values for each of the triad components is shown to the right of each of the tables. In Table 5 the RTR values for the three trace metals were used to calculate means ($n = 3$), which were then used along with the RTRs for the organic compound classes to calculate overall mean RTR values ($n = 4$).

For the three metals and three organic compound classes shown, several of the sites in the Hylebos Waterway were the most contaminated (Table 5). Station 22 in the turning basin was especially highly contaminated with aromatic hydrocarbons and PCBs. Compared to the other stations, those in City Waterway were moderately contaminated and those in Blair Waterway were slightly contaminated. The concentrations of the selected chemicals was relatively uniform within the Carr Inlet reference area.

The bioassay data (Table 6) indicated that the sediments at station 11 in Blair Waterway were most toxic, i.e., had highest mortality in amphipods and oyster larvae and highest abnormal development in oyster larvae. Stations 22 and 23 in Hylebos Waterway were also relatively toxic, compared to the stations in Blair Waterway and Carr Inlet. The

TABLE 5 RTR Values for Contaminants Quantified at 20 Stations in Commencement Bay and Carr Inlet[a]

Site	Station	Total LMWH	Total HMWH	Total PCBs	Cd	Cu	Hg	Mean RTR
Upper Hyle-	12	56.1	174.1	15.7	26.4	22.9	9.2	66.3
bos Water-	14	47.9	290.8	12.0	20.0	18.2	6.6	91.4
way	17	56.1	321.5	24.0	3.6	32.6	6.0	103.9
Hylebos	22	154.8	524.0	286.0	3.6	38.2	10.0	245.5
Turning	23	132.8	240.7	214.0	2.7	23.5	8.0	149.7
Basin	24	49.8	196.3	36.0	2.7	30.7	9.8	74.1
Blair	11	17.7	20.3	2.4	2.7	6.6	1.4	11.0
Waterway	12	30.6	63.6	10.0	2.7	11.8	1.6	27.4
	13	28.1	45.6	3.1	2.7	10.1	4.0	20.6
	21	34.8	41.2	1.0	2.7	8.6	2.6	20.4
City	11	148.1	228.6	2.1	4.5	24.8	10.6	98.0
Waterway	13	97.1	148.5	20.0	5.4	29.8	22.0	71.2
	16	109.7	84.2	5.1	5.4	25.3	2.2	52.5
	17	132.6	174.7	7.1	5.4	26.9	2.2	81.5
	20	91.0	83.1	2.7	4.5	25.3	4.8	47.1
	22	158.7	203.3	4.6	1.4	6.4	4.4	92.7
Carr	11	0.9	1.1	1.0	0.9	0.8	1.0	1.0
Inlet	12	1.1	1.2	1.0	0.9	1.1	2.0	1.1
	13	0.5	0.4	1.0	0.9	0.8	1.0	0.7
	14	1.5	1.3	1.0	1.4	1.3	0.6	1.2

NOTE:
[a]RTR values were calculated by dividing individual station values by the mean for Carr Inlet.

SOURCE: Tetra Tech Inc., 1985.

chemical data indicated a distinct difference in chemical concentrations between the Hylebos Turning Basin and the adjacent Upper Hylebos Waterway, whereas the bioassay data indicated only a small difference between the two areas. The chemical data indicated that station 11 in City Waterway was moderately contaminated (mean RTR of 98), whereas the bioassay data indicated it was the most toxic among the selected stations.

The benthos data (Table 7) indicated that small differences in total abundance and species richness occurred at most stations relative

TABLE 6 RTR Values for Sediment Bioassays Performed with Samples from 19 Stations in Commencement Bay and Carr Inlet[a]

Site	Station	Percent amphipod mortality	Percent relative oyster larvae mortality	Percent oyster larvae abnormality	Mean RTR
Upper Hylebos Waterway	12	1.2	0.8	3.5	1.8
	14	0.7	1.2	1.9	1.3
	17	1.4	1.5	3.1	2.0
Hylebos Turning Basin	22	2.7	1.6	3.0	2.4
	24	1.4	1.1	1.9	1.5
Blair Waterway	11	1.0	0.8	1.3	1.0
	12	1.5	1.2	1.4	1.4
	13	1.4	1.3	1.6	1.4
	21	1.1	1.1	1.8	1.3
City Waterway	11	3.9	1.5	4.8	3.4
	13	1.5	1.5	2.5	1.8
	16	1.1	1.5	2.5	1.7
	17	1.3	1.8	1.7	1.6
	20	2.3	1.5	2.5	2.1
	22	1.4	1.0	1.5	1.3
Carr Inlet	11	1.9	1.0	1.3	1.4
	12	0.8	0.9	0.8	0.8
	13	0.5	1.0	0.8	0.8
	14	0.8	1.0	1.1	1.0

NOTE:
[a]RTR values were determined by dividing individual values by the mean for Carr Inlet.

SOURCE: Tetra Tech, Inc., 1985.

to the Carr Inlet mean, but that major differences in amphipod abundance were observed at most stations. The mean RTR values were influenced mainly by the values from the amphipod abundance RTR values. Either none, one, or two amphipods were found in most of the samples, whereas a mean of 71.2 per sample were found in Carr Inlet. The exceptions were stations 22 and 24 in Hylebos Waterway, station 11 in City Waterway and Station 13 in Blair Waterway where more amphipods were encountered. Based upon the three selected measures of benthos in

TABLE 7 RTR Values for Benthic Community Indices Measured at 20 Stations in Commencement Bay and Carr Inlet[a]

Site	Station	1/Total abundance	1/Species richness	1/Amphipod abundance	Mean RTR
Upper Hylebos Waterway	12	0.3	1.7	71.2	24.4
	14	0.6	1.5	71.2	24.4
	17	0.6	1.9	71.4	24.6
Hylebos Turning Basin	22	1.0	2.2	11.9	5.0
	23	7.2	4.9	71.2	27.8
	24	0.6	1.8	23.8	8.7
Blair Waterway	11	0.2	1.2	71.4	24.3
	12	nd	nd	nd	
	13	0.3	1.3	23.8	8.5
	21	0.4	1.5	71.2	24.4
City Waterway	11	0.1	3.8	35.7	13.2
	13	2.6	2.3	71.4	25.4
	16	2.3	3.2	71.2	25.6
	17	0.3	1.2	71.4	24.3
	20	0.4	1.2	71.2	24.3
	22	0.4	1.1	71.2	24.3
Carr Inlet	11	0.6	0.8	1.0	0.8
	12	1.1	0.8	5.1	2.3
	13	1.4	1.2	1.9	1.5
	14	1.3	1.2	0.4	1.0

NOTE:
[a] RTR values were determined by dividing individual values by the mean for Carr Inlet.

SOURCE: Tetra Tech, Inc., 1985.

Table 7, Hylebos-23, City-13, and City-16 had the most highly altered communities. Total abundance and species richness were low at Hylebos-23 and there were no amphipods there.

   Tetra Tech, Inc. (1985) identified a number of correlations among the chemistry, bioassay, and benthos results in the full data set. Among the subset of data summarized here, there are several interesting patterns. Station Hylebos-23 had highly altered benthos (low abundance, low species richness, devoid of amphipods), was highly contaminated (mainly with aromatic hydrocarbons and PCBs) and was relatively highly toxic. Whereas station 22 in Hylebos Waterway was the second most toxic among the 20 stations, it had the least altered benthos. Curiously, it had the highest amphipod abundance. Therefore, based upon the triad of measures at the 20 selected stations, station 22 in Hylebos Waterway was the most contaminated, and very toxic to amphipods and oyster larvae, but had minimally altered benthos relative to the conditions in the reference site. The Blair Waterway stations were minimally contaminated, slightly toxic, and had minimally altered benthos. Station City-11 at the head of the waterway was the most toxic and the most contaminated of the City Waterway stations, and had low species richness, but had high total abundance and a mean of two amphipods in the benthos grabs. Overall, the City Waterway stations were moderately contaminated (mainly with aromatic hydrocarbons), were moderately toxic (station 11 was the most toxic of the 20), and had relatively highly altered benthos (often species poor and without amphipods).

Triaxial plots of RTR values for three of the stations are illustrated in Figure 3. Note three unique scales are used to plot the values on the same figure. As was observed with the triaxial plots of San Francisco Bay data, very little parallelism is indicated among the

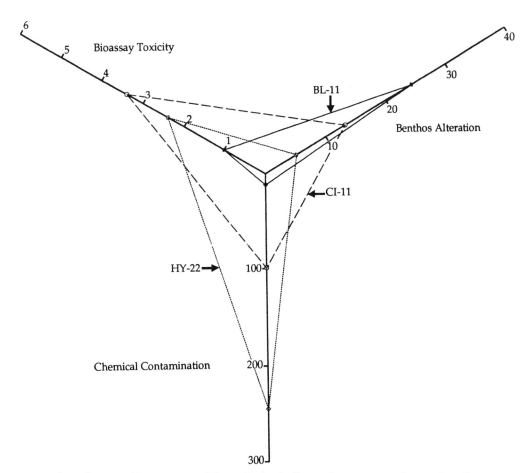

FIGURE 3 The Sediment Quality Triad for three stations in Commencement Bay waterways, based upon RTR values from Tables 5, 6, and 7. Each of the axes has a unique scale.

three triad components among stations. Contaminant concentrations were highest at Hylebos-22, toxicity was highest at City-11, and benthos were most altered at Blair-11. Given these data, it would be difficult to rank the three stations; all three should be classified as having low sediment quality, but based upon different measures.

Selected bioassay and chemical data from all 56 stations are compared in Table 8 to illustrate the levels of contamination associated with three means and ranges in toxicity. The toxicity data chosen for this example are those from the amphipod bioassay. The data from the chemical analyses are those for three organic compound classes, four trace metals, and total organic carbon. Sediment texture data are also listed as percent fines (silt + clay). Three ranges in amphipod mortality were established following an examination of the frequency distribution of the results: 1.0 to 3.8 dead out of 20, 4.0 to 7.4 dead, and 10.4 to 20 dead. Most of the sediments that killed a mean of 4 or more amphipods out of 20 were significantly different than the Carr Inlet stations at $p < 0.05$ (Tetra Tech, Inc., 1985). The mean contaminant concentrations were usually highest in those samples with highest toxicity. For example, the mean concentration of low-molecular-weight aromatic hydrocabons was 6.98 ppm in the most toxic samples and 1.60 ppm in the least toxic samples. However, this pattern in co-occurrence did not obtain with the PCB data. The largest difference in chemical concentration between the least toxic and the most toxic stations was with copper where a 33-fold difference was apparent. The largest differences in chemical concentrations usually occurred between the intermediate toxicity and highly toxic stations. This pattern matches the largest difference in mean mortality; i.e., between the intermediate group of stations and the high-toxicity group of stations.

The data in Table 8 indicate very large standard deviations and ranges in chemical concentrations among stations that had the least, intermediate, and the highest toxicity. For example, among the five stations that were most toxic to amphipods, the standard deviation in the cadmium concentration was nearly twice the mean, and the range among those five stations (0.8 to 184 ppm) nearly covered the range for all 56 stations. While the mean total organic carbon increased with increased mean amphipod mortality, the ranges within each group of stations were also very high. The percent of the sediments composed of fines was lowest among the most toxic stations.

Based upon the full set of data from this study (Tetra Tech, Inc., 1985) and from previous studies in the Commencement Bay waterways, areas have been classified and ranked for remedial actions. Included among the highest priority problem areas were the Upper Hylebos Waterway, the Hylebos Turning Basin, and Upper City Waterway. Remedial action planning is proceeding under the direction of the Washington Department of Ecology (Dave Bradley, WDOE, personal communication).

CONCLUSIONS AND RECOMMENDATIONS

The Sediment Quality Triad is a concept for use in classifying sediment quality that relies upon synoptic measures of chemical

TABLE 8 Means, Standard Deviations, and Ranges in Contaminant Concentrations in Sediments from 56 Stations in Commencement Bay and Carr Inlet Associated with Three Means (Standard Deviations and Ranges) in Mortality of Rhepoxynius abronius.

	Rhepoxynius abronius mortality[a]	LMWPAH ppm	HMWPAH ppm	PCBs ppm	Cd ppm	Cu ppm	Zn ppm	Hg ppm	TOC %	Fines %
Mean	15.7	6.98	9.79	0.02	41.7	2820	941	11.2	7.3	33.4
S.D.	3.9	8.44	12.82	0.01	79.8	4881	1373	22.8	6.2	22.8
Range n=5	10.4–20	0.68–21.39	0.73–30.89	0.01–0.03	0.8–184	32–11400	29–3320	0.1–52	0.6–16	3–66
Mean	5.2	2.03	6.18	0.30	2.8	117.8	211.4	0.3	2.9	66.2
S.D.	1.0	1.32	6.44	0.61	2.2	98.1	341.6	0.2	3.2	27.2
Range n=21	4.0–7.4	0.03–4.80	0.06–30.00	0.01–2.00	0.3–9.6	5–385	15–1620	0.1–1.1	0.4–15.1	4–88
Mean	2.8	1.60	4.86	0.08	2.3	86	108.6	0.21	2.6	51.1
S.D.	0.8	1.41	4.80	0.10	1.3	70	80.1	0.14	2.3	27.5
Range n=30	1.0–3.8	0.02–5.85	0.02–18.40	0.003–0.42	0.4–6.0	5–311	15–268	0.03–0.6	0.2–10.9	1–91

NOTE:
[a] Number dead out of 20 animals

SOURCE: Tetra Tech Inc., 1985.

contamination and measures of biological effects. The three components of the triad provide measures of contamination, toxicity, and resident benthos community structure. Data resulting from the three measures can be used to descriptively compare sediment quality among sampling stations, to classify or rank the relative quality of sediment sampling stations, to determine the spatial extent of poor sediment quality, to characterize putative uncontaminated reference conditions, and to estimate the contaminant levels associated with ranges in biological effects. This concept has been used in a number of assessments and surveys performed in various regions of the United States. Overall, a total of 300 to 400 stations have been tested thus far. The triad of measures provides a powerful preponderance of evidence of sediment quality and has been effective in identifying those areas where sediments are not only contaminated, but also elicit damaging biological effects. This approach is complementary to the bioassay, equilibrium partitioning, and AET approaches described elsewhere in this volume.

Available data from studies in which the triad concept has been used usually indicate a general, overall pattern of co-occurrence between chemical contamination and biological effects. However, this pattern has a relatively large degree of uncertainty and variation on a station-to-station basis. There is often a large range in chemical concentrations among groups of stations with similar toxicity and benthos alterations. Also, sediments that demonstrate high biological effects often have relatively high concentrations of complex mixtures of chemicals, precluding the identification of individual chemicals as the etiological agents. Therefore, caution must be used in setting absolute standards or criteria, based solely upon field effects-based sediment values. Some estimate of uncertainty must accompany any such standards or criteria. Descriptive interpretations of resulting data are needed to identify ecological relationships between controlling physical/chemical parameters and biological variables and to classify sediments.

Studies of potentially polluted areas with the triad concept are needed to assess and estimate the extent of poor sediment quality that is biologically damaging. New approaches to treating and evaluating the resulting data are needed. The RTR approach has certain weaknesses and could be improved. Short-cut methods for acquiring chemical, toxicity, and benthos data are needed to reduce costs. An effort to pool data from triad studies from many parts of the country is needed to determine if there is agreement in the concentrations of sediment-associated contaminants that co-occur with measures of biological effects. These concentrations, in turn, should be compared with those determined to be toxic in spiked sediment bioassays and to exceed water quality standards through the theoretical, equilibrium-partitioning approach.

REFERENCES

Battelle Pacific Northwest Laboratories. 1986. Reconnaissance Survey of Eight Bays in Puget Sound, Vols. 1 and 2. Prepared for U.S. EPA

Region 10. Seattle, Wash.: Battelle. 231 pp.

Chapman, P. M. 1986. Sediment quality criteria from the Sediment Quality Triad: An example. Envir. Toxicol. Chem. 5:957-964.

Chapman, P. M., R. N. Dexter, S. F. Cross, and D. G. Mitchell. 1986. A field trial of the Sediment Quality Triad in San Francisco Bay. NOAA Technical Memorandum NOS OMA 25. National Oceanic and Atmospheric Administration, Rockville, Md. 134 pp.

Chapman, P. M., R. N. Dexter, and E. R. Long. 1987. Synoptic measures of sediment contamination, toxicity, and infaunal community composition (the Sediment Quality Triad) in San Francisco Bay. Mar. Ecol. Prog. Series 37:75-96.

Gentile, J. H., K. J. Scott, S. Lussier, M. Redmond. 1985. Application of Laboratory Population Responses for Evaluating the Effects of Dredged Material. Field Verification Program Tech. Rpt. D-85-8. U.S. EPA/ACOE Final Report. Narragansett, RI: U.S. EPA. 72 pp.

Long, E. R. and P. M. Chapman. 1985. A Sediment Quality Triad: Measures of sediment contamination, toxicity and infaunal community composition in Puget Sound. Mar. Pollut. Bull. 16:405-415.

Long, E. R., D. MacDonald, M. B. Matta, K. VanNess, M. Buchman, and H. Harris. 1988. Status and Trends in Concentrations of Contaminants and Measures of Biological Stress in San Francisco Bay. NOAA Tech. Memo. NOS/OMA 41. Rockville, Md: Ocean Assessments Division. 265 pp.

Rogerson, P. F., S. C. Schimmel, G. Hoffman. 1985. Chemical and Biological Characterization of Black Rock Harbor Dredged Material. Field Verification Program Tech. Rpt. D-85-9. U.S. EPA/ACOE Final Report. Narragansett, RI: U.S. EPA. 110 pp.

Schiewe, M. H., E. G. Hawk, D. I. Actor, and M. M. Krahn. 1985. Use of a bacterial bioluminescence assay to assess toxicity of contaminated marine sediments. Can. J. Fish. Aquat. Sci. 42:1244-1248.

Stober, Q. J. and K. K. Chew. 1984. Renton Sewage Treatment Plant Project: Seahurst Baseline Study. Final Report prepared for Seattle METRO. Fisheries Research Institute, University of Washington, Seattle, WA.

Swartz, R. C., W. A. DeBen, K. A. Sercu, and J. O. Lamberson. 1982. Sediment toxicity and the distribution of amphipods in Commencement Bay, Washington, USA. Mar. Poll. Bull. 13:359-364.

Swartz, R. C., F. A. Cole, D. W. Schults, and W. A. DeBen. 1986. Ecological changes in the Southern California Bight near a large sewage outfall: benthic conditions in 1980 and 1983. Mar. Ecol. Prog. Series 31:1-13.

Tetra Tech, Inc. 1985. Commencement Bay Nearshore/Tideflats Remedial Investigation. Vol. 1. Prepared for Washington State Dept. of Ecology, U.S. Environmental Protection Agency, Region 10. Final Report EPA 910/9-85-134b. Olympia: WDOE.

A REVIEW OF THE DATA SUPPORTING THE EQUILIBRIUM PARTITIONING APPROACH TO ESTABLISHING SEDIMENT QUALITY CRITERIA

Dominic M. Di Toro
Manhattan College

The development of sediment quality criteria has been underway for some time, and a number of approaches have been suggested (see Chapman, 1987 for a review). The discussion that follows presents the data and interpretation that support the equilibrium partitioning approach (Pavlou and Weston, 1983; Pavlou, 1987) adopted for establishing sediment quality criteria by the U.S. Environmental Protection Agency (EPA) Criteria and Standards Division. The acknowledgment section of this paper lists the many contributors to this effort. In this regard the author should be viewed as a spokesman for workers involved.

Perhaps the first question to be answered is "why not use the existing procedure for the development of water quality criteria?" After all, water quality criteria have demonstrated their utility; have been reviewed by independent scientific groups; and--most important--a methodology has been developed (Stephan et al., 1985) that presents the supporting logic, establishes the minimum toxicological data set required to develop a criteria, and specifies the numerical procedures to calculate resulting criteria values. A natural extension would be to apply these methods directly to sediments.

One reason water quality criteria have practical utility is that they are based on straightforward measurements for most chemicals, either total concentration or the recently proposed weak acid extractions for certain metals, and they appear to be broadly applicable. The experience with site-specific modifications of the national water quality criteria have demonstrated that the "water effect ratio" has averaged 3.5 (Spehar and Carlson, 1984; Carlson et al., 1986). The implication is that subtle differences in water chemistry are not an overwhelming impediment to nationally applicable, numerical water quality criteria.

The primary impediment to direct application of the water quality paradigm to sediment quality criteria is the use of total sediment chemical concentration as a measure of bioavailable, or even potentially bioavailable, concentration. This is not supported by the available data (see, for example, Luoma, 1983). A review of recent experiments is presented below. Different sediments can differ by factors of ten or more in toxicity for the same total chemical concentration of a toxicant. This is a severe obstacle since, without some quantitative estimate of the bioavailable chemical concentration in a sediment, it is impossible to evaluate its toxicity based on chemical measurements.

This is true regardless of the methodology used to assess biological impact, be it field data sets comprising benthic biological and chemical sampling or laboratory toxicity experiments. Without a unique relationship between the chemical measurement and the biological endpoints, which applies across the range of sediment properties that affect bioavailability, the cause and effect linkage is not supportable. If the same total chemical concentration is ten times more toxic in one sediment than another, how does one set a universal sediment quality criteria that depends only on the total sediment chemical concentration? Some attempt must be made to address the issue of bioavailability. Further it appears that any sediment quality criteria methodology that depends on chemical measurements in the sediment must face this issue as well. It is not unique to the equilibrium partitioning methodology.

BIOAVAILABILITY AND PORE WATER CONCENTRATION

The observation that provided the key insight to the solution of the problem of quantifying the bioavailability of chemicals in sediments was that the dose-response curve for the biological effect of concern could be correlated not to the total sediment chemical concentration (μg chemical/g sediment) but to the interstitial water (i.e., pore water) concentration (μg chemical/liter pore water). Since this observation is a critical part of the logic behind the equilibrium partitioning approach to sediment criteria, a substantial amount of data has been assembled to support this observation. The data are presented in a uniform fashion in Figures 1 through 5. The biological response variable--survival rate, growth rate, body burden--is plotted versus the total sediment concentration in the top panel and versus the measured pore water concentration in the bottom panel.

The kepone experiments, Figure 1, are particularly dramatic (Adams et al., 1985; Ziegenfuss et al., 1986). Consider first the top panels. For the low organic carbon sediment (0.09 percent) the fiftieth percentile total kepone concentration for both *C. tentans* mortality LC_{50} and growth rate reduction EC_{50} are < 1 μg/g. By contrast the 1.5 percent organic carbon sediment EC_{50} and LC_{50} are approximately 8 and 10 μg/g respectively. The high organic carbon sediment (12 percent) has still higher LC_{50} and EC_{50}s on a total sediment kepone concentration basis (42 and 49 μg/g). However, as shown in the bottom panels, essentially all the data collapse into a single curve when the pore water concentrations are used as the correlating concentrations. Possible reasons for this observation will be discussed below. It is important at this stage only to realize that on a pore water basis the biological responses are essentially the same for the three different sediments: the EC_{50} = 23 μg/liter and LC_{50} = 28 μg/liter, whereas when they are evaluated on a total sediment kepone basis they exhibit an almost 50-fold range in kepone toxicity. Figure 2 presents similar data for fluoranthene and cadmium and the marine amphipod *Rhepoxynius* (Kemp and Swartz, 1986; Swartz et al.,

1987). The results of the fluoranthene experiments parallel those for kepone. The lowest organic carbon fraction sediments, 0.2 percent exhibits the lowest LC_{50} on a total sediment concentration basis (3.1 µg/g) and as the organic carbon concentration increases the LC_{50}s increase (6.7 and 11 µg/g). On a pore water basis, however, the data collapse to a single dose response curve. The cadmium experiments were done using constant pore water concentrations and a sediment amended with varying quantities of organic carbon. The unamended and 0.25 percent additional organic carbon exhibit essentially similar responses. However the 1 and 2 percent amended sediments had much higher LC_{50}s. Using the pore water concentrations again collapses the data into one dose response curve.

Figure 3 presents data for DDT and endrin and the freshwater amphipod *Hyalella* (Nebeker and Schuytema, 1988). The response is again similar to that observed above. On a total sediment concentration basis, the organism response is different for the different sediments. On a pore water basis, however, the dose responses are similar.

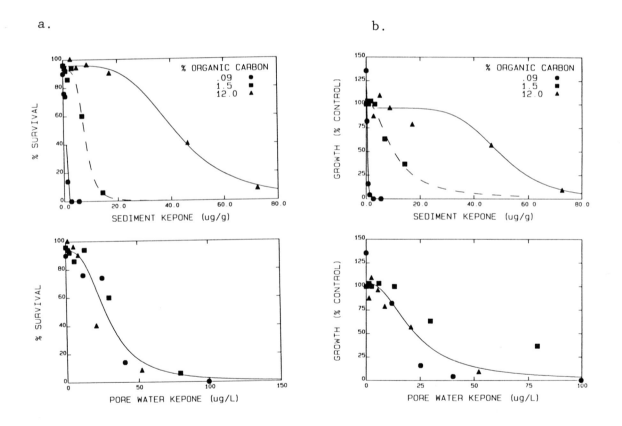

FIGURE 1 Acute (a) and chronic (b) toxicity of kepone to *Chironomus tentans*. SOURCE: Adams et al., 1983.

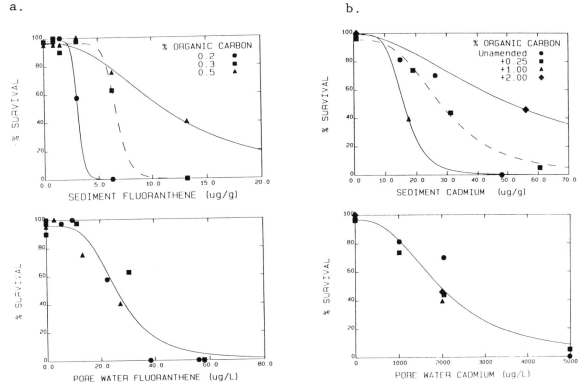

FIGURE 2 Acute toxicity of fluoranthene (a) and cadmium (b) to *Rhepoxynius abronius*. SOURCE: Swartz et al., 1987.

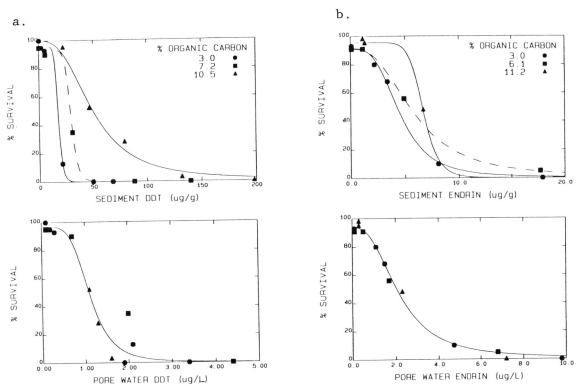

FIGURE 3 Acute toxicity of DDT (a) and endrin (b) to *Hyalella*.
SOURCE: Nebeker and Schuytema, 1988.

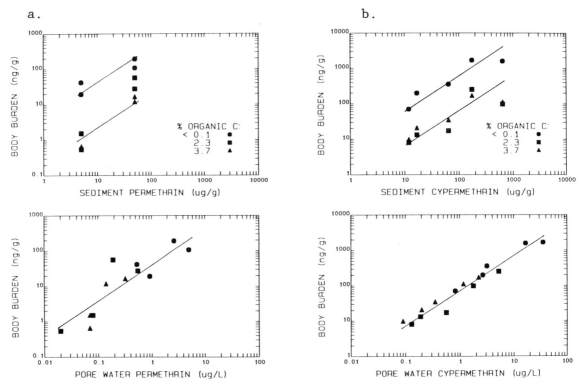

FIGURE 4 Bioaccumulation of permethrin (a) and cypermethrin (b) in *Chironomus tentans*. SOURCE: Muir et al., 1985.

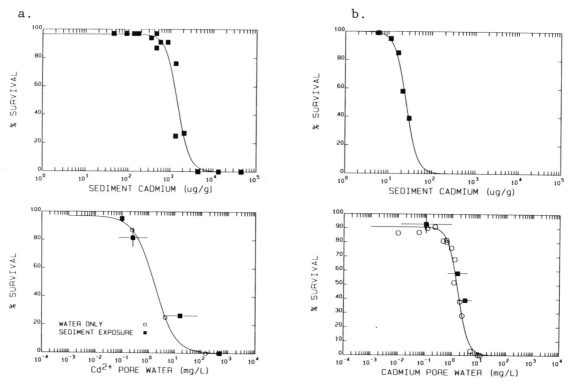

FIGURE 5 Acute toxicity of cadmium to *Ampelisca* (a) and *Rhepoxynius* (b). SOURCE: Scott and DiToro, 1988; Swartz et al., 1985.

Data for another biological endpoint, namely organism body burden, are examined in Figure 4. Two synthetic pyrethroids, cypermethrin and permethrin, and *C. tentans* were used (Muir et al., 1985). Three sediments, one of which was laboratory grade sand, were employed. The bioaccumulation from the sand was approximately an order of magnitude higher than the organic carbon containing sediment for both cypermethrin and permethrin (top panels). On a pore water basis, however, the bioaccumulation appeared to be linear (the lines are slope = 1) and independent of sediment type (bottom panels).

Two sets of data are compared in Figure 5. which make an additional point. They compare the response of *Rhepoxynius* (Swartz et al., 1985) and *Ampelisca* (Scott and Di Toro, 1988) to cadmium in seawater-only exposures and to measured pore water concentrations in sediment exposures (lower panels). Note that the responses are the same with or without the sediment present. The dose response curves using total cadmium concentrations are also shown (top panels). It is interesting to note that two organisms show essentially the same sensitivity to cadmium. Yet the total cadmium LC_{50}s differ by almost two orders of magnitude (25 and 2000 µg/g respectively) for the different sediments.

These observations--that organism dose response curves for different sediments can be collapsed into one curve if pore water is considered as the dose concentration--can be interpreted in a number of ways. However, from a purely empirical point of view it suggests that if it were possible to either measure the pore water concentration of a chemical, or to predict it from the total sediment concentration and the relevant sediment properties, then that concentration could be used to quantify the chemical dose the sediment would deliver to an organism. Thus one is lead to examine the state of the art with respect to predicting the partitioning of chemicals between the solid and liquid phase. This is examined in the next section.

PARTITIONING OF CHEMICALS

A discussion of modeling sorption to particles is best organized by classes of chemicals. For nonpolar hydrophobic organic chemicals sorbing to natural soils and sediment particles, a number of empirical models have been suggested (see Karickhoff, 1984 for an excellent review). The characteristic that indexes the hydrophobicity of the chemical is the octanol-water partition coefficient, K_{ow}. The important particle property is the mass fraction of organic carbon, f_{oc}. For particles with $f_{oc} > 0.5$ percent the organic carbon appears to be the predominate sorption phase. The only other important environmental variable appears to be the particle concentration itself (O'Connor and Connolly, 1980). For the reversible (or labile) component of sorption, a model has been proposed that predicts the partition coefficient of nonpolar hydrophobic chemicals over a range of nearly seven orders of magnitude with a log_{10} standard error of 0.38 (Di Toro, 1985).

For this class of chemicals the partitioning problem appears to be solved. If c_{pore} is the aqueous pore water concentration (µg chemical/liter pore water), and r is the solid-phase concentration (µg chemical/g sediment), then defining the partition coefficient K_p as

$$r = K_p c_{pore}, \qquad (1)$$

then for sediment-pore water partitioning, K_p is given almost exactly by[1]

$$K_p = f_{oc} K_{ow}. \qquad (2)$$

What is important about this equation is that the partition coefficient for this class of chemicals is linear in the organic carbon fraction f_{oc}. As a consequence, the relationship between solid-phase concentration r and pore water concentration c_{pore} can be written

$$r = f_{oc} K_{ow} c_{pore}, \qquad (3)$$

or:

$$\frac{r}{f_{oc}} = K_{ow} c_{pore}. \qquad (4)$$

If we define

$$r_{oc} = \frac{r}{f_{oc}} \qquad (5)$$

as the organic carbon normalized sediment concentration (μg chemical/g organic carbon) then

$$r_{oc} = K_{ow} c_{pore}. \qquad (6)$$

Hence we arrive at the following conclusion: for a specific chemical with fixed K_{ow} the organic carbon normalized total sediment concentration r_{oc} is proportional to the pore water concentration c_{pore}.

ORGANIC CARBON NORMALIZATION

From the above discussion we conclude that if a dose-response curve correlates to pore water concentration, it should correlate equally well to organic carbon-normalized total chemical concentration independent of sediment properties. Of course this only applies to nonpolar hydrophobic organic chemicals, since the rationale is based on a partitioning theory for these chemicals. Figures 6 through 8 present these comparisons. The lower panels present the response-total sediment concentration data which is organic carbon normalized (μg chemical/g organic carbon). The top panels repeat the response-pore water concentration plots. Note that in all cases the correlation is

---

[1] The exact equation is $\log_{10} K_p = \log_{10} f_{oc} + 0.0028 + 0.983 * \log_{10} K_{ow}$ (Di Toro, 1985).

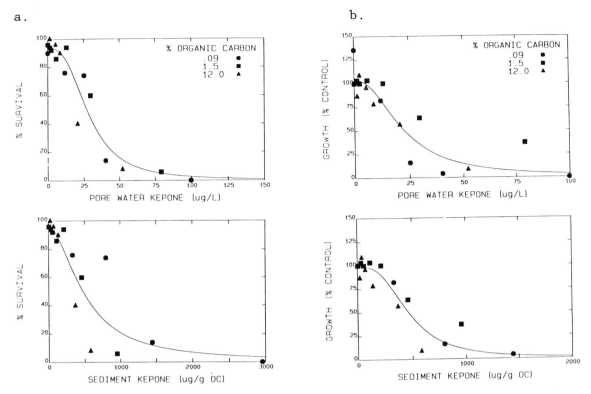

FIGURE 6 Acute (a) and chronic (b) toxicity of kepone to *Chironomus tentans*. SOURCE: Adams et al., 1983.

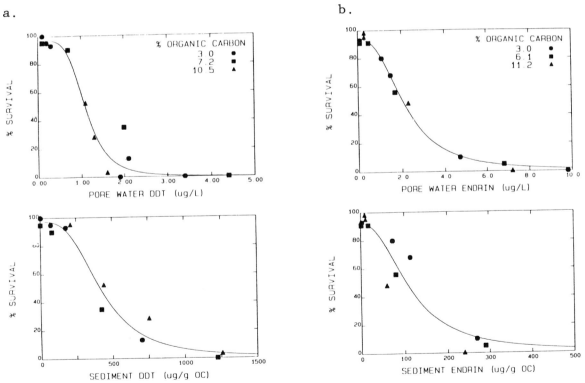

FIGURE 7 Acute toxicity of DDT (a) and endrin (b) to *Hyalella*. SOURCE: Nebeker and Schuytema, 1988.

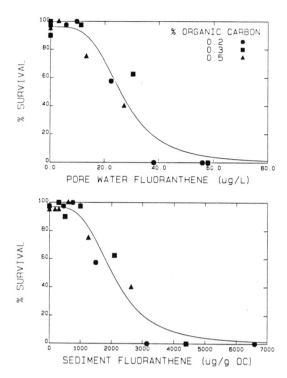

FIGURE 8 Acute toxicity of fluoranthene to *Rhepoxynius abronius*. SOURCE: Swartz et al., 1987.

essentially the same whether pore water or organic carbon-normalized sediment concentrations are used for the chemical dose. This implies that either of these concentrations can be used. Since it is much more convenient to measure total sediment concentration and organic carbon fraction the latter seems more practical.

EQUILIBRIUM PARTITIONING APPROACH TO SEDIMENT CRITERIA

The evidence presented above suggests that the pore water concentration correlates to the biological responses examined. Hence the biological effect levels generated in toxicity experiments can be associated with the pore water concentrations. The procedure to set a sediment quality criteria for a chemical would be to perform a series of toxicity tests using benthic plants and animals, and establish the range of effect concentrations based on the pore water concentrations. The procedures set forth in the national guidelines (Stephan et al., 1985) could be used directly. If it turned out that the most sensitive benthic and pelagic species have similar sensitivity, then this would be equivalent to requiring that the pore water concentration be at the water quality criteria concentration. Hence the sediment quality criteria, r_{SQC} would be calculated using the water quality criteria concentration c_{WQC}, and the partition coefficient K_p, as follows:

$$r_{SQC} = K_p c_{WQC} \tag{7}$$

where r_{SQC} would be the total sediment concentration that is in equilibrium with the pore water at the criteria level concentration. It is from the equilibrium requirement that this approach is termed the "equilibrium partitioning" method. With remarkable foresight this approach was suggested for establishing sediment quality criteria by Pavlou and Weston (1983) before the evidence discussed above was available. For nonpolar hydrophobic organic chemicals, $K_p = f_{oc} K_{ow}$, so that

$$r_{SQC} = f_{oc} K_{ow} c_{WQC}. \qquad (8)$$

Therefore, in order to compute a sediment quality criteria, r_{SQC}, for a particular chemical and sediment, one needs to know (1) the water quality criteria, c_{WQC}, (2) the octanol-water partition coefficient of the chemical, K_{ow}, and (3) the organic carbon fraction of the sediment, f_{oc}. A more easily remembered quantity is the organic carbon-normalized sediment quality criteria $r_{oc,SQC}$, where

$$r_{oc,SQC} = \frac{r_{SQC}}{f_{oc}} = K_{ow} c_{WQC}. \qquad (9)$$

This quantity is sediment independent, to the extent that organic carbon is the sole determinant of hydrophobic partitioning. It has been suggested that the limit of applicability is $f_{oc} \geq 0.5$ percent (Karickhoff, 1984). However, the fluoranthene data presented above suggest that it might even apply to sediments with lower organic carbon fractions. These procedures have been used to generate interim sediment criteria values (Cowan and Di Toro, 1988).

THEORETICAL SPECULATIONS

The data presented above raise a number of interesting issues. The most surprising result is that the biological effects examined appear to correlate to the interstitial water concentration, independent of sediment type. This has been interpreted to mean that exposure is primarily via the pore water, however the data correlate equally well to the organic carbon-normalized sediment concentration. This may just reflect the validity of the organic carbon-based, hydrophobic chemical partitioning model.

However, another interpretation is possible. Consider the hypothesis that the chemical potential (or fugacity) of a chemical controls its biological activity. For a chemical dissolved in pore water at concentration c_{pore}, the chemical potential, μ_{pore} is

$$\mu_{pore} = \mu_o + RT \ln(c_{pore}), \qquad (10)$$

where μ_o is the standard state chemical potential, and RT is the product of the universal gas constant and absolute temperature (Stumm and Morgan, 1970). For a chemical dissolved in organic carbon--

assuming that particle organic carbon can be characterized as a homogeneous phase--its chemical potential is

$$\mu_{oc} = \mu_o + RT\ln(r_{oc}), \qquad (11)$$

where r_{oc} is the weight fraction of chemical in organic carbon. If the pore water is in equilibrium with the sediment organic carbon, then

$$\mu_{pore} = \mu_{oc}. \qquad (12)$$

The chemical potential that the organism experiences from either route of exposure is the same. Hence, so long as the sediment is in equilibrium with pore water, the route of exposure is immaterial. Further if chemical potential (or equivalently, fugacity) is proportional to biological effects then the issue becomes "in which phase is μ most directly measured?"

Pore water concentration is an obvious suggestion. However, it is necessary that chemical complexed to colloidal--or as it is loosely called, dissolved organic carbon (DOC)--be a small fraction of the total measured concentration. If the partition coefficient for DOC is on the order of K_{ow} (Landrum, 1987), then the requirement is

$$c_{DOC}K_{ow} \ll 1. \qquad (13)$$

For $\log_{10} K_{ow} > 5$ this may not be the case, and c_{pore} would not be a valid measure of the free chemical concentration.

Total sediment concentration normalized by sediment organic carbon fraction is a second obvious choice. Note that this measurement is not affected by DOC complexing since that is affecting the distribution within the aqueous phase but not the validity of Equation 11. The only requirement is that sediment organic carbon be the only sediment phase that contains significant amounts of the chemical. At $f_{oc} < 0.5$ percent this may no longer be the case.

SEDIMENT QUALITY CRITERIA FOR METALS

The equilibrium partitioning methodology for establishing sediment quality criteria requires that effects concentration be determined in an accessible phase and that the chemical potential of the chemical be computed. The experiments presented above suggest that pore water concentrations of cadmium correlate to biological effect. A substantial number of water column experiments point to the fact that biological effects can be correlated to the divalent metal activity (Me^{2+}) (Sunda and Guillard, 1976; Sunda et al., 1978; Anderson and Morel, 1978; Zamuda and Sunda, 1982). Hence the required partitioning model should predict (Me^{2+}) in the pore water.

Models are available for cation and anion sorption to metal oxides in laboratory systems (see the articles in Stumm [1987] for recent summaries). The models for natural particles are less well developed. Since the ability to predict partition coefficients is required if the pore water metal concentrations are to be inferred from the total concentration, some practical model is necessary.

A start in this direction was made during a recent conference (see Di Toro et al., 1987). A more formal presentation is available (Jenne et al., 1986). The basic idea is that instead of only one site of sorption, organic carbon, as is assumed for nonpolar hydrophobic chemicals, three sites of sorption are considered. In oxic soils and sediments these have been identified as particulate organic carbon (POC) and the oxides of iron and manganese (Jenne, 1968, 1977; Luoma and Bryan, 1981; Oakley et al., 1982). They are important because they have a large sorptive capacity. Further they appear as coatings on the particles and occlude the other mineral components. Thus they provide the primary sites for sorption of metals and they restrict the importance of the clay and other mineral components of soils and sediments.

In addition to the sites of adsorption it is necessary to quantify the fraction of total sediment metal that is chemically interacting with the pore water. A substantial effort has been expended over the years in attempting to determine the "bioavailable" portion of trace metals in soils and sediments using chemical extractions (see Jenne [1987] for a review and recommended procedure). The use of a relatively mild reductant (hydroxylamine hydrochloride), which dissolves the Fe and Mn oxides and liberates the sorbed metals, is recommended. The reported results using extracted iron normalization (Tessier et al., 1984) have been very encouraging. Similar results using an acid extraction have been found for arsenic in *Nereis,* a deposit feeding polychaete, and *Macoma,* a deposit-feeding bivalve (Langston, 1980); and copper in aquatic plants (Campbell et al., 1985). For mercury body burdens in various benthic species (Langston, 1986), a strong correlation exists between the sediment concentration normalized by organic matter content.

The most direct evidence for the utility of the extraction-phase normalization procedure, however, is from simultaneous observations of interstitial water and sediment metal concentrations. Initial data of this type (Tessier et al., 1985; Johnson, 1986) suggest that the extraction partitioning methodology can be used to establish metals criteria in a way that directly addresses the bioavailability problem.

APPLICATION TO MARINE SEDIMENTS

The model presented above is directed at the oxic layer in freshwater sediments. It is not clear whether a similar model can be developed for estuarine and marine sediments. Certainly the role of metal sulfides must be explicit in the formulation. Interestingly, some data from estuaries (Langston, 1980, 1986) indicate iron normalization may also apply. The application of the equilibrium partitioning approach to these sediments is in the formative stage.

ACKNOWLEDGMENTS

The EPA project director of the EPA Criteria and Standards Division's sediment quality criteria effort is Christopher Zarba. The organizations and personal involved are (in alphabetical order) the Battelle Pacific Northwest Laboratory, Christina Cowan (Project Leader); Everett Jenne; Drexel University, Herbert Allen; Envirosphere, Spyros Pavlou; EPA ERL Corvallis, Richard Swartz; EPA ERL Duluth, Nelson Thomas; EPA ERL Narragansett, David Hansen, John Scott; Manhattan College, John Mahony. Many other people have contributed ideas and data to the effort. The author particularly thanks the scientists that provided raw data and results prior to publication for inclusion in this review.

REFERENCES

Adams, W. J., Kimerle, R. A. and Mosher, R. G. 1985. Aquatic safety assessment of chemicals sorbed to sediments. In Aquatic Toxicology and Hazard Assessment: Seventh Symposium, R. D. Cardwell, R. Purdy and R. C. Bahner, eds. Philadelphia: American Society for Testing and Materials. Pp. 429-453.

Anderson, D. M. and Morel, F. M. M. 1978. Copper sensitivity of *Gonyaulax tamarensis*. Limnol. Oceanogr. 23:283-295.

Campbell, P. G. C., A. Tessier, M. Bisson, and R. Bougie, 1985. Accumulation of copper and zinc in the yellow lily, *Nuphar variegatum*: relationships to metal partitioning in the adjacent lake sediments. Can. J. Fish. Aquat. Sci. 42:23-32.

Carlson, A. R., H. Nelson, and D. Hammermeister. 1986. Development and validation of site-specific water quality criteria for copper. Environ. Toxicol. and Chem. 5:997-1012.

Chapman, G. A. 1987. Establishing sediment criteria for chemicals--regulatory perspective. In Fate and Effects of Sediment--Bound Chemicals in Aquatic Systems, K. L. Dickson, A. W. Maki, and W. A. Brungs, eds. New York: Pergamon Pres. Pp. 355-376.

Cowan, C. E. and D. M. Di Toro. 1988. Interim Sediment Criteria Values for Nonpolar Hydrophobic Compounds. Richland, Washington: Battelle Pacific Northwest Laboratories.

Di Toro, D. M. 1985. A particle interaction model of reversible organic chemical sorption. Chemosphere 14(10):1503-1538.

Di Toro, D. M., F. Harrison, E. Jenne, S. Karickhoff, and W. Lick. 1987. Synopsis of discussion session 2: Environmental fate and compartmentalization. In Fate and Effects of Sediment-Bound Chemicals in Aquatic Systems, K. L. Dickson, A. W. Maki, and W. A. Brungs, eds. New York: Pergamon Press. Pp. 136-147.

Jenne, E. A. 1968. Controls on Mn, Fe, Co, Ni, Cu, and Zn concentrations in soils and water--the significant role of hydrous Mn and Fe oxides. In Advances in Chemistry. Washington, D.C.: American Chemical Society. Pp. 337-387.

Jenne, E. A. 1977. Trace element sorption by sediments and soil--sites and processes. In Symposium on Molybdenum in the Environment, Vol.

2, W. Chappell and K. Petersen, eds. New York: M. Dekker, Inc. Pp. 425-553.

Jenne, E. A. 1987. Sediment Quality Criteria for Metals: II Review of Methods for Quantitative Determination of Important Adsorbents and Sorbed Metals in Sediments. Prepared for EPA Criteria and Standards Division, Washington, D.C. Richland, Wash.: Battelle, Pacific Northwest Laboratories. Pp. 1-38.

Jenne, E. A., D. M. Di Toro, H. E. Allen, and C. S. Zarba, 1986. An activity-based model for developing sediment criteria for metals: A new approach. Chemicals in the Environment, International Conference, Lisbon, Portugal.

Johnson, C. A. 1986. The regulation of trace element concentrations in river and estuarine waters contaminated with acid mine drainage: The adsorption of Cu and Zn on amorphous Fe oxyhydroxides. Geochim. et Coschim. Acta 50:2433-2438.

Karickhoff, S. W. 1984. Organic pollutant sorption in aquatic systems. J. Hydraulic Div. ASCE 110(6):707-735.

Kemp, P. F. and Swartz, R. C. 1986. Acute toxicity of interstitial and particle-mound cadmium to a marine infaunal amphipod. Submitted to J. Exp. Marine Biol. Ecol.

Landrum, P. F., S. R. Nihart, B. J. Eadie, and L. R. Herche. 1987. Reduction in bioavailability of organic contaminants to the amphipod *Pontoporeia hoyi* by dissolved organic matter of sediment interstitial water. Environ. Toxicol. and Chem. 6:11-20.

Langston, W. J. 1980. Arsenic in U.K. estuarine sediments and its availability to benthic organisms. J. Mar. Biol. Assoc. U.K. 60:869-881.

Langston, W. J. 1986. Metals in sediments and benthic organisms in the Mersey estuary. Estuar. Coast. and Shelf Sci. 23:239-261.

Luoma, S. N. 1983. Bioavailability of trace methals to aquatic organisms--a review. Sci. Total Environ. 28:1-22.

Luoma, S. N. and G. W. Bryan. 1981. A statistical assessment of the form of trace metals in oxidized sediments employing chemical extractants. Sci. Total Environ. 17:165-196.

Nebeker, A. and G. Schuytema. 1988. DDT/Endrin Results. Evaluation of Carbon Normalization Theory. Report. Corvallis, Ore.: U.S. EPA Environmental Research Laboratory.

O'Connor, D. J. and J. Connolly. 1980. The effect of concentration of adsorbing solids on the partition coefficient. Water resources 14:1517-1523.

Oakley, S. M., K. J. Williamson, and P. O. Nelson. 1980. The Geochemical Partitioning and Bioavailability of Trace Metals in Marine Sediments. Corvallis: Oregon State University Water Resources Institute. Pp. 1-84.

Pavlou, S. P. and D. P. Weston. 1983. Initial Evaluation of Alternatives for Development of Sediment Related Criteria for Toxic Contaminants in Marine Waters (Puget Sound). Phase I. Development of Conceptual Framework. Bellevue, Wash.: JRB Associates.

Pavlou, S. P. 1987. The use of the equilibrium partitioning approach in determining safe levels of contaminants in marine sediments. In Fate and Effects of Sediment-Bound Chemicals in Aquatic Systems, K. L. Dickson, A. W. Maki, and W. A. Brungs, eds. New York: Pergamon Press. Pp. 388-412.

Scott, J. and D. M. Di Toro. 1988. Preliminary Experimental Results. EPA Environmental Research Laboratory, Narragansett, R.I. Manhattan College, Bronx, N.Y.

Spehar, R. L. and A. R. Carlson. 1984. Derivation of site-specific water quality criteria for cadmium and the St. Louis River basin, Duluth, Minnesota. Environ. Toxicol. and Chem. 3:651-655.

Stephan, C. E., D. I. Mount, D. J. Hansen, J. H. Gentile, G. A. Chapman, and W. A. Brungs. 1985. Guidelines for Deriving Numerical National Water Quality Criteria for the Protection of Aquatic Organisms and their uses. PB85-227049. Washington, D.C.: U.S. EPA. Pp. 1-98.

Stumm, W. 1987. Aquatic Surface Chemistry: Chemical Processes at the Particle-Water Interface. New York: John Wiley & Sons.

Stumm, W. and J. J. Morgan. 1970. Aquatic Chemistry. New York: Wiley-Interscience.

Sunda, W. and R. R. L. Guillard. 1976. The relationship between cupric ion activity and the toxicity of copper to phytoplankton. J. Mar. Res. 34:511-529.

Sunda, W. G., D. W. Engel, and R. M. Thuotte. 1978. Effect of chemical speciation of toxicity of cadmium to grass shrimp, *Palaemonetes pugio*: Importance to free cadmium ion. Environ. Sci. Tech. 12:409-413.

Swartz, R. C., D. W. Schults, T. H. DeWitt, G. R. Ditsworth, and J. O. Lamberson. 1987. Toxicity of fluoranthene in sediment to marine amphipods: A test of the equilibrium partitioning approach to sediment quality criteria. Presented at the 8th Annual Meeting, Society for Environmental Toxicology and Chemistry, Pensacola, Florida.

Tessier, A., P. G. C. Campbell, J. C. Auclair, and M. Bisson, 1984. Relationships between the partitioning of trace metals in sediments and their accumulation in the tissues of the freshwater mollusc *Elliptio complanata* in a mining area. Can. J. Fish. Aquat. Sci. 41:1463-1472.

Tessier, A., F. Rapin, and R. Carignan. 1985. Trace metals in oxic lake sediments: Possible adsorption onto iron oxyhydroxides. Geochim. et Cosmochim. Acta 49:183-194.

Zamuda, C. D. and W. G. Sunda. 1982. Bioavailability of dissolved copper to the American oyster *Crassostrea virginica*. I. Importance of chemical speciation. Mar. Biol. 66:77-82.

Ziegenfuss, P. S., W. J. Renaudette, and W. J. Adams. 1986. Methodology for assessing the acute toxicity of chemicals sorbed to sediments: Testing the equilibrium partitioning theory. In Aquatic Toxicology and Environmental Fate, T. J. Poston and R. Purdy, eds. Philadelphia: American Society for Testing and Materials. Pp. 479-493

MARINE SEDIMENT TOXICITY TESTS

Richard C. Swartz
U.S. Environmental Protection Agency

ABSTRACT

Sediment toxicity tests have been developed on the basis of virtually all levels of biological organization from subcellular through model ecosystems. Rapid, cost-effective techniques based on acute exposures are often used in research and regulatory programs to determine the spatial and temporal distribution of sediment toxicity, and the relative toxicity of individual chemicals and complex wastes spiked into sediment. Sediment toxicity tests are part of several comprehensive methods for generating sediment quality criteria. Major research needs include test methods for chronic exposures, field validation of acute toxicity tests and the geochemical integrity of test materials, the relation between toxicity and the bioavailability/partitioning of contaminants in different sediment phases, models of toxicological interactions between sediment contaminants, and sediment wasteload allocation models.

INTRODUCTION

Marine pollution often results in the chemical contamination of the seabed and detrimental effects on benthic communities. The initial development of sediment toxicity tests in the early 1970s reflected an increasing research and regulatory interest in methods of documenting benthic degradation (Gannon and Beeton, 1971; Hoss et al., 1974; Hanson, 1974; Cardwell et al., 1976; Lee and Mariani, 1976). The Ocean Dumping Regulations promulgated by the U.S. Environmental Protection Agency (EPA) in 1977 included sediment toxicity tests in the evaluation of applications for dredged material disposal permits. Methods for solid-phase bioassays to be used in conjunction with the Ocean Dumping Regulations were published by EPA and the U.S. Army Corps of Engineers (COE) in 1977 (U.S. EPA/COE, 1977). Research on sediment toxicity tests and their regulatory applications has expanded greatly since 1977. This paper reviews marine sediment toxicity tests with respect to the variety of methods that are available, and their effectiveness, relevance to remedial actions, and applications in research and regulatory programs.

SEDIMENT TOXICITY TESTS--A SUMMARY REVIEW

Sediment toxicity tests have been applied at virtually all levels of biological organization ranging from inhibition of enzymatic activity at the subcellular level to alterations of the structure and function of macrobenthic assemblages in experimental ecosystems (Table 1). However, only a few of these methods are commonly used to assess sediment toxicity. The original EPA/COE (1977) solid-phase bioassay simulates a dredged material disposal operation in a 20 liter or larger exposure chamber. A crustacean, infaunal bivalve, and infaunal polychaete must be included among the test species. Twenty individuals of each species are placed on a 3-cm deep layer of clean sediment, allowed to acclimate for 48 hours, and then covered by a 1.5-cm deep layer of test sediment. Controls are covered by a 1.5-cm layer of clean sediment. Five replicates are prepared for the control and each test sediment. The primary response criterion is survival after 10 days relative to controls. This procedure, or modifications of it, has been used by COE to evaluate applications for dredged material disposal permits.

The amphipod acute sediment toxicity test is technically well-developed and widely applied, especially on the Pacific coast of the United States. This method evolved from the EPA/COE (1977) bioassay method after early research with the solid-phase test showed that amphipods were consistently more sensitive to polluted sediment than other major benthic taxa (Swartz et al., 1979). The typical experimental design for the amphipod test includes 5 replicates for each sediment treatment. Each replicate consists of 20 individual amphipods placed in a 1 liter beaker containing a 2-cm deep layer of test sediment and 825 ml of overlying water, at a salinity of 28 ppt (parts per thousand) for marine tests. The exposure system is static, aerated, and maintained at a constant temperature, usually 15°C. At the initiation of the test, the amphipods quickly swim to the bottom and burrow completely into the sediment. In the absence of stress, they remain buried during the 10-day exposure period. There are three response criteria: mortality after 10 days, ability of survivors to bury in clean sediment, and emergence of amphipods during the exposure. Typical control treatments include clean sediment from the amphipod collection site, sediment with the same particle-size distribution as the test material, carrier control for spiked chemicals (e.g., acetone), and a positive response control based on the effects of a chemical with known amphipod toxicity (e.g., cadmium). The method was originally developed for the phoxocephalid amphipod, *Rhepoxynius abronius*, but has been used with a variety of other marine, estuarine, and freshwater amphipod genera including *Eohaustorius*, *Corophium*, *Grandidierella*, *Ampelisca*, *Hyalella*, and *Pontoporeia*. Three detailed descriptions of the acute amphipod test are available (EVS Consultants, Inc. and Tetra Tech, Inc., 1986; Swartz et al., 1985b; Reish and Lemay, 1988), and Subcommittee E47.03 of the American Society for Testing and Materials (ASTM) is presently adapting the procedure as an ASTM standard method. The literature on the amphipod test includes interlaboratory (Mearns et al., 1986), intermethod (Williams et al., 1986), and interspecies

TABLE 1

Biological organization	Response criterion	References
Biochemistry	Enzyme induction, Microtox	Lee et al, 1979; Schiewe et al., 1985; Reichert et al., 1985; Varanasi et al., 1985; E.V.S. Consultants, Inc. and Tetra Tech, Inc., 1986; Williams et al., 1986; Tetra Tech, Inc., 1986a; PTI Environmental Services, 1988; Geisy et al., 1988.
Cell	Chromosome damage	E.V.S. Consultants, Inc. and Tetra Tech, Inc., 1986; Chapman et al., 1982; Landolt and Kocan, 1984; Landolt et al., 1984; Long and Chapman, 1985; Chapman, 1986
Development	Larval abnormalities	Hoss et al., 1974; Cardwell et al., 1976; E.V.S. Consultants, Inc. and Tetra Tech, Inc., 1986; Williams et al., 1986; Tetra Tech, Inc., 1985, 1986a; PTI Environmental Services, 1988; Long and Chapman, 1985; Chapman and Morgan, 1983; Chapman et al., 1987.
Physiology	Respiration, osmoregulation	Chapman et al., 1982; Long and Chapman, 1985; Chapman, 1986, 1987; Kehoe, 1983; Alden and Butt, 1987.
Behavior	Burrowing, feeding, sediment and predator avoidance	Chapman et al., 1987; Rubinstein, 1979; McGreer, 1979; Pearson et al., 1981, 1984; Olla and Bejda, 1983; Mohlenberg and Kiorboe, 1983; Phelps et al., 1983; Olla et al., 1984, 1988; Oakden et al., 1984a, 1984b; Swartz et al., 1985b, 1986b; Mearns et al., 1986; Clark and Patrick, 1987;
Reproduction	Fertilization, fecundity	Chapman et al., 1983, 1987; Nimmo et al., 1982;
Pathology	Fin erosion	Hargis et al., 1984
Individual	Mortality	Lee and Mariani, 1976; E.V.S. Consultants, Inc. and Tetra Tech, Inc., 1986; Tetra Tech, Inc., 1985, 1986a, 1986b; PTI Environmental Service, 1988; Long and Chapman, 1985; Chapman, 1986; Chapman et al., 1987; Oakden, 1984a; Swartz et al., 1979, 1982, 1984, 1985a, 1985b, 1986a, 1986b; Mearns et al., 1986; Shuba et al., 1978; Tsai et al., 1979; Peddicord, 1980; Tatem, 1980; McLeese and Metcalfe, 1980; McLeese et al., 1982; Alden and Young, 1982; Ott, 1986; Reish and Lemay, 1988; Breteler et al., 1988; DeWitt et al., 1988.
Population	Life cycle, "r"	Chapman et al., 1987; Chapman and Fink, 1984; Tietjen and Lee, 1984
Community	Structure, function, recolonization	Hansen, 1974; Tagatz and Tobia, 1978; Hansen and Tagatz, 1980; Rubinstein et al., 1980; Elmgren et al., 1980; Grassle et al., 1981; Oviatt et al., 1982, 1984; Perez, 1983; Bauer et al., 1988

(Swartz et al., 1979) comparisons, field validation (Swartz et al., 1982, 1985b, 1986a), field toxicity surveys (Chapman et al., 1982; Tetra Tech, Inc., 1985, 1986b; Swartz et al., 1979, 1985b; Breteler et al., 1988), bioassays of the toxicity of specific chemicals or complex wastes (Reichert et al., 1985; Varanasi et al., 1985; Oakden et al., 1984a, 1984b; Swartz et al., 1986, 1984), and development of sediment quality criteria (Tetra Tech, Inc., 1986a; PTI Environmental Services, 1988; Long and Chapman, 1985; Chapman, 1986; Chapman et al., 1987).

Two other frequently used sediment toxicity tests are based on the development of bivalve larvae and inhibition of bacterial bioluminescence (Microtox). These methods lack the direct ecological relevance of the amphipod test, but may be equally or more sensitive to sediment contaminants (Williams et al., 1986). Descriptions of standard methods are available for both tests (E.V.S. Consultants, Inc. and Tetra Tech, Inc., 1986; Chapman and Morgan, 1983). Both Pacific oysters *(Crassostrea gigas)* and blue mussels *(Mytilus edulis)* are used in the larval test. Response criteria are survival and abnormal shell development of larvae exposed for 48 hours to a suspension of 20 g, wet weight, of sediment in 1 liter of filtered, sterilized, 28 ppt seawater. The bivalve larvae test has been used primarily to document the distribution of sediment toxicity (Cardwell et al., 1976; Williams et al., 1986; Long and Chapman, 1985; Chapman and Morgan, 1983; Chapman et al., 1987; Tetra Tech, Inc., 1985) and to develop sediment quality criteria (Tetra Tech, Inc., 1986a; PTI Environmental Services, 1988). The Microtox technique measures the inhibition of light emission by the luminescent bacterium *(Photobacterium phosphoreum)* exposed for 15 minutes to either organic or saline sediment extracts (E.V.S. Consultants, Inc. and Tetra Tech, Inc., 1986). Schiewe et al. (1985) demonstrated a significant relation between the extract concentration causing a 50 percent reduction in luminescence and the concentrations of classes of organic chemicals. In comparative studies, the Microtox assay registered a larger proportion of positive responses than lethality tests with *Rhepoxynius abronius* (Williams et al., 1986) or the freshwater cladoceran, *Daphnia magna* (Giesy et al., 1988). Because of uncertainty about the bioavailability of extracted chemicals and the irrelevance of bacterial luminescence to benthic ecosystems, the greater sensitivity of the Microtox test may reflect chemical contamination rather than a potential for ecological degradation. Microtox has also been used to examine the distribution of sediment toxicity (Schiewe et al., 1985; Williams et al., 1986; Giesy et al., 1988) and to develop sediment quality criteria (Tetra Tech, Inc., 1986a; PTI Environmental Services, 1988). Experimental designs for sediment toxicity tests with bivalve larvae and bacterial luminescence are similar to those described above for acute amphipod tests.

Most of the sediment toxicity tests cited in Table 1 are not routinely used in sediment toxicity surveys or permit application reviews. These include methods to assess the effects of contaminated sediment on complex biological phenomena including predator-prey interactions (Pearson et al., 1981), the intrinsic rate of population growth ("r") (Tietjen and Lee, 1984), recruitment of benthic assemblages from planktonic eggs and larvae (Hansen and Tagatz, 1980), and nutrient flux in recovering benthic mesocosms (Oviatt et al., 1984). These more sophisticated techniques generally have a relatively high cost in time, expertise, and resources. Their utility lies in evaluating higher level ecological impacts when equivocal results are obtained from the more standard toxicity tests.

EFFECTIVENESS OF SEDIMENT TOXICITY TESTS

There are some important limitations and advantages of sediment toxicity tests (Table 2). Bioavailability and toxicity of sediment contaminants can be greatly altered by collection, handling, and storage of sediment samples. Freezing and long storage can usually be avoided (U.S. EPA/COE, 1977; E.V.S. Consultants, Inc. and Tetra Tech, Inc., 1986; Swartz et al., 1985b). However, sediment samples are routinely mixed, sieved, or extracted with poorly understood effects on geochemical properties. Similarly, chemicals experimentally spiked into sediment in the laboratory may not be bioavailable in the same way as "naturally" contaminated sediment. Research is needed to compare sediment geochemistry in the field with that of sediments used in toxicity tests.

TABLE 2 Limitations and Advantages of Sediment Toxicity Tests

---

Limitations

- Sediment collection, handling and storage may alter bioavailability.
- Results may reflect test conditions other than chemical toxicity.
- Route of exposure can be uncertain.
- Field validation is needed for sediment spiking methods.
- Few comparisons of methods and species.
- Few chronic methods.
- Inherent limitations of lab tests to predict ecological events.
- Tests applied to field samples can't discriminate effects of individual chemicals.

Advantages

- Provide a direct benthic, biological impact assessment.
- Legal and scientific precedence; some standard methods.
- Tests applied to field samples reflect cumulative effects of all contaminants.
- Tests applied to spiked chemicals provide unequivocal analysis of causal relations.
- Sediment toxicity tests can be applied to all chemicals of concern.
- Only method available to examine contaminant interactions.
- Limited expertise or special equipment is required.
- Methods are rapid and cost-effective.
- Toxicity tests are amenable to field validation.

Toxic effects of natural sediment features beyond the tolerance limits of test species can sometimes be confused with contaminant effects. Information such as the salinity (Swartz et al., 1985b) and sediment particle size requirements (Ott, 1986; DeWitt et al., 1988) known for *Rhepoxynius abronius* should be developed for other test species. The broad tolerance of some benthic taxa to natural sediment features often extends to contaminant effects (Olla et al., 1988), and is not a proper justification of the use of certain pelecypods and polychaetes in sediment toxicity tests.

Uncertainty about the route of exposure can obfuscate toxicity results, especially when epibenthic or pelagic organisms are used as test species. Most burrowing species have direct exposures to sediment particles and interstitial water. However, if the primary exposure is through the overlying water, the degree of exposure is determined by the mechanisms controlling transport across the sediment-water interface. Relative toxicity may then be determined by factors such as sediment bioturbation, rather than absolute contamination.

Although many sediment toxicity tests have been developed, there are no standard chronic methods and few comparisons of species or methods. EPA's Region 10 Office of Puget Sound, is currently comparing the relative sensitivity of 13 acute and chronic test methods. Previous methods comparisons have shown a general concordance of acute tests in identifying the most and least contaminated sediment samples, although concordance is less at intermediate levels of contamination. Different toxicity tests may be particularly sensitive to different kinds of chemicals and, therefore, no single method will necessarily meet all requirements of sediment toxicity surveys (Swartz et al., 1985). For these reasons, many investigations now employ several test methods (Williams et al., 1986; Chapman et al., 1982, 1987; Long and Chapman, 1985; Tetra Tech, Inc., 1985).

There is an inherent inability of simple, acute laboratory tests to predict or reflect ecological events. For example, in a field validation of the acute amphipod test along the sediment pollution gradient on the Palos Verdes Shelf, off California, there was generally a good correspondence between sediment contamination, toxicity, and benthic community degradation (Swartz et al., 1985, 1986). However, at one site intermediate between areas of major and minor impacts, there were substantial contamination and biological perturbations, but no acute amphipod toxicity. These simple tests often are not sensitive to the long-term events that effect chronic toxicity and ecological succession.

Sediment bioassays determine the cumulative toxicity of all chemicals in samples collected from the field. This is a major advantage over other analytical methods because many chemicals and their toxicological interactions are unknown or unmeasured. Conversely, this sensitivity to cumulative effects makes it impossible to attribute toxicity to specific chemicals on the sole basis of bioassay results on field sediments. Sediment toxicity tests should be part of a comprehensive analysis of sediment quality that also includes chemical, geological, and biological assessments (Swartz et al., 1985). This is the basic concept of the very effective benthic assessment method often called the "Sediment Quality Triad" (Long and Chapman, 1985; Chapman, 1986; Chapman et al., 1987; Swartz et al., 1982, 1985, 1986).

Causal relations are unequivocal when toxicity tests are applied to unpolluted sediment spiked with individual chemicals or complex wastes. The spiking method can be applied to any chemical of concern. It also offers the only experimental procedure for examining interactions between sediment contaminants, an important problem that has not yet received much attention (Oakden et al., 1984; Samolloff et al., 1983; Plesha et al., 1988; Swartz et al., 1988).

A major advantage of most sediment toxicity tests is that they require limited expertise or equipment and are rapid and cost-effective. Bioassay results are usually available within two weeks of sample collection. Analyses of macrobenthos and chemical samples from the same survey typically require months for completion at much higher costs for equipment and expertise. Toxicity tests can quickly and inexpensively locate "hot spots" where more comprehensive assessments can be focused.

APPLICATIONS OF SEDIMENT TOXICITY TESTS

Sediment toxicity tests have a variety of applications in research and regulatory programs (Table 3). They are used principally to determine patterns of toxicity in the field and quantify the toxicity of materials spiked into sediment. Field surveys can examine the distribution of toxicity in space, time, or depth in the sediment (Williams et al., 1986; Giesy et al., 1988; Chapman et al., 1982, 1983, 1987; Long and Chapman, 1985; Chapman, 1986; Chapman and Morgan, 1983; Tetra Tech, Inc., 1985, 1986b; Alden and Butt, 1987; Tsai et al., 1979; Swartz et al., 1982, 1985b, 1986a; Alden and Young, 1982; Breteler et al., 1988; Chapman and Fink, 1984). Relative sediment toxicity is presently used in a variety of impact assessments, disposal permit decisions, and monitoring programs. Wasteload allocation models that combine sediment toxicity distributions with particle/contaminant transport, deposition, and resuspension models are currently being developed. Examination of the vertical distribution of toxicity in a sediment core reflects the historic pattern of contamination in depositional environments. Such data are particularly relevant to remedial investigations that consider capping and "no action" alternatives. The toxicity of field-collected sediment is also used as a research variable for comparisons with biological effects and geochemical sediment characteristics (Long and Chapman, 1985; Chapman, 1986; Tetra Tech, Inc., 1985, 1986b; Chapman et al., 1987; Swartz et al., 1982, 1985b, 1986a).

The sediment spiking method can be used to determine the toxicity of individual chemicals or complex wastes like sewage effluents, sludges and drilling fluids (Hansen, 1974; Reichert et al., 1985; Varanasi et al., 1985; Pearson et al., 1981; Olla and Bejda, 1983; Olla et al., 1984, 1988; Oakden et al., 1984a, 1984b; Swartz et al., 1984, 1986b; McLeese and Metcalfe, 1980; McLeese et al., 1982; Ott, 1986; Tagatz and Tobia, 1978; Bauer et al., 1988; Clark and Patrick, 1987;

Hansen and Tagatz, 1980; Rubinstein et al., 1980; Elmgren et al., 1980; Grassle et al., 1981; Oviatt et al., 1982, 1984; Perez, 1983). Contaminated sediment can also be mixed into clean sediment to determine an LC_{50} or other measure of effects (Swartz et al., 1989). This procedure can determine the relative toxicity of field-collected samples that cause 100 percent mortality of test specimens. Layering, rather than spiking, could be used to test the effectiveness of proposed sediment capping materials in experimental designs similar to the original EPA/COE (1977) solid-phase bioassay.

TABLE 3 Research and Regulatory Applications of Sediment Toxicity Tests

Field sediment	Spatial distribution of toxicity
	Temporal distribution of toxicity
	Depth distribution of toxicity
	Dilution--LC_{50} in clean sediment
Spiked sediment	Single chemical
	LC_{50}
	safe concentration
	sediment quality criterion
	Multiple chemicals
	joint action
	interaction models
	Complex wastes
	sewage
	sludge
	drilling fluids
	dredged material
Sediment features	Salinity
	Particle-size distribution
	Organic carbon concentration
Research variable	Relation to benthic community structure, function, sediment conditions
Comprehensive sediment evaluation methods	Apparent Effects Threshold
	Sediment Quality Triad
Regulatory applications	Dredged material permit decisions
	Environmental impact assessment
	Wasteload allocation
	Remedial action alternatives
	Sediment quality criteria

Sediment spiking provides a toxicological approach to the development of numerical sediment quality criteria. Safe concentrations of chemicals in sediment can be estimated from dose-response relations by the same rationale used to generate water quality criteria. Sediment toxicity tests are also part of other methods of developing sediment quality criteria, (e.g., Apparent Effects Threshold [Tetra Tech, Inc., 1986a, 1986b; PTI Environmental Services, 1988], Sediment Quality Triad [Chapman, 1986]), and are being used to validate criteria based on the equilibrium partitioning method. There is a close agreement between estimates of safe sediment concentrations of fluoranthene, after organic carbon (OC) normalization, based on the methods of equilibrium partitioning--1,330 μg/g OC (using the chronic lowest observed effect level [U.S. EPA, 1980]); the amphipod Apparent Effects Threshold--891 μg/g OC [Tetra Tech, Inc., 1986b]; and fluoranthene sediment toxicity tests--817 μg/g OC (10-day LC_{50} for *Rhepoxynius abronius*; Swartz, unpublished data). Preliminary research based on the toxicological approach indicates that a simple additivity model can predict interactions between sediment contaminants (Swartz et al., 1988).

CONCLUSIONS AND RESEARCH RECOMMENDATIONS

Acute sediment toxicity tests are well-developed and have become an integral part of benthic ecosystem impact assessments. There is a broad range of test methods with a variety of biological response criteria. Standard methods have been established for acute toxicity tests that are rapid and cost-effective. They are principally applied to determine spatial/temporal patterns of toxicity, and the relative toxicity of individual chemicals and complex wastes spiked into clean sediment. These methods are used in a variety of regulatory programs including dredged material disposal permits, sediment quality criteria, wasteload allocations, and remedial actions at sites of major sediment contamination. Benthic impact assessments are most effective when toxicity tests are combined with biological, chemical, and geological indicators of sediment degradation.

Future research should focus on development of standard methods for chronic sediment toxicity tests and field validation of acute sediment toxicity tests. A toxicological data base should be established for selected chemicals and sensitive infaunal species. Issues concerning the bioavailability of contaminants in different sediment phases, and the toxicological interactions of sediment contaminants must be resolved as part of the development of sediment quality criteria. Wasteload allocation models should incorporate sediment quality criteria or toxicity predictions with models of the transport, deposition and resuspension of sediment particles and contaminants.

ACKNOWLEDGMENTS

I thank Janet Lamberson and Steve Ferraro for their reviews of this manuscript. Contribution Number N-068, U.S. EPA Environmental Research Laboratory, Narragansett, Rhode Island and Newport, Oregon.

REFERENCES

Alden, R. W., III, and A. J. Butt. 1987. Statistical classification of the toxicity and polynuclear aromatic hydrocarbon contamination of sediments from a highly industrialized seaport. Environ. Toxicol. Chem. 6:673-684.

Alden, R. W., III and R. J. Young, Jr. 1982. Open ocean disposal of materials dredged from a highly industrialized estuary: An evaluation of potential lethal effects. Arch. Environ. Contam. Toxicol. 11:567-576.

Bauer, J. E., R. P. Kerr, M. F. Bautista, C. J. Decker, and D. C. Capone. 1988. Stimulation of microbial activities and polycyclic aromatic hydrocarbon degradation in marine sediments inhabited by *Capitella capitata*. Mar. Environ. Res. 25:63-84.

Breteler, R. J., K. J. Scott, and S. P. Sheperd. 1988. Application of a new sediment toxicity test using marine amphipods, *Ampelisca abdita*, to San Francisco Bay sediments. Submitted to ASTM Twelfth Symposium on Aquatic Toxicology and Hazard Assessment, Sparks, Nevada, April 24-26, 1988.

Cardwell, R. D., C. E. Woelke, M. I. Carr, and E. W. Sanborn. 1976. Sediment and elutriate toxicity to oyster larvae. In Proceedings of the Special Conference on Dredging and its environmental effects, P. A. Krenkel, J. Harrison and J. C. Burdick, III, eds., New York: American Society of Civil Engineers.

Chapman, P. M. 1987. Oligochaete respiration as a measure of sediment toxicity in Puget Sound, Washington. Hydrobiologia 155:249-258.

Chapman, P. M. 1986. Sediment quality criteria from the sediment quality triad: An example. Environ. Toxicol. Chem. 5:957-964.

Chapman, P. M., R. N. Dexter, and E. R. Long. 1987. Synoptic measures of sediment contamination, toxicity and infaunal community composition (the Sediment Quality Triad) in San Francisco Bay. Mar. Ecol. Prog. Ser. 37:75-96.

Chapman, P. M. and R. Fink. 1984. Effects of Puget Sound sediments and their elutriates on the life cycle of *Capitella capitata*. Bull. Environ. Contam. Toxicol. 33:451-459.

Chapman, P. M. and J. D. Morgan. 1983. Sediment bioassays with oyster larvae. Bull. Environ. Contam. Toxicol. 31:438-444.

Chapman, P. M., D. R. Munday, J. Morgan, R. Fink, R. M. Kocan, M. L. Landolt, and R. N. Dexter. 1983. Survey of Biological Effects of Toxicants upon Puget Sound Biota. II. Tests of Reproductive Impairment. Technical Report NOS 102 OMS 1. Rockville, Md.: National Oceanic and Atmospheric Administration.

Chapman, P. M., G. A. Vigers, M. A. Farrell, R. N. Dexter, E. A. Quinlan, R. M. Kocan, and M. Landolt. 1982. Survey of biological effects of toxicants upon Puget Sound biota. 1. Broad-scale toxicity survey. NOAA Technical Memorandum OMPA-25. Boulder, Colo.

Clark, J. R. and J. M. Patrick, Jr. 1987. Toxicity of sediment-incorporated drilling fluids. Mar. Poll. Bull. 18:600-603.

DeWitt, T. H., G. R. Ditsworth, and R. C. Swartz. 1988. Effects of natural sediment features on the phoxocephalid amphipod, *Rhepoxynius abronius*: Implications for sediment toxicity bioassays. Mar. Environ. Res. 25:99-124.

Elmgren, R., G. A. Vargo, J. F. Grassle, J. P. Grassle, D. R. Heinle, G. Langlois, and S. L. Vargo. 1980. Trophic interactions in experimental marine ecosystems perturbed by oil. In Microcosms in Ecological Research, Symposium Series 52. J. P. Giesy, Jr., ed. Washington, D.C.: U.S. Department of Energy. Pp. 779-800.

E.V.S. Consultants, Inc. and Tetra Tech, Inc. 1986. Recommended Protocols for Conducting Laboratory Bioassays on Puget Sound Sediments. Seattle, Wash.: E.V.S. Consultants, Inc.

Gannon, J. E. and A. M. Beeton. 1971. Procedures for determining the effects of dredged sediments on biota-benthos viability and sediment selectivity tests. J. Water Poll. Contr. Fed. 43:392-398.

Giesy, J. P., R. L. Graney, J. L. Newsted, C. J. Rosiu, and A. Benda. 1988. Comparison of three sediment bioassay methods using Detroit River sediments. Environ. Toxicol. Chem. 7:483-498.

Grassle, J. F., R. Elmgren, and J. P. Grassle. 1981. Response of benthic communities in MERL experimental ecosystems to low level chronic additions of No. 2 fuel oil. Mar. Environ. Res. 4:279-297.

Hansen, D. J. 1974. Aroclor 1254: Effect on composition of developing estuarine animal communities in the laboratory. Mar. Sci. 18:19-33.

Hansen, D. J. and M. E. Tagatz. 1980. A laboratory test for assessing impacts of substances on developing communities of benthic estuarine organisms. In Aquatic Toxicology, J. G. Eaton, P. R. Parrish, and A. C. Hendricks, eds. STP 707. Philadelphia: American Society for Testing and Materials. pp 40-57.

Hargis, W. J., Jr., M. H. Roberts, and D. E. Zwerner. 1984. Effects of contaminated sediments and sediment-exposed effluent water on an estuarine fish: Acute toxicity. Mar. Environ. Res. 14:337-354.

Hoss, D. E., L. C. Coston, and W. E. Schaaf. 1974. Effects of seawater extracts of sediments from Charleston Harbor, SC, on larval estuarine fishes. Estuar. Coastal. Mar. Sci. 2:323-328.

Kehoe, D. M. 1983. Effects of Grays Harbor estuary sediment on the osmoregulatory ability of Coho Salmon smolts (*Oncorhynchus kisutch*). Bull. Environ. Contam. Toxicol. 30:522-529.

Landolt, M. L. and R. M. Kocan. 1984. Lethal and sublethal effects of marine sediment extracts on fish cells and chromosomes. Helgolander Meersunters. 37:479-491.

Landolt, M. L., R. M. Kocan, and R. N. Dexter. 1984. Anaphase aberrations in cultured fish cells as a bioassay of marine sediments. Mar. Environ. Res. 14:497-498.

Lee, G. F. and G. M. Mariani. 1976. Evaluation of the significance of waterway sediment-associated contaminants on water quality at the dredged material disposal site. In Aquatic Toxicology and Hazard Evaluation, F. L. Mayer and J. L. Hamelink, eds. ASTM STP 634.

Philadelphia: American Society for Testing and Materials. Pp. 196-213.

Lee, R. F., S. C. Singer, K. R. Tenore, W. S. Gardner, and R. M. Philpot. 1979. Detoxification system in polychaete worms: Importance in the degradation of sediment hydrocarbons. In Marine Pollution: Functional Responses, W. B. Vernberg, A. Calabrese, F. P. Thurberg and F. J. Vernberg, eds. New York: Academic Press. Pp. 23-37.

Long, E. R. and P. M. Chapman. 1985. A sediment quality triad: Measures of sediment contamination, toxicity, and infaunal community composition in Puget Sound. Mar. Poll. Bull. 16:405-415.

McGreer, E. R. 1979. Sublethal effects of heavy metal contaminated sediments on the bivalve *Macoma balthica* (L.). Mar. Poll. Bull. 10:259-262.

McLeese, D. W. and C. D. Metcalfe. 1980. Toxicities of eight organochlorine compounds in sediment and seawater to *Crangon septemspinosa*. Bull. Environ. Contam. Toxicol. 25:921-928.

McLeese, D. W., L. E. Burridge, and J. Van Dinter. 1982. Toxicities of five organochlorine compounds in water and sediment to *Nereis virens*. Bull. Environ. Contam. Toxicol. 28:216-220.

Mearns, A. J., R. C. Swartz, J. M. Cummins, P. A. Dinnel, P. Plesha, and P. M. Chapman. 1986. Inter-laboratory comparison of a sediment toxicity test using the marine amphipod, *Rhepoxynius abronius*. Mar. Environ. Res. 18:13-37.

Mohlenberg, F. and T. Kiorboe. 1983. Burrowing and avoidance behaviour in marine organisms exposed to pesticide-contaminated sediment. Mar. Poll. Bull. 14:57-60.

Nimmo, D. R., T. L. Hamaker, E. Mathews, and W. T. Young. 1982. The long-term effects of suspended particulates on survival and reproduction of the mysid shrimp, *Mysidopsis bahia*, in the laboratory. In Ecological Stress and the New York Bight: Science and Management, G. F. Mayer, ed. Columbia, S.C.: Estuarine Research Federation. Pp. 413-422.

Oakden, J. M., J. S. Oliver, and A. R. Flegal. 1984a. EDTA chelation and zinc antagonism with cadmium in sediment: Effects on the behavior and mortality of two infaunal amphipods. Mar. Biol. 84:125-130.

Oakden, J. M., J. S. Oliver, and A. R. Flegal. 1984b. Behavioral responses of a phoxocephalid amphipod to organic enrichment and trace metals in sediment. Mar. Ecol. Prog. Ser. 14:253-257.

Olla, B. L. and A. J. Bejda. 1983. Effects of oiled sediment on the burrowing behaviour of the hard clam, *Mercenaria mercenaria*. Mar. Environ. Res. 9:183-193.

Olla, B. L., A. J. Bejda, A. L. Studholme, and W. H. Pearson. 1984. Sublethal effects of oiled sediment on the sand worm, *Nereis (Neanthes) virens*: Induced changes in burrowing and emergence. Mar. Environ. Res. 13:121-139.

Olla, B. L., V. B. Estelle, R. C. Swartz, G. Braun, and A. L. Studholme. 1988. Responses of polychaetes to cadmium-contaminated sediment: Comparison of uptake and behavior. Environ. Toxicol. Chem. 7:587-592.

Ott, F. S. 1986. Amphipod sediment bioassays: Effect of grain size, cadmium, methodology, and variations in animal sensitivity on interpretation of experimental data. Ph.D. Dissertation, Univ. of Washington, Seattle.

Oviatt, C., J. Frithsen, J. Gearing, and P. Gearing. 1982. Low chronic additions of No. 2 fuel oil: Chemical behavior, biological impact and recovery in a simulated estuarine environment. Mar. Ecol. Prog. Ser. 9:121-136.

Oviatt, C. A., M. E. Q. Pilson, S. W. Nixon, J. B. Frithsen, D. T. Rudnick, J. R. Kelly, J. F. Grassle, and J. P. Grassle. 1984. Recovery of a polluted estuarine ecosystem: A mesocosm experiment. Mar. Ecol. Prog. Ser. 16:203-217.

Pearson, W. H., D. L. Woodruff, P. C. Sugarman, and B. L. Olla. 1984. The burrowing behavior of sand lance, *Ammodytes hexapterus*: Effects of oil-contaminated sediment. Mar. Environ. Res. 11:17-32.

Pearson, W. H., D. L. Woodruff, P. C. Sugarman and B. L. Olla. 1981. Effects of oiled sediment on predation on the little neck clam, *Protothaca staminea*, by the Dungeness crab, *Cancer magister*. Estuar. Coastal Shelf Sci. 13:445-454.

Peddicord, R. K. 1980. Direct effects of suspended sediments on aquatic organisms. In Contaminants and Sediments, Vol. 1, R. A. Baker, ed. Ann Arbor, Mich.: Ann Arbor Science Publishers. Pp. 501-536.

Perez, K. T. 1983. Environmental assessment of a phthalate ester, di-(2-ethylhexyl) phthalate (DEHP), derived from a marine microcosm. In Aquatic Toxicology and Hazard Assessment, W. E. Bishop, R. D. Cardwell, and B. B. Heidolph, eds. STP 801. Philadelphia: American Society for Testing and Materials. Pp. 180-191.

Plesha, P. D., J. E. Stein, M. H. Schiewe, B. B. McCain, and U. Varanasi. 1988. Toxicity of marine sediments supplemented with model mixtures of chlorinated and aromatic hydrocarbons to the infaunal amphipod, *Rhepoxynius abronius*. In press. Mar. Environ. Res.

Phelps, H. L., J. T. Hardy, W. H. Pearson, and C. W. Apts. 1983. Clam burrowing behavior: Inhibition by copper-enriched sediment. Mar. Poll. Bull. 14:452-455.

PTI Environmental Services. 1988. Sediment Quality Values Refinement: Tasks 3 and 5--1988 Update and Evaluation of the Puget Sound AET. PTI Contract C717-01. Bellevue, Wash: PTI Environmental Services.

Reichert, W. L., B.-T. L. Eberhart, and U. Varanasi. 1985. Exposure of two species of deposit-feeding amphipods to sediment associated [^{3}H]Benzo[a]pyrene: uptake, metabolism and covalent binding to tissue macromolecules. Aquat. Toxicol. 6:45-56.

Reish, D. J. and J. A. Lemay. 1988. Bioassay manual for dredged materials. Contract DACW-09-83R-005. Los Angeles: U.S. Army Corps of Engineers.

Rubinstein, N. I. 1979. A benthic bioassay using time-lapse photography to measure the effect of toxicants on the feeding behavior of lugworms (Polychaeta: Arenicolidae). In Marine Pollution: Functional Responses, W.B. Vernberg, A. Calabrese, F.P. Thurberg, and F. J. Vernberg, eds. New York: Academic Press. Pp. 341-351.

Rubinstein, N. I., C. N. D'Asaro, C. Sommers, and F. G. Wilkes. 1980. The effects of contaminated sediments on representative estuarine species and developing benthic communities. In Contaminants and

Sediments, Vol. 1, R. A. Baker, ed. Ann Arbor: Ann Arbor Science Pub. Pp. 445-461.

Samolloff, M. R., J. Bell, D. A. Birkholz, G. R. B. Webster, E. G. Arnott, R. Pulak, and A. Madrid. 1983. Combined bioassay-chemical fractionation scheme for the determination and ranking of toxic chemicals in sediments. Environ. Sci. Technol. 17:329-334.

Schiewe, M. H., E. G. Hawk, D. I. Actor, and M. M. Krahn. 1985. Use of bacterial bioluminescence assay to assess toxicity of contaminated marine sediments. Can. J. Fish. Aquat. Sci. 42:1244-1248.

Shuba, P. J., H. E. Tatem, and J. H. Carroll. 1978. Biological Assessment Methods to Predict the Impact of Open-water Disposal of Dredged Material. Technical Report D-78-50. Vicksburg, Miss.: U.S. Army Waterways Experiment Station.

Swartz, R. C., F. A. Cole, D. W. Schults, and W. A. DeBen. 1986a. Ecological changes on the Palos Verdes Shelf near a large sewage outfall: 1980-1983. Mar. Ecol. Prog. Ser. 31:1-13.

Swartz, R. C., W. A. DeBen, and F. A. Cole. 1979. A bioassay for the toxicity of sediment to the marine macrobenthos. J. Water Poll. Control Fed. 51:944-950.

Swartz, R. C., W. A. DeBen, K. A. Sercu, and J. O. Lamberson. 1982. Sediment toxicity and the distribution of amphipods in Commencement Bay, Washington, USA. Mar. Poll. Bull. 13:359-364.

Swartz, R. C., G. R. Ditsworth, D. W. Schults, and J. O. Lamberson. 1986b. Sediment toxicity to a marine infaunal amphipod: Cadmium and its interaction with sewage sludge. Mar. Environ. Res. 18:133-153.

Swartz, R. C., P. F. Kemp, D. W. Schults, G. R. Ditsworth, and R. J. Ozretich. 1989. Toxicity of sediment from Eagle Harbor, Washington to the infaunal amphipod, *Rhepoxynius abronius*. In press. Environ. Toxicol. Chem. 8(3).

Swartz, R. C., P. F. Kemp, D. W. Schults, and J. O. Lamberson. 1988. Effects of mixtures of sediment contaminants on the marine infaunal amphipod, *Rhepoxynius abronius*. Environ. Toxicol. Chem. 7:1013-1020.

Swartz, R. C., D. W. Schults, G. R. Ditsworth, and W. A. DeBen. 1984. Toxicity of sewage sludge to *Rhepoxynius abronius*, a marine benthic amphipod. Arch. Environ. Contamination Toxicol. 13:207-216.

Swartz, R. C., D. W. Schults, G. R. Ditsworth, W. A. DeBen, and F. A. Cole. 1985a. Phoxocephalid amphipod bioassay for marine sediment toxicity. In Aquatic Toxicology and Hazard Assessment: Proceedings of the Seventh Annual Symposium, R. D. Cardwell, R. Purdy, and R. C. Bahner, eds. STP 854. Philadephia: American Society for Testing and Materials. Pp. 284-307.

Swartz, R. C., D. W. Schults, G. R. Ditsworth, W. A. DeBen, and F. A. Cole. 1985b. Sediment toxicity, contamination, and macrobenthic communities near a large sewage outfall. In Validation and Predictability of Laboratory Methods for Assessing the Fate and Effects of Contaminants in Aquatic Ecosystems, T. P. Boyle, ed. STP 865. Philadelphia: American Society for Testing and Materials. Pp. 152-175.

Tagatz, M. E. and M. Tobia. 1978. Effects of barite ($BaSO_4$) on development of estuarine communities. Estuar. Coastal Mar. Sci. 7:401-407.

Tatem, H. E. 1980. Exposure of benthic and epibenthic estuarine animals to mercury and contaminated sediment. In Contaminants and Sediments, Vol. 1, R. A. Baker, ed. Ann Arbor, Mich.: Ann Arbor Science Publishers. Pp. 537-549.

Tetra Tech, Inc. 1986a. Development of Sediment Quality Values for Puget Sound. DACW67-85-0029, Work Order 0001C, TC3090-02; Task 6 Final Report. Bellevue, Wash.: Tetra Tech, Inc.

Tetra Tech, Inc. 1986b. Eagle Harbor preliminary investigation. Final Report EGHB-2, TC-3025-03. Bellevue, Wash.: Tetra Tech, Inc.

Tetra Tech, Inc. 1985. Commencement Bay Nearshore/Tideflats Remedial Investigation. TC-3752, Final Report, EPA-910/9-85-134b. Bellevue, Wash.: Tetra Tech, Inc.

Tietjen, J. H. and J. J. Lee. 1984. The use of free-living nematodes as a bioassay for estuarine sediments. Mar. Environ. Res. 11:233-251.

Tsai, C., J. Welch, K. Chang, J. Shaefer, and L. E. Cronin. 1979. Bioassay of Baltimore Harbor sediments. Estuaries 2:141-153.

U.S. Environmental Protection Agency. 1980. Water Quality Criteria for Fluoranthene. Washington, D.C.: U. S. EPA.

U.S. Environmental Protection Agency/U.S. Army Corps of Engineers. 1977. Ecological Evaluation of Proposed Discharge of Dredged Material into Ocean Waters. Vicksburg, Miss.: U. S. Army Engineer Waterways Experiment Station.

Varanasi, U., W. L. Reichert, J. E. Stein, D. W. Brown, and H. R. Sanborn. 1985. Bioavailability and biotransformation of aromatic hydrocarbons in benthic organisms exposed to sediment from an urban estuary. Environ. Sci. Tech.. 19:836-841.

Williams, L. G., P. M. Chapman and T. C. Ginn. 1986. A comparative evaluation of sediment toxicity using bacterial luminescence, oyster embryo, and amphipod sediment bioassays. Mar. Environ. Res. 19: 225-249.

SIGNIFICANCE OF CONTAMINATION

EFFECTS OF CONTAMINATED SEDIMENTS ON MARINE BENTHIC BIOTA AND COMMUNITIES

K. John Scott
Science Applications International Corporation

ABSTRACT

Our understanding of the effects of contaminants on benthic organisms lags well behind that for water column species because of the way in which sediments mediate bioavailability and because test protocols using infaunal organism are still in the developmental stage. Although quantitative analyses of benthic communities continue to be a primary assessment tool, their interpretation as to contaminant effects remains difficult, especially under conditions of moderate contamination. It is, therefore, important to determine how sediment contaminants affect lower levels of biological organization. Contaminant effects have been described for subcellular, cellular, tissue, whole organism, and population level systems in benthic organisms. The routine application of these responses has been limited, however. There is a critical research need to develop test protocols that address chronic effects and bioaccumulation in both short- and long-lived benthic organisms. Such methodologies would allow for the effective interpretation of changes in benthic communities in response to contamination and the significance of the community in food chain transfer of sediment contaminants.

INTRODUCTION

The degree of sediment contamination in our nation's coastal waters has become the subject of intensive research and monitoring programs during the last 10 years. The role that benthic systems play in the sequestering and redistribution of contaminants is now well recognized (Baker, 1980; Dickson et al., 1987). National and international efforts have sought to document the degree of contamination in sediments and in the associated fauna as a means to rank coastal sites for remedial action (NOAA, 1988; see Marine Pollution Bulletin baselines for examples). Regulatory actions will hopefully control further source inputs to these systems; however, with the ever increasing localization of population centers along the coastline, determining the contribution of nonpoint sources to sediment contamination will remain the biggest challenge.

Research on the effects of contaminants on benthic organisms lags well behind that for water column species because of the way in which sediments mediate contaminant bioavailability and because test protocols using a wide range of infaunal organisms are still in the developmental stage. The Environmental Protection Agency's (EPA) sediment quality criteria program is beginning to address the relationship between sediment properties, contaminant availability, and subsequent biological effects (Zarba, 1988). Approaches to developing these criteria rely heavily on acute responses and sediment-water equilibrium partitioning theory. Sediment criteria that depend on acute responses will not, however, provide information necessary to assess chronic, long-term effects that regulate an organism's growth, reproduction, and subsequent function in the benthic community.

Although test methods have been developed to assess the effects of water column and sediment associated contaminants on the development of benthic communities (Hansen and Tagatz, 1980), most evaluations of community responses to sediment contaminants are the result of field benthic and contaminant surveys. The implications of chemical-specific effects on species abundances and community structure are thus largely correlational. Compounding the problem of determining chronic, population level effects is the lack of chronic test methodologies using benthic species. These tests will be essential to the prediction of contaminant effects on population abundances and for the interpretation of the significance of community changes.

The focus of this paper is to briefly review the types of responses that sediment contaminants have been shown to elicit in macrobenthic communities. Effects at this level of biological organization, however, integrate individual organism and population responses throughout the biological hierarchy.

As shown in Figure 1, contaminants can induce effects at any level of biological organization from biochemical or pathological effects to reproduction and population interaction effects (Sheehan, 1984). The time scales of these effects can vary from hours to years. Therefore, information on the type of response and the life history of the target organism(s) are necessary for making predictions at the community level.

To understand community responses to sediment contaminants, therefore, a brief review of effects at several levels of the biological hierarchy will be provided. This discussion will follow a summary of the extent of contamination in natural sediments and the bioaccumulation of contaminants in benthic organisms.

EXTENT OF CONTAMINATION

Sediments

The most comprehensive assessment of sediment contamination in U.S. coastal waters has been conducted by the National Oceanic and Atmospheric Administration's (NOAA) Status and Trends (NS & T) program. Results for benthic and mussel watch sediment surveillance studies from

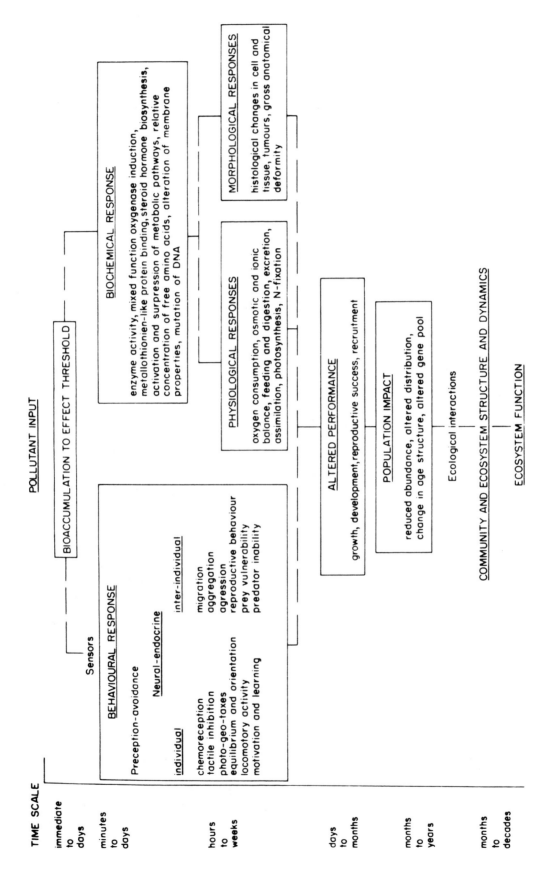

FIGURE 1 A conceptual chronology of effects following exposure to toxicants. SOURCE: Sheehan, 1984.

177 sites are presented in NOAA (1988). The authors conclude that silver, tin, mercury, and all of the organic compounds are the most representative of anthropogenic input and that, although there are exceptions, sediment contamination is most prevalent around urban centers. These studies provide a range of contaminant concentrations for sites not directly influenced by point sources, and thereby reflect basin- or baywide conditions. For the most part, benthic communities at the NOAA NS & T sites would be expected to exhibit a range of responses consistent with this range of contamination. Where contaminant concentrations are at moderate levels, these communities would not be drastically impoverished in either numbers of species or species abundances.

Point-source inputs have resulted in many cases in which sediments are much more contaminated than those cited above. Some examples are

- New Bedford Harbor, Massachusetts, where PCB discharges have caused sediment PCB concentrations to be in the part per thousand range (Weaver, 1984);
- Eagle Harbor, Washington, where extremely high aromatic hydrocarbon concentrations (120,000 ng/g) resulted from a creosote wood treating plant (Malins et al., 1985);
- Black Rock Harbor, Connecticut, where multiple source inputs have elevated concentrations of PAHs, PCBs, Cd, Cu, Pb and Zn higher than at any NS & T site (Rogerson et al., 1985).

The identification of "hot spots" will surely continue as a result of a rising public and regulatory concern about sediment contamination. At sites such as these, the benthic communities are drastically altered and are commonly dominated by contaminant-tolerant, opportunistic species assemblages.

TISSUES

The bioaccumulation of contaminants from sediment matrices is a complex problem that is presently receiving considerable attention in the research and regulatory communities. An understanding of uptake rates of contaminants and the processes regulating bioaccumulation is especially relevant to the establishment of sediment quality criteria. A significant body of literature exists on tissue residue levels in marine organisms based on laboratory and field studies where comparable data are available for sediment concentrations.

Most of this literature deals with filter-feeding species, such as the mussel, *Mytilus edulis,* or with epibenthic forms and bottom-feeding fish that are not in the most intimate contact with the sediments. To assess the mechanisms of contaminant bioaccumulation from sediments, and the subsequent food chain transfer, there is a critical need for studies using infaunal and deposit-feeding organisms. In fact, there is a real technological limitation in the determination of bioaccumulation processes in one of the most important components of benthic communities--the small, fast growing, highly productive opportunistic species of polychaetes, bivalves, and crustaceans. These taxa are known to be important food sources for demersal fish (Becker and Chew, 1987). This limitation is due to contaminant detection limits and sample-size constraints associated with these small organisms, many of which are only retained on mesh sizes smaller than 300 μm. We also know very little about contaminant effects in this group.

A general review of metal residues in aquatic organisms can be found in Prosi (1979). This reviewer and others have observed that there is considerable variation in metal accumulation, even in individuals of the same species found in the same locale (Bradford and Luoma, 1980). This variability primarily results from the differential ability of various species and/or life stages to accumulate, store, and metabolize metal compounds. For example, Bryan (1976) found that the polychaetes *Nereis* and *Nephtys* can regulate the uptake of iron, manganese, and zinc, but cannot regulate cadmium, copper, silver, or lead. Crustaceans also are able to regulate copper, manganese, and zinc, but not cadmium (Bryan, 1976). The ability to detoxify metal compounds (Jenkins and Brown, 1984) may account in large part for the fact that metal bioaccumulation levels are generally found to correspond closely the ambient sediment concentrations. Jenkins and Brown (1984) have found that the binding capacity of the organism's metallothionein protein pool may be directly related to the metal tissue residues that cause toxic effects. This hypothesis suggests that metals are gradually accumulated to the point where the metallothionein pool is saturated and further uptake "spills over," causing toxicity. The implication of these findings is that there is a small difference between concentrations that have no apparent effect and those causing toxicity. It thus appears that, from a bioaccumulation standpoint, metals do not appear to be a significant problem because toxicity would occur before the bioaccumulation of elevated tissue residues.

On the other hand, the accumulation of organic compounds do pose a serious problem because of the affinity of many PAHs and chlorinated organics for animal lipid pools. Although many marine organisms have the ability to metabolize these organic compounds via mixed function oxidase (MFO) systems (Lee, 1984), bioaccumulation, persistence, and the potential for food chain transfer of organics is much more prevalent than for metals. The uptake of PAHs from water has been well documented, however, uptake from sediments by deposit feeders is not well understood. Available evidence indicates that PAHs are tightly bound to the organic fraction of the sediments and are relatively unavailable for bioaccumulation (Neff, 1985). Any accumulation would thus result from PAH desorption from particles to the interstitial water for uptake across the integument or gills. Further compounding the interpretation of PAH residue data is the fact that these contaminants appear to be rapidly metabolized and detoxified by MFO systems (Stegeman, 1981).

Organochlorine compounds such as PCBs and pesticides tend to be much more persistent in benthic systems (Nimmo, 1985). The potential for food chain transfer from sediments to deposit-feeding invertebrates to demersal fish has been documented by Goerke et al. (1979) and Young and Mearns (1978). The level of accumulation of non-polar chlorinated organics in deposit feeders appears to be a function of the organic carbon content of the sediment and the lipid resources of the organism (Lake et al., 1987). This hypothesis has recently been tested by Lake et al. (in prep.) using field-collected sediments with a range of sediment PCB and total organic carbon concentrations. Accumulation factors for deposit feeders, when normalized for sediment TOC and organism lipid content, ranged from 2.3 to 7.3. The development and application of this equilibrium partitioning approach is an important component of EPA's program to set sediment quality criteria.

SIGNIFICANCE OF CONTAMINATION TO THE BENTHOS BACKGROUND

There are many examples from the literature of changes in benthic communities due to contaminated sediments. The most dramatic and clear-cut community changes have been demonstrated for sediments contaminated by oil spills (Sanders et al., 1980; Jacobs, 1980), where there are fairly dramatic shifts in species composition due to immediate acute mortalities and changing recruitment and survivorship patterns. Other examples have been described for sediments with gradients in metal concentrations (Rygg, 1986). The most common cases of community effects, however, are those documented for sewage outfalls and other types of organic enrichment gradients (Pearson and Rosenberg, 1978; Stainken, 1984; Swartz et al., 1985b, 1986; Stull, 1986). Because of the overwhelming multiple effects that organic enrichment may have on community structure (BOD, COD, pH, H_2S, CH_4) it is often difficult to ascribe community changes to specific contaminants, which may be correlated with a gradient of total organic carbon.

The integrated response of the community is most often an after-the-fact assessment of community effects (Sheehan, 1984). Although it is important to understand the contaminant related processes operating at the community level, it is equally critical to determine the mechanisms occurring at the individual and population level which are ultimately expressed in a community effect. An appreciation of these mechanisms is directly related to a regulatory evaluation of contaminant effects, as most applicable tests must be conducted at the lower levels of biological organization.

In the discussion that follows, I will attempt to illustrate how biological effects at lower levels of organization may influence integrated types of responses in benthic populations and communities. I will supplement selected observations from the literature with those drawn from our experience in the Field Verification Program (FVP), a joint effort of the Corps of Engineers (COE) and EPA to evaluate the utility of a variety of biological endpoints as to their effectiveness in predicting the biological effects resulting from the disposal of contaminated dredged materials. The biological responses evaluated in this program included genetic, pathological, physiological, reproductive, population and community endpoints.

The sediments that were used in the FVP were dredged from Black Rock Harbor, Connecticut, and disposed of at the COE's dredged material disposal site in Central Long Island Sound. These sediments were contaminated with PAHs (sum = 142,000 ng/g dry), PCBs (A1254 = 6,400 ng/g dry), copper (2,900 µg/g dry), chromium (1,480 µg/g dry), and cadmium (24 µg/g dry). Field and laboratory measurements of biological effects were determined over a five-year period, which included both pre- and post-disposal phases. A synthesis of the results of the FVP is provided in Gentile et al. (1988).

INDIVIDUAL RESPONSES

Biochemical

As discussed earlier, exposure to metals may result in metal binding by proteins and PAH exposure can result in oxidation by mixed function oxidase systems. In their work on metal induction of metallothionein and very low-molecular-weight ligands, Jenkins and Sanders (1986) have found correlations between the increased accumulation of cadmium in these pools and growth and reproduction in the polychaete *Neanthes*. The authors have also observed this relationship for growth in crab larvae. Exposure to crude oil has been shown to induce mixed function oxidase activity in two benthic polychaetes, *Capitella capitata* and *Nereis virens*. There is some indication that this response may contribute to the relative tolerance of these species, as measured by acute mortality, to oiled sediments. Fries and Lee (1984) suggest that induction of the MFO system in *Nereis* due to oil exposure slows growth and prevents reproduction because of an interference with the normal production of gonadotrophic and ovulation-inhibiting hormones.

Genetic

Genetic studies using benthic invertebrates are rare; a genetic response has only recently been developed by Pesch et al. (1981), who described the response, sister chromatid exchange (SCE), in the polychaete *Neanthes arenaceodenta*. This genetic endpoint was evaluated in the FVP using another polychaete, *Nephtys incisa*. Exposure to Black Rock Harbor sediments did cause an increase in SCE frequency in both laboratory and field exposed worms (Pesch et al., 1988). The long-term consequences of this genetic response is unknown; however, tumor induction or some other mutagenicity may result. It is clear that these effects are important as genetic polymorphism will enhance adaption to stressed environments. Grassle and Grassle (1977) have demonstrated that the opportunistic, contaminant-tolerant polychaete *Capitella capitata* consists of at least six sympatric sibling species.

In a separately funded study conducted by the EPA for the National Cancer Institute (NCI), chemical fractions of these same Black Rock Harbor sediments were subjected to three genetic screening assays: the Ames test, the metabolic cooperation assay, and SCE (Gardner et al., 1987). The latter two tests were conducted using the Chinese hamster V79 lung fibroblasts. The results of all three assays established that these sediments exhibited some form of genotoxicity. Further evidence of tumor promotion will be discussed below.

Pathologies

Myers and Hendricks (1985) surveyed the existing literature on pathological studies with aquatic organisms and found the state of our pathological mechanisms in this group to be far behind the mammalian counterpart. Among aquatic species, the least amount of work has been done on marine infauna. There are several reports that investigated the carcinogenic effects of petroleum hydrocarbons on neoplasia induction in bivalves exposed to oil (Brown et al., 1977; Yevich and Barszcz, 1977; Harshbarger et al., 1979). Data are beginning to accumulate for bottom-dwelling fish (Malins et al., 1984) that indicate a close association between fish diseases and heavily contaminated sediments. Aromatic hydrocarbons are the most commonly implicated chemicals in these studies.

In the FVP, the pathological responses to Black Rock Harbor sediments were examined in four infaunal species: the tube-building amphipod *Ampelisca abdita*; the polychaetes *Nephtys incisa* and *Neanthes arenaceodentata*; and the bivalve *Yoldia limatula* (Yevich et al., 1986). In *A. abdita*, Black Rock Harbor sediments caused necrosis of gill epithelia and a loss of normal gill architecture. This species also exhibited atrophied mucous cells and tube glands. These data are consistent with observations that *Ampelisca* builds shorter and more poorly constructed tubes in sediments containing high concentrations of BRH material. Histopathological changes were noted in the epidermis and parapodial muscle of both polychaetes. Mucous-secreting cells were also affected in *Neanthes*. No pathological effects were observed in *Yoldia*, which was probably due to a lack of exposure because it would not burrow into Black Rock Harbor sediments.

Oysters and flounder exposed to Black Rock Harbor sediments in the NCI study referred to above showed a high degree of pathological abnormalities (Gardner et al., 1987). Neoplastic tumors in the oyster were found primarily in the renal excretory tissues, with some tumors found in the gill, gonad, gastrointestinal, heart, and neural tissues. Neoplastic lesions also developed in the kidney, pancreas and oral epithelial surfaces of winter flounder exposed to these sediments.

Physiology

Physiological measurements, when integrated and expressed as energy balance, have been shown to be sensitive indicators of stress (Bayne, 1975). This approach has been modified by Gilfillan et al. (1976), who reported a reduced carbon flux in the clam *Mya arenaria* following exposure to petroleum. Roesijadi and Anderson (1979) developed a condition index that varied in concert with the abundance of free amino acids in *Macoma inquinata*, also exposed to oil-contaminated sediments.

Bioenergetic responses of the polychaetes *N. incisa* and *N.*

arenaceodenta were evaluated with the Black Rock Harbor sediments (Johns et al., 1985; Johns and Gutjahr-Gobell, 1988). Laboratory exposures to mixtures of Black Rock Harbor and control sediments caused reductions in net growth efficiency and scope for growth in each species, respectively. Respiration and excretion rates in *Nephtys* collected from the disposal site were lower than the same rates in worms collected from control areas. Responses of this nature may explain why the eventual growth and population size of this species was impacted at the FVP disposal site.

Behavior

A behavioral response is, for mobile organisms, the first line of defense against exposure to a heavily contaminated sediment. Avoidance of noxious sediments is also a common response of infaunal benthic invertebrates (Swartz, 1987) and this parameter has been incorporated into sediment toxicity tests with amphipods (Swartz et al., 1985a). Other behavioral responses in benthic invertebrates are generally difficult to describe because of observational problems associated with the sedimentary medium.

Behavioral effects were observed in the studies with Black Rock Harbor sediments. The amphipod *Ampelisca* showed a distinct emergence response to these sediments. The bivalve *Yoldia* failed to burrow into any Black Rock Harbor mixtures greater than 50 percent and, when pushed into the sediments, would not feed. The burrowing activities of *Nephtys* were also altered, such that it would only create its burrows in sediment layers without Black Rock Harbor material. The implications of changes in burrowing behavior are twofold. Inability to rebury into contaminated sediments makes the organism susceptible to predation (Pearson et al., 1981). It also disrupts regular feeding behavior for deposit feeders affecting ontogenetic growth and the normal organism-sediment interactions.

Growth and Reproduction

Growth and reproduction are two very closely linked processes, especially as related to individual maturation rates and attainment of reproductive age or size. All of the responses discussed above can significantly affect growth rates by impairing feeding mechanisms, food assimilation and the conversion of energy resources into body tissues. Regardless of the mechanism, the end result will be slower growth and maturation rates.

The results of several studies examining the effects of heavy metals and petroleum hydrocarbons on reproduction in benthic polychaetes have been summarized by Reish (1980). In these water-only exposures, all species showed reproductive effects to all compounds; however, the relationship of these effects to sediment contamination is uncertain because the organisms were not exposed to the contaminants in sediments. Similarly, Jenkins and Mason (1988) have demonstrated growth and reproductive effects in *Neanthes* as a result of water column cadmium exposures.

A series of chronic experiments were conducted for the FVP using *A. abdita* exposed to suspended Black Rock Harbor sediments (Gentile et al., 1985, 1987; Scott and Redmond, in press). Long-term exposures to these sediments caused significant reduction in the mean size of this amphipod in all Black Rock Harbor treatments. The most critical growth reduction occurred with the female amphipods. Because they grew slower and matured later, time-specific fecundity was reduced. Exposed females also produced fewer eggs because they were smaller. Growth in the polychaete *N. incisa*, was also impaired by exposure to Black Rock Harbor bedded sediments, probably as a result of its inability to create burrows and effectively feed in these sediments (Johns et al., 1985).

Survival

Acute mortalities resulting from exposure to contaminated sediments under laboratory conditions have been well documented (Swartz, 1987). This response continues to be one of the primary tools used to evaluate the relative toxicity of sediments. In general, crustaceans--particularly amphipods--have been shown to be the most sensitive group. This was the case in the FVP where *A. abdita's* acute and chronic responses were the most sensitive of 11 species tested.

Chronic survival studies with benthic species and contaminated sediments are rare because of a paucity of chronic test methods. Survival of benthic species are usually inferred from population abundance comparisons among contaminated and noncontaminated field sites. Differential survival among the species in the community is one of the primary mechanisms leading to shifts in community dominance and diversity.

POPULATION AND COMMUNITY LEVEL RESPONSES

The Successional Paradigm

The response of soft-bottom benthic communities to pollutant stress was extensively reviewed by Pearson and Rosenberg (1978) who suggested a model for community recolonization and succession following disturbances due to anthropogenic influences. Rhoads et al. (1978), in the evaluation of community changes following dredged material disposal, recognized similarities in the types of species that were dominant at various stages of the recolonization process. Based upon the work of McCall (1977) and Pearson and Rosenberg (1978), they identified two distinct stages of succession, each typified by species with certain life history characteristics.

The stages of succession and their relation to sediment processes are illustrated in Figure 2. The first stage is characterized by small tube-dwelling polychaetes or oligochaetes who are short lived, opportunistic species. They are either suspension or deposit feeders and they have little contact with deeper subsurface sediments. These species are progressively replaced by an intermediate assemblage consisting of shallow-dwelling bivalves or tubicolous amphipods. The final stage, or equilibrium assemblage, is dominated by infaunal deposit feeders who are large, long-lived, and feed deep in the sediment. As can be seen in Figure 2, this progressive infaunalization increases sediment bioturbation and particle mixing and, hence, pore water exchange at the sediment- water interface.

Population and Community Responses to Contamination

Rates of population growth were predicted for two infaunal species in the FVP. Assessments of *Nephtys* population growth were

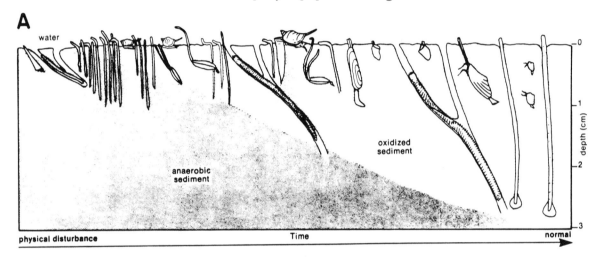

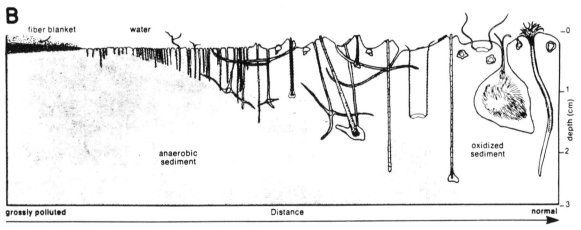

FIGURE 2 A. Development of organism-sediment relationships over time following a physical disturbance in Long Island Sound. B. Organism-sediment relationships associated with pollution gradient due to pulp mill effluent (Pearson and Rosenberg, 1978). SOURCE: Rhoads and Germano, 1982.

conducted by Zajac and Whitlatch (in press) using a size-classified population model and field-collected data on abundance and individual size. Combined with laboratory-derived estimates of growth rates, the model predicted depressed population growth on the FVP disposal mound. Although *Ampelisca* was not an abundant species at the site, a similar analysis was made using the laboratory data described above. Intrinsic rates of growth, "r," calculated as the difference between estimated birth and death rates, were less than zero for all Black Rock Harbor exposures (1 to 4 mg/liter). These values indicate that the population would become extinct with continued exposure to these sediments (Scott and Redmond, 1988).

As described above, the responses of individuals to sediment contamination can easily be translated into population responses in terms of abundance and distribution. Studies in Loch Creren, Scotland by Pearson (1981) are typical of the shifts in species dominance one might expect to result from organic enrichment. He examined community structure in a gradient away from an outfall of an alginate factory, which created a localized, highly anaerobic deposit. Figures 3 and 4 show changes in total density, biomass, species richness and species composition along this gradient. Species dominance shifted from opportunistic polychaetes near the outfall to longer-lived bivalves, polychaetes, and ophiuroids as distance from the outfall increased. Swartz et al. (1985b) observed the same shift in species type with distance from a sewage outfall on the California coast. These authors numerically classified the stations in the gradient, using both species composition and contaminant concentrations, and found that each analysis grouped the stations similarly. These data suggest a high degree of correlation between the two parameters and the possibility that contaminants, as well as other factors associated with organic enrichment, may be responsible for community effects. In a summary of benthic community data from the New York Bight, Boesch (1982) described similar transitions in species dominance which were related to organic enrichment. The opportunist *Capitella capitata* was the dominant species closest to the sewage disposal site. Amphipods were virtually absent from this zone. A similar pattern of increased opportunists and decreased amphipod abundance has been described in benthic communities responding to oil spills (Sanders et al., 1980; Cabioch et al., 1978). Swartz et al. (1982) have also demonstrated a relationship between sediment toxicity and amphipod distributions.

Beyond this enrichment zone described for the New York Bight, community density and species diversity increase and composition of the community is typical of those in muddy fine sands in the region. As was seen in Swartz et al. (1985b), unusually high densities were found in a transition zone, which Boesch attributed to the exclusion of predators due to oxygen or toxic stress. A major impact on benthic systems resulting from organic enrichment is the increase in sediment oxygen demand. With the input of suspended solids there is a concomitant increase in abundance of opportunistic suspension feeders to the exclusion of bioturbating deposit feeders. The result is a migration of the sulfide zone (redox-potential discontinuity layer, RPD) closer to the sediment-water interface.

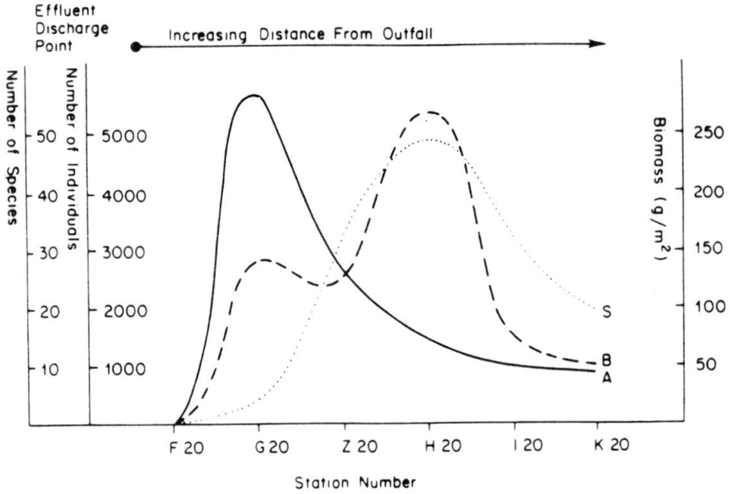

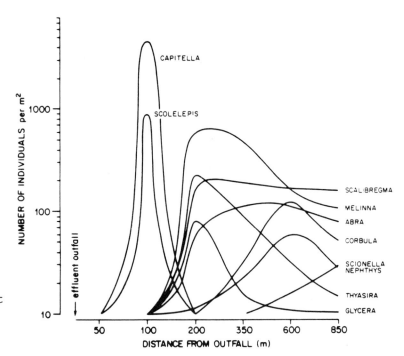

FIGURE 3 Species-abundance-biomass diagram of faunal change along a gradient of organic enrichment in Loch Creren. SOURCE: Pearson, 1981.

FIGURE 4 Abundance changes of some dominant species along an enrichment gradient in Loch Creren. SOURCE: Pearson, 1981.

The following discussion will describe, in some detail, our studies on the benthic community recolonization process resulting from the disposal of Black Rock Harbor sediments at the FVP disposal site (Scott et al., 1987). The recolonization process was measured by documenting the rate of recolonization of the disposal site and comparing this with the ambient (control) community. The parameters used to describe recolonization and convergence with the predisposal system were species numbers, abundance of numerically dominant species, degree of infaunalization (successional stage), and depth of biogenic mixing of the bottom sediments (another measure of infaunalization). The latter two parameters were described using the REMOTS^R interface camera (Rhoads and Germano, 1982).

The benthic community, sampled at four stations at the FVP site on an easterly transect, prior to disposal (baseline) and off the disposal mound for six months following disposal, was dominated by a subsurface infaunal deposit-feeding assemblage consisting of the protobranch bivalves *Nucula annulata* and *Yoldia limatula* and the polychaete worm *Nephtys incisa*. All three of these organisms are Stage III taxa (sensu Rhoads and Germano, 1982) that have mean life spans greatly exceeding one year. Reproduction may take place two or more times per year. They are important in bioturbating the sediment column to a depth of approximately 4 cm; this biogenic mixing controls both pore water and solid-phase chemistry. This subsurface deposit-feeding assemblage was overlain by near-surface populations of the deposit-feeding polychaetes *Mediomastus ambiseta* and the suspension-feeding mactrid bivalve *Mulinia lateralis*. These latter two species are well known members of Stage I series representing opportunistic adaptive strategies. Mean life spans are less than one year and reproduction of the population occurs several times per year. The sympatric association of opportunistic colonizers (Stage I taxa) with longer-lived species (Stage III taxa) is common in estuaries and embayments. This assemblage type has been previously described for the CLIS silt-clay facies (Sanders, 1956; Michael, 1975) and for the silt-clay basinal facies of Buzzards Bay (Sanders, 1958).

The temporal pattern of recolonization consisted of two separate processes operating at different time scales. The first process was the immediate recolonization of the dredged material mound, which occurred during the first six months following disposal. Short-lived, relatively tolerant, early colonizing species, *Polydora* and *Mulinia*, populated the mound in significant densities, and in some cases were most abundant on the mound apex. This phase of the recolonization of the FVP site was not unlike that seen for other disturbed sites within Long Island Sound and elsewhere (Rhoads et al., 1978).

The greater abundances on the mound are not surprising since many early colonizers thrive in disturbed bottoms where the surface sediments contain high inventories of labile organic matter (Rice and Rhoads, 1988). In defaunated habitats (McCall, 1977), neither competition for space nor the biologically mediated geochemical conditions of the sediment would pose problems for recruitment. In fact, the appearance of sedimentary sulfides at the surface may stimulate settlement in some opportunistic species (Cuomo, 1985).

The second component of the recovery process, which may begin concurrently with the initial colonization, is the progressive development of subsurface bioturbation associated with the re-establishment of the long-lived species. The time scale of this process may be on the order of one to two years or more. It was not until 19 months after disposal that head-down feeding voids were observed on REMOTS^R images at the FVP site mound stations, even though the major frequency mode of the Biological Mixing Depth (BMD) at those stations had converged with that of reference site within one year. The mound station continued to have a significantly lower BMD even though head-down deposit feeders from the ambient community were among the recolonizers.

Both *Nephtys* and *Yoldia* experienced some recolonization of the mound stations during the first six months following disposal. Gradual recruitment of these species occurred during the next 18 months, until population densities approached those at the reference station. The recolonization pattern for *Yoldia* and *Nephtys* showed that the mound was recovering and converging with the ambient seafloor. The size structure of the *Nephtys* population was, however, not similar to that of off-mound stations in that these worms were significantly smaller (Zajac and Whitlatch, in press). The protobranch *Nucula* did not recolonize the mound in significant numbers. At the mound apex the lack of recruitment was likely related to the presence of a large sand fraction (1 to 3 cm deep). *Nucula* is also absent from other sand-covered disposal mounds in CLIS. Recent analyses of grab samples collected at a mound station without a sand lens show that *Nucula* have been unable to colonize the fine sediments at this station, indicating a toxic effect on this species (Scott, unpublished data).

The failure of the ambient Stage III assemblage (*Nephtys*, *Nucula*, and *Yoldia*) to become fully established on the mound after two years may have been due to grain-size effects, as other deep bioturbating organisms were present, although in low densities. It may be that the sand lens on the surface of the mound was effectively capping the subsurface contaminated BRH sediments. As a result, when the headdown feeders grew to a size where feeding depths penetrated the subsurface contaminated silts, feeding activities and survival may have been impaired. The physiological and pathological studies with *Nephtys*, and the behavioral observations on both *Nephtys* and *Yoldia* would support this conclusion.

IMPLICATIONS OF CHANGES IN COMMUNITY STRUCTURE

Alterations in the species composition, abundance, and diversity of benthic communities have two primary affects on the broader ecosystem:

1. influences on higher trophic levels and habitat resource value, and
2. influences on sediment processes and biogeochemical cycling.

These processes have been summarized in Swartz and Lee (1980), Lee and Swartz (1980), and are described in Table 1.

Boesch (1982) has reviewed the trophic interactions between soft-bottom benthic communities and bottom-feeding fish in the New York Bight. In examining changes in community structure in silt-clay facies, it is clear that responses to disturbance during the colonizing phase involve a shift from subsurface, deposit-feeding species to those inhabiting and feeding on surface sediments and suspensions. It has been shown that dense populations of these tolerant species can limit recruitment of other organisms by direct predation on settling larvae.

TABLE 1 Benthic Ecosystem Attributes Associated with Pioneering and Late Stage Series

System attribute	Successional stage	
	Early (Stage I)	Late (Stage III)
Secondary production	High potential for r-selected taxa	Lower potential for K-selected taxa
Prey availability	High--prey are concentrated near surface	Lower--infauna are deep burrowing[a]
Potential for food-web contamination	Highest for suspended or recently sedimented particulates. Body burdens may be low related to short mean life spans	Highest for deeply buried contaminents. Longer mean life spans may lead to significant body burdens
Contaminant/nutrient recycling	Limited to solutes in ≤ 3 cm	Solutes exchanged over distances to 20 cm or deeper
Potential for bottom-water hypoxia	High--storage systems for labile detritus	Low--a recycling or "purging" system

NOTE:
[a]Nonlethal predation of distal ends of siphons or caudal segments may be important for some predator species.

SOURCE: Rhoads and Germano, 1986.

These early stages are typically characterized as having high abundance, biomass, and secondary production.

This assemblage, predominating the sediment surface, is much more available to bottomfish as a food source (Becker and Chew, 1987). Because this group of species is the first to colonize contaminated sediments, the potential for food web contamination needs to be addressed. The small size, rapid turnover, and high production rates make these organisms an excellent food source for fast growing juvenile fish. The short life spans of colonizing species would indicate that bioaccumulation under these conditions may not be a problem, but that has yet to be determined.

The distribution of sediment contaminants and nutrients is also altered by shifts in community structure and composition. Shallow-burrowing forms will only affect transport over the top few centimeters and, as the RPD rebounds toward the sediment-water interface, anoxic conditions in the deeper layers would more tightly bind some contaminants. As colonization progresses, this condition would undergo reversal.

Another effect of changes in species composition is an alteration in habit structure (Lee and Swartz, 1980), particularly in the surface sediments. Formation of dense tube mats can significantly influence hydrodynamics and surface flow characteristics of the boundary layer which have been demonstrated to increase sedimentation rates (Rhoads and Boyer, 1982). Changes in the grain-size distribution can also result from extensive pelletization of the surface sediments (Cuomo and Rhoads, 1987). These factors can restructure the composition of surface sediments which may ultimately affect larval recruitment of benthic species.

CONCLUSIONS AND RECOMMENDATIONS

The effects of sediment contamination on changes in benthic communities have been largely determined through correlational analyses. There are few laboratory studies directly linking a sediment contaminant to community effects (e.g., Tagatz, 1983), and we are left with the problem of attempting to interpret community change and the relative importance of sediment contaminants as a cause of that change. Sediment physical properties and biological interactions must also be factored into these interpretations.

Benthic communities show responses to sediment contamination under severe contaminant stress. Locations exhibiting these conditions are prime candidates for remedial action. The larger issue requiring attention, however, is the long-term, low-level contaminant input to coastal systems and the resultant subtle changes in species composition and abundance. Given our inability to discriminate between contaminant effects and natural variability, these types of changes are most likely going unnoticed.

The major contaminant inputs and sediment contaminant impacts occur on coastal systems, many of which are in estuaries. In their review of stress effects in benthic communities, Boesch and Rosenberg (1981) suggest that estuarine organisms are more resistant to stress than are those in more stable environments, e.g., the deep-sea. They relate this observation to the ability of estuarine and nearshore fauna to tolerate a wide range of environmental factors, such as salinity, temperature, and suspended solids. This tolerance has selected for a suite of species that are opportunistic, have high turnover rates, and contribute to the extremely dynamic nature of nearshore biological systems. The variable recruitment and distribution patterns of these species contribute to the difficulty in understanding contaminant effects. It appears that these opportunists are capable of adapting to contaminant stress, but the costs to the organism, in terms of long-term population consequences, are largely unknown. The significance of bioaccumulation in this group and, subsequently, in the food chain transfer of contaminants is also unclear.

The effects of the long-term degradation of benthic systems and the ability of benthic organisms to adapt to this chronic stress is a critical factor in the management of point- and nonpoint-source discharges. As such, there is a pressing need for the development of laboratory test systems to evaluate this chronic stress in benthic organisms. This effort should concentrate on chronic level effects and on bioaccumulation studies with both opportunistic and longer-lived deposit feeders. The methods should have the ability to predict such effects under field conditions and be able to predict linkages between lower and higher levels of biological organization. As the data base on these effects grows, our ability to interpret subtle community changes will increase dramatically.

ACKNOWLEDGMENTS

Support for the preparation of this paper was partially provided under EPA contract #68-03-3529 to Science Applications International Corporation. Dr. Donald Rhoads' (SAIC) review of the manuscript is appreciated. The contents of the manuscript do not necessarily reflect the views or policies of EPA. Nor does the mention of trade names constitute endorsement or recommendation for use by EPA.

REFERENCES

Baker, R. A. 1980. Contaminants and Sediments. Ann Arbor, Mich.: Ann Arbor Science Publishers.

Bayne, B. L. 1975. Cellular and physiological measures of pollution effect. Mar. Poll. Bull. 16:127-128.

Becker, D. S. and K. K. Chew. 1987. Predation on *Capitella* spp. by small-mouthed pleuronectids in Puget Sound, Washington. Fish. Bull. 85:471-479.

Boesch. D. F. 1982. Ecosystem consequences of alterations of benthic community structure and function in the New York Bight region. In Ecological stress in the New York Bight: Science and Management, G. F. Mayer, ed. Columbia, South Carolina: Estuarine Research Federation. Pp. 543-568.

Boesch, D. F. and R. Rosenberg. 1981. Response to stress in marine benthic communities. In Stress Effects on Natural Systems, G. W. Barrett and R. Rosenberg, eds. New York: John Wiley and Sons. Pp. 179-200.

Bradford, W. L. and S. N. Luoma. 1980. Some perspectives on heavy metal concentrations in shellfish and sediment in San Francisco Bay, California. In Contaminants and sediments: Volume 2, R. A. Baker, ed. Ann Arbor, Mich.: Ann Arbor Science Publishers. Pp. 501-532.

Brown, R. S., R. E. Wolke, S. B. Saila, and C. W. Brown. 1977. Prevalence of neoplasia in 10 New England populations of the soft shell clam *Mya arenaria*). Ann. N.Y. Acad. Sci. (1977b) 298:522-534.

Bryan, G. W. 1976. Some effects of heavy metal tolerance in aquatic organims. In Effects of Pollutants on Aquatic Organisms, A. P. M. Lockwood, ed. Cambridge, England: Cambridge University Press. Pp. 7-34.

Cabioch, L., J.-C. Dauvin, and F. Gentile. 1978. Preliminary observations on pollution of the sea bed and disturbance of sublittoral communities in Northern Brittany by oil from the *Amoco Cadiz*. Mar. Poll. Bull. 9:303-307.

Cuomo, M. C. 1985. Sulfide as a larval settlement cue for *Capitella* sp. I. Biogeochemistry 1:181-196.

Cuomo, M. C. and D. C. Rhoads. 1987. Biogenic sedimentary fabrics associated with pioneering Polychaete assemblages: modern and ancient. J. Sed. Petrol. 57:537-543.

Dickson, K. L., M. A. Maki, and W. A. Brungs. 1987. Fate and Effects of Sediment Bound Chemicals in Aquatic Systems. New York: Pergamon Press.

Fries, C. R. and R. F. Lee. 1984. Pollutant effects on the mixed function oxygenase (MFO) and reproductive systems of the marine polychaete *Nereis virens*. Mar. Biol. 79:187-193.

Gardner, G. R., P. P. Yevich, A. R. Malcolm, and R. P. Pruell. 1987. Carcinogenic effects of Black Rock Harbor sediment on American oysters and winter flounder. Project report to the National Cancer Institute. ERL-Narragansett contribution #901.

Gentile, J. H., K. J. Scott, S. M. Lussier, and M. S. Redmond. 1985. Application of Laboratory Population Responses for Evaluating the Effects of Dredged Material. Technical Report D-85-8. Vicksburg, Miss.: U.S. Army Engineer Waterways Experiment Station.

Gentile, J. H., K. J. Scott, S. M. Lussier, and M. S. Redmond. 1987. The Assessment of Black Rock Harbor Dredged Material Impacts on Laboratory Population Responses. Technical Report D-87-3. Vicksburg, Miss.: U.S. Army Engineer Waterways Experiment Station.

Gentile, J. H., G. G. Pesch, J. Lake, P. P. Yevich, G. Zaroogian, P. Rogerson, J. Paul, W. Galloway, K. J. Scott, W. Nelson, D. M. Johns, and W. Munns. 1988. Applicability and Field Verification of Predictive Methodologies for Aquatic Dredged Material Disposal. Technical Report D-88-5. Vicksburg, Miss.: U.S. Army Engineer Waterways Experiment Station.

Gilfillan, E. S., D. Mayo, S. Hanson, D. Donovan, and L. C. Jiang. 1976. Reduction in carbon flux in *Mya arenaria* caused by a spill of No. 6 fuel oil. Mar. Biol. 37:115-123.

Goerke, H., G. Eder, K. Weber, and W. Ernst. 1979. Patterns of organochlorine residues in animals of different trophic levels from the Weser Estuary. Mar. Poll. Bull. 10:127-132.

Grassle, J. F. and J. P. Grassle. 1977. Temporal adaptations in sibling species of *Capitella*. In Ecology of Marine Benthos, B. C. Coull, ed. Columbia, S.C.: University of South Carolina Press. Pp. 177-189.

Hansen, D. J. and M. E. Tagatz. 1980. A laboratory test for assessing impacts of substances on developing communities of benthic estuarine organisms. In Aquatic Toxicology, J. G. Eaton, P. R. Parrish, and A. C. Hendricks, eds. STP 707. Philadelphia: American Society for Testing and Materials. Pp. 40-57.

Harshbarger, J. C., S. V. Otto, and S. C. Chang. 1979. Proliferative disorders in *Crassostrea virginica* and *Mya arenaria* from the Chesapeake Bay and intranuclear virus-like inclusions in *Mya arenaria* with germinomas from a Maine oil spill site. Halitos 8:243-248.

Jacobs, R. P. 1980. Effects of the *Amoco Cadiz* oil spill on the seagrass community at Roscoff with special reference to the benthic infauna. Mar. Ecol. Prog. Ser. 2:207-212.

Jenkins, K. D. and D. A. Brown. 1984. Determining biological significance of contaminant bioaccumulation. In Concepts in Marine Pollution Measurements, H. H. White, ed. College Park, Md.: Maryland Sea Grant. Pp. 354-375.

Jenkins, K. D. and B. M. Sanders. 1986. Relationships between free cadmion ion activity in sea water, cadmion accumulation and subcellular distribution, and growth in polychaetes. Environ. Health Persp. 65:205-210.

Jenkins, K. D. and A. Z. Mason. 1988. Relationships between subcellular distributions of cadmion and perturbations in reproduction in the polychaete *Neanthes arenaceodentata*. Aq. Toxicol. 12:229-244.

Johns, D. M., R. Gutjahr-Gobell, and P. Schauer. 1985. Use of Bioenergetics to Investigate the Impact of Dredged Material on Benthic Species: A Laboratory Study with Polychaetes and Black Rock Harbor Material. Technical Report D-85-7. Vicksburg, Miss.: U.S. Army Engineer Waterways Experiment Station.

Johns, D. M. and R. Gutjahr-Gobell. 1988. Bioenergetic Effects of Black Rock Harbor Dredged Material on the Polychaete *Nephtys incisa*: A Field Verification. Technical Report D-88-3. Vicksburg, Miss.: U.S. Army Engineer Waterways Experiment Station.

Lake, J. L., N. Rubinstein, and S. Pavignano. 1987. Predicting bioaccumulation: Development of a simple partitioning model for use as a screening tool for regulating ocean disposal of wastes. In Fate and Effects of Sediment Bound Chemicals in Aquatic Systems, K. L. Dickson, A. W. Maki, and W.A. Brungs, eds. New York: Pergamon Press. Pp. 151-166.

Lake, J. L., N. Rubinstein, H. Lee II, C. A. Lake, J. Heltshe, and S. Pavignano. In prep. Equilibrium partitioning and bioaccumulation of sediment associated contaminants by infaunal organisms.

Lee, H. L. and R. C. Swartz. 1980. Biological processes affecting the distribution of pollutants in marine sediments. Part II. Biodeposition and bioturbation. In Contaminants and Sediments, Volume 2, R. A. Baker, ed. Ann Arbor, Mich.: Ann Arbor Science Publishers. Pp. 555-606.

Lee, R. F. 1984. Factors affecting bioaccumulation of organic pollutants by marine animals. In Concepts in Marine Pollution Measurements, H. White, ed. College Park, Md.: Maryland Sea Grant Publication. Pp. 339-354.

Malins, D. C., B. B. McCain, D. W. Brown, S. L. Chan, M. S. Myers, J. T. Landhal, P. G. Prohaska, A. J. Friedman, L. D. Rhodes, D. G. Burrows, W. D. Grolund, and H. O. Hodgens. 1984. Chemical pollutants in sediments and diseases of bottom dwelling fish in Puget Sound, Washington. Environ. Sci. Technol. 18:705-713.

Malins, D. C., M. M. Krahn, M. S. Myers, L. D. Rhodes, D. W. Brown, C. A. Krone, B. B. McCain, and S. L. Chan. 1985. Toxic chemicals in sediments and biota from a creosote-polluted harbor: Relationships with hepatic neoplasms and other hepatic lesions in English sole (*Paraphrys vetulus*). Carcinogenesis 6:1463-1469.

McCall, P. L. 1977. Community patterns and adaptive strategies of the infaunal benthos of Long Island Sound. J. Mar. Res. 35:221-266.

Meyers, T. R. and J. D. Hendricks. 1985. Histopathology. In Fundamentals of Aquatic Toxicology, G. M. Rand and S. R. Petrocelli, eds. New York: Hemisphere Publishing Corporation. Pp. 283-331.

Michael, A. D. 1975. Structure and stability in three marine benthic communities in southern New England. In Brookhaven Symposium on the Effects of Energy Related Activities on the Outer Continental Shelf, E. Morowitz, ed. Pp. 109-125.

Neff, J. M. 1985. Polycyclic aromatic hydrocarbons. In Fundamentals of Aquatic Toxicology, G. M. Rand and S. R. Petrocelli, eds. New York: Hemisphere Publishing Corporation. Pp. 416-454.

Nimmo, D. R. 1985. Pesticides. In Fundamentals of Aquatic Toxicology, G. M. Rand and S. R. Petrocelli, eds. New York: Hemisphere Publishing Corporation. Pp. 335-373.

National Oceanic and Atmospheric Administration (NOAA). 1988. A summary of data on chemical contaminants in sediments collected during 1984, 1985, 1986, and 1987. NRC Symposium Proceedings.

Pavlou, S. P. 1987. The use of the equilibrium partitioning approach in determining safe levels of contaminants in marine sediments. In Fate and Effects of Sediment Bound Chemicals in Aquatic Systems, K. L. Dickson, A. W. Maki, and W. A. Brungs, eds. New York: Pergamon Press. Pp. 388-412.

Pearson, T. H. 1981. Stress and catastrophe in marine benthic ecosystems. In Stress Effects on Natural Ecosystems, G. W. Barrett and R. Rosenberg, eds. New York: John Wiley and Sons. Pp. 201-214.

Pearson, T. H. and R. Rosenberg. 1978. Macrobenthic succession in relation to organic enrichment and pollution of the marine environment. Oceanogr. Mar. Biol. Ann. Rev. 16:229-311.

Pearson, W. H., D. L. Woodruff, P. C. Sugarman, and B. L. Olla. 1981. Effects of oiled sediment on predation on the littleneck clam, *Protothaca staminea*, by the Dungeness crab, *Cancer magister*. Est. Coast Shelf Sci. 13:445-454.

Pesch, G., C. E. Pesch, and A. R. Malcolm. 1981. *Neanthes arenaceodentata*, a cytogenetic model for marine genetic toxicology. Aq. Toxicol. 1:301-311.

Pesch, G., C. E. Pesch, A. R. Malcolm, P. F. Rogerson, and G. R. Gardner. 1987. Sister Chromatid Exchange in Marine Polychaetes Exposed to Black Rock Harbor Sediments. Technical Report D-87-5. Vicksburg, Miss.: U.S. Army Engineer Waterways Experiment Station.

Prosi, F. 1979. Heavy metals in aquatic organisms. In Metal Pollution in the Aquatic Environment, U. Forstner and G. T. Wittman, eds. New York: Springer Verlag. Pp. 271-318.

Reish, D. J. 1980. The effect of different pollutants on ecologically important polychaete worms. EPA Ecological Research Series 600/3-80-053.

Rhoads, D. C., P. L. McCall, and J. Y. Yingst. 1978. Disturbance and production on the estuarine seafloor. Amer. Sci. 66:577-586.

Rhoads, D. C. and J. D. Germano. 1982. Characterization of organism-sediment relations using sediment profile imaging: An efficient method of remote ecological monitoring of the seafloor (REMOTS System). Mar. Ecol. Progr. Ser. 8:115-128.

Rhoads D. C. and L. F. Boyer. 1982. The effects of marine benthos on physical properties of sediments. In Animal-Sediment Relations, P. L. McCall and M. J. Tevesz, eds. New York: Plenum Press Geobiology Series. Pp. 3-52.

Rhoads, D. C. and J. D. Germano. 1986. Interpreting long-term changes in benthic community structure: A new protocol. Hydrobiol. 142:291-308.

Rice D. L. and D. C. Rhoads. In press. Early diagenesis of organic matter and the nutritional value of sediment. In Ecology of Deposit-Feeding. G. Lopez, ed. New York: Elsevier.

Rogerson, P. F., S. C. Schimmel, and G. Hoffman. 1985. Chemical and Biological Characterization of Black Rock Harbor Dredged Material. Technical Report D-85-9. Vicksburg, Miss.: U.S. Army Engineer Waterways Experiment Station.

Roesiijadi, G. and J. W. Anderson. 1979. Condition index and free amino acid content of *Macoma inquinata* exposed to oil-contaminated marine sediments. In Marine Pollution: Functional Responses, W. B. Vernberg, A. Calabrese, F. P. Thurberg, and F. J. Vernberg, eds. New York: Academic Press. Pp. 69-84.

Rygg, B. 1986. Heavy metal pollution and log-normal distribution of individuals among species in benthic communities. Mar. Poll. Bull. 17:31-36.

Sanders, H. L. 1956. Oceanography of Long Island Sound 1952-1954, X. Biology of marine bottom communities. Bull. Bingham Oceanogr. Coll. 15:345-414.

Sanders, H. L. 1958. Benthic studies in Buzzards Bay, I. Animal-sediment relationships. Limnol. Oceanogr. 38:265-380.

Sanders, H. L., J. F. Grassle, G. R. Hampson, L. S. Morse, S. Garner-Price, and C. C. Jones. 1980. Anatomy of an oil spill: Long term effects from the grounding of the barge *Florida* off West Falmouth, Massachusetts. J. Mar. Res. 38:265-380.

Scott, J., D. Rhoads, J. Rosen, S. Pratt, and J. Gentile. 1987. The Impact of Open-water Disposal of Black Rock Harbor Dredged Material on Benthic Recolonization at the FVP Site. Technical Report D-87-4. Vicksburg, Miss.: U.S. Army Engineer Waterways Experiment Station.

Scott, K. J., and M. S. Redmond. In press. The effects of a contaminated dredged material on laboratory populations of the tubicolous amphipod, *Ampelisca abdita*. In Aquatic Toxicology and Hazard Assessment: 12th Volume, U. M. Cowgill and L. R. Williams, eds., STP 1027. Philadelphia: American Society for Testing and Materials.

Sheehan, P. J. 1984. Effects on individuals and populations. In Effects of Pollutants at the Ecosystem Level, P. J. Sheehan, D. R. Miller, G. C. Butler, and P. Bourdeau, eds. New York: John Wiley & Sons, Ltd. Pp. 23-50.

Stainken, D. 1984. Organic pollution and the macrobenthos of Raritan Bay. Environ. Toxicol. Chem. 3:95-111.

Stegeman, J. J. 1981. Polynuclear aromatic hydrocarbons and their metabolism in the marine environment. In Polycyclic Hydrocarbons and Cancer, Volume 3, H. Gelboin and P. O. Ts'O, eds. New York: Academic Press. Pp. 1-60.

Stull, J. K., C. I. Haycock, R. W. Smith, and D. B. Montagne. 1986. Long-term changes in the benthic community on the coastal shelf off Palos Verdes, Southern California. Mar. Biol. 91:539-551.

Swartz, R. C. and H. F. Lee. 1980. Biological processes affecting the distribution of pollutants in marine sediments. Part 1. Accumulation, trophic transfer, biodegradation and migration. In Contaminants and sediments, Volume 2, R. A. Baker, ed. Ann Arbor, Mich.: Ann Arbor Science Publishers. Pp. 533-553.

Swartz, R. C., W. A. DeBen, J. K. Jones, J. O. Lamberson, and F. A. Cole. 1985a. Phoxocephalid amphipod bioassay for marine sediment toxicity. In Aquatic Toxicology and Hazard Assessment: Seventh Symposium, R. D. Cardwell, R. Purdy, and R. C. Bahner, eds. STP 854. Philadelphia: American Society for Testing and Materials Pp. 284-307.

Swartz, R. C., D. W. Schults, G. R. Ditsworth, W. A. DeBen, and F. A. Cole. 1985b. Sediment toxicity, contamination, and macrobenthic communities near a large sewage outfall. In Validation and Predictability of Laboratory Methods for Assessing the Fate and Effects of Contaminants in Aquatic Ecosystems, T. P. Boyle, ed. STP 865. Philadelphia: American Society for Testing and Materials. Pp. 152-175.

Swartz, R. C., F. A. Cole. D. W. Schults, and W. A. DeBen. 1986. Ecological changes in the Southern California Bight near a large sewage outfall: Benthic conditions in 1980 and 1983. Mar. Ecol. Progr. Ser. 31:1-13.

Swartz, R. C. 1987. Toxicological methods for determining the effects of contaminated sediment on marine organisms. In Fate and Effects of Sediment Bound Chemicals in Aquatic Systems, K. L. Dickson, A. W. Maki, and W. A. Brungs, eds. New York: Pergamon Press. Pp. 183-198.

Tagatz, M. E., G. R. Plaia, C. H. Deans, and E. M. Lores. 1983. Toxicity of creosote-contaminated sediment to field- and laboratory-colonized estuarine benthic communities. Environ. Toxicol. Chem. 2:441-450.

Weaver, G. 1984. PCB contamination in and around New Bedford, Mass. Environ. Sci. Technol. 18:22-27.

Yevich, P. P. and C. A. Barszcz. 1977. Neoplasia in soft-shell clams *Mya arenaria*) collected from oil-impacted sites. Ann. N.Y. Acad. Sci. 298:409426.

Yevich, P. P., C. A. Yevich, K. J. Scott, M. Redmond, D. Black, P. Schauer, and C. E. Pesch. 1986. Histopathological Effects of Black Rock Harbor Dredged Material on Marine Organisms: A Laboratory Investigation. Technical Report D-86-1. Vicksburg, Miss.: U.S. Army Engineer Waterways Experiment Station.

Young, D. R. and A. J. Mearns. 1978. Pollutant flow through food webs. In Southern California Coastal Water Research Project, 1978 Annual Report, El Segundo, California. Pp. 185-202.

Zajac, R. and R. Whitlatch. In press. Population ecology of the polychaete *Nephtys incisa* in southern New England waters and the effects of disturbance. Estuaries.

Zarba, C. 1988. National perspective on sediment quality. NRC Symposium Proceedings.

SEDIMENT CONTAMINATION AND MARINE ECOSYSTEMS: POTENTIAL RISKS TO HUMAN HEALTH

Donald C. Malins
Pacific Northwest Research Foundation

ABSTRACT

It is recognized that exposure of aquatic organisms to contaminated sediments results in the bioaccumulation of toxic chemicals (Capuzzo et al., 1988). Toxic responses may occur at the biochemical-cellular, organismal, population, and community levels and range from metabolic impairment to changes in community structure and function (Capuzzo et al., 1988; Buhler and Williams, 1988). The toxic insults are not limited to the initial organisms impacted, but may extend throughout the food web and include the human consumer of seafood.

Contamination of the sediment is especially significant because of the host of benthic species that inhabit the ocean floor--species that serve as initial contaminant reservoirs and are food for a variety of organisms. Thus, the sediment is the starting point for the transfer of toxic chemicals through wide expanses of the food web (Malins et al., 1984; U.S. EPA, 1985).

Many gaps exist in our understanding of the far ranging effects of sediment contamination on the myriad organisms that inhabit rivers, estuaries, and coastal areas. They vary from a limited understanding of synergistic/antagonistic interactions to a shallow perspective of chronic effects. Knowledge about mechanisms that mediate chemical accumulations, metabolic changes and biological effects have been especially elusive. Yet, we know far more about the impacts of sediment contamination on aquatic species than on the human consumer. Simply stated, our understanding of *events* and *processes* that lead to potential human health effects from the consumption of contaminated seafood are virtually unknown, as is the extent of the impact on human populations (Swain, 1988; Friberg, 1988).

Many of the chemicals that contaminate fish and shellfish in polluted environments are transferred to humans through the diet (Malins et al., 1986). Thus, humans are logically viewed as an intimate part of marine food webs. Metabolically resistant (refractory) chemicals, such as PCBs, DDT derivatives, and other halogenated compounds are readily transferred from the sediments to benthic species, such as worms, clams. and bottom-feeding fish

(Malins et al., 1984). Thus, the potential exists for their transfer and bioconcentration through the food web.

Compounds that are actively metabolized, such as the aromatic hydrocarbons, are not readily transferred through aquatic food webs, although they do accumulate in organisms, such as shellfish, that have a limited ability to metabolize them (Malins et al., 1986). Thus, with fish the readily metabolized compounds may be of less concern for the human consumer than the refractory compounds, some of which are known to accumulate in the edible muscle (MacLeod et al., 1981; Romberg et al., 1984; Table 1). The same conclusion cannot be drawn with respect to shellfish contaminated with xenobiotics. Overall, however, it is important to remember, as indicated, that little is known about the propensity for humans to accumulate and bioconcentrate through the diet the thousands of different parent chemicals and metabolites arising from contamination of sediments (Malins et al., 1986). Also, only a paucity of information exists about the nature and extent of the human health effects.

Having broadly delineated some of the complex problems that can be attributed to one of the "original sins" of chemical contamination--pollution of the sediments--several specific questions will now be addressed:

TABLE 1 Concentration (ppm, Wet Weight) of PCBs, DDT and AHs in Edible Tissues of Striped Bass and Salmon

	PCBs	ΣDDT	ΣAHs
Striped bass (Hudson River, New York)	7.00	1.01	t[a]
Striped bass (Montauk, Long Island)	0.80	0.11	t
Striped bass (Orient Point, New York Bight)	3.00	0.73	t
Chinook salmon (Denny Way, Seattle)	1.35	0.01	PHN[b]
Chinook salmon (Richmond Beach, Seattle)	0.23	0.01	PHN

NOTES:
[a]Trace
[b]Phenanthrene identified

SOURCE: MacLeod et al., 1981 and Romberg et al., 1984.

1. What is actually known about the dietary transfer to humans of toxic chemicals from contaminated fish and shell fish?
2. What are the human health implications?
3. How can the gaps in knowledge be filled?

TRANSFER OF TOXIC CHEMICALS TO HUMAN POPULATIONS

An almost singular emphasis has been placed on PCBs as a "model" for considering the exposure of humans to contaminants through marine food webs (U.S. EPA, 1985; Swain, 1988). These studies suggest that refractory organic compounds have the potential for being transferred to human populations through consumption of contaminated sea food (U.S. EPA, 1985; Swain, 1988). For example, a study was conducted on the exposure of humans to PCBs through the consumption of fish from Lake Michigan (Humphrey, 1976). In the 18 counties that border Lake Michigan, 381,000 licensed sports fishermen caught 14 million pounds of trout and salmon annually. As a group, the fishermen and their families consumed 36.6 pounds of fish per year (Humphrey, 1976, 1983), which is over three times the national average. Adults were used in a matched cohort study (MacLeod et al., 1981; Romberg et al., 1984) in which samples of serum from each group were analyzed for PCBs, then the results were compared with data obtained from interviews with each individual involved in the study. The findings generally revealed an increase in PCB serum levels with increased fish consumption (Humphrey, 1983; Table 2). These data also provided evidence that the Michigan residents were exposed through the diet to levels of PCBs significantly above those for the average population (Swain, 1988).

TABLE 2 PCBs in Human Serum as a Function of Fish Consumption and Geographic Location of Fish Source

Source of fish	Amount consumed (kg)[a]	Sample N	Serum PCB (μg/kg) Range	Mean	Median
No source	0	29	ND[b]-41	17.3	15
Lake St. Clair	5-66.8	15	ND-38	19.4	17
Lake Michigan	0-2.73	39	ND-41	18.5	20
Lake Michigan	10.91-118	90	25-366	72.7	56

NOTES:
[a] Presumably kg/yr, though not so specified.
[b] ND = Not detected; detection limit specified as <5μg/kg.

SOURCE: Humphrey, 1983.

It was reasoned that additional groups especially at risk were pregnant women and their unborn and newborn offspring. Thus, a longitudinal study was designed to assess the impact of contaminated fish consumption on these groups (Jacobson et al., 1983). Briefly, the study revealed that infants were exposed to PCBs in utero, as well as postpartum via the breast milk, when the mothers consumed contaminated fish.

The above findings are not surprising. Evidence with rodents exposed to PCBs revealed essentially the same potential for bioaccumulation (see U.S. EPA, 1985). In addition, a recent study with seals (Reijnders, 1988) showed that the consumption of PCB-contaminated fish resulted in substantial accumulations of these compounds. Comparable results were also obtained with mink fed PCB-contaminated fish (Reijnders, 1986).

The findings with the PCBs add an additional dimension to the concern expressed after the Minamata, Japan, mercury poisoning incident in the 1950s (Takeuchi, 1972). In this case, fish were shown to be the source of methylmercury exposure in humans seriously afflicted with neurological and other damage (Takeuchi, 1972). Unfortunately, little or no information exists on the transfer to human populations of the wide variety of xenobiotics that exist together with the PCBs in edible tissues of aquatic life exposed to pollutants. It is not known, for

TABLE 3 DDT[a] concentrations (μg/gm, ppm Wet Weight) in Uncooked and Pan Fried White Croaker (*Genyonemus lineatus*) Fillets

Composite number	Percent original weight	DDT (μg/g)		
		Uncooked	Pan fried	Pan fried normalized
1	27.8	0.202	0.247	0.069
2	35.7	0.167	0.184	0.066
3	35.2	1.110	0.844	0.297
4	33.2	1.070	0.531	0.176
5	31.8	0.410	0.506	0.161
Mean ± SE	32.7	0.57±0.20	0.46±0.12	0.15±0.04
Mean percent loss of DDT due to frying				74

NOTE:
[a] Refers to DDT, DDD, and DDE.

SOURCE: Puffer et al., 1982.

example, whether synergistic or antagonistic interactions play an important part in the disposition in humans of xenobiotics derived from contaminated seafood. Moreover, little is known about the accumulation of metabolites in edible tissues, especially those compounds that are not detected by conventional analytical techniques. Although, as stated, there are indications that metabolites of aromatic hydrocarbons may not accumulate to a significant degree in the edible tissue of fish (Malins et al., 1987), substantive information on possible contamination from many other compounds simply does not exist (Swain, 1988). Clearly of significance is the finding that pan-fried fish tend to have a significantly lower concentration of DDT derivatives than the uncooked fish (Puffer et al., 1982; Table 3); however, the influence of cooking on the concentrations of other contaminants is virtually unknown.

THE HUMAN HEALTH IMPLICATIONS

Considerations of human health effects from the consumption of contaminated seafood have focused, for the most part, on the PCBs; however, some concern has been expressed about inorganic compounds (e.g., arsenic) that accumulate in the muscle of fish (Friberg, 1988). Some studies, for example, point to a possible threat from arsenic (Friberg, 1988). While most of the arsenic in seafood is in the form of arsenobetaine, which is considered relatively atoxic, "extreme consumption" of seafood may give rise to an intake of several hundred micrograms of inorganic arsenic per day--an exposure, which over a lifetime, may be associated with a "significant increase in skin cancer" (Friberg, 1988). Unfortunately, related studies that focus tightly on cause-effect relationships have yet to be conducted. The daily intake of tin through the consumption of seafood is not particularly high; however, more studies need to be conducted on the potential toxicity of trimethyltin resulting from biochemical alkylation reactions. Also, despite regulations pertaining to methylmercury, groups having a "high" fish intake, or an intake of fish with a "high" methylmercury content, may exceed established tolerance levels. In this regard, a special concern exists about pregnant women (Friberg, 1988).

Results from the relatively large amount of research on PCBs suggests that these compounds may pose a significant problem for the consumer of fish from polluted areas. In Lake Michigan studies (Jacobson et al., 1983, 1984; Fein et al., 1984), for example, effects observed among infants born to mothers in "high fish consumption" categories included delays in developmental maturation at birth (Fein et al., 1984). The infants were also smaller in physical size, and had a reduced head circumference and neuromuscular maturity (Table 4). They also exhibited an altered lability of state, increased startle reflexes, and were classified by physicians to be within the "worrisome" neonatal category (Jacobson et al., 1984, 1988). Swain (1988) makes the point that these observations suggest "an effect of contaminants upon the centers of higher integration in infants secondarily exposed via maternal circulation." A study conducted with rats (Hertzler and Daly, 1985) supports the conclusion that PCBs derived from fish indeed have

TABLE 4 Adjusted Birth Size and Gestational Age Measures by Overall Contaminated Fish Consumption and Cord Serum PCB Level[a]

	Overall contaminated fish consumption[b]			Cord serum PCB level[c]		
	Non-fish eaters (n=71)	Fish eaters (n=242)	P	>3 ng/mL (n=166)	≤3 ng/mL (n=75)	P
Birth weight (kg)	3.66±0.54	3.47±0.53	<0.05	3.57±0.54	3.41±0.54	<0.05
Head circumference (cm)	35.48±1.36	34.92±1.31	<0.01	35.28±1.18	34.63±1.19	<0.001
Gestational age based on last menstrual period (wk)	40.82±3.07	40.31±2.97		41.03±3.01	39.77±3.06	<0.05
Gestational age (Ballard examination) (wk)	39.85±1.42	39.15±1.40	<0.01	39.41±1.40	39.47±1.41	
Neuromuscular	19.96±2.48	18.52±2.44	<0.001	19.95±2.42	19.00±2.40	
Physical maturity	17.13±2.25	16.67±2.19		16.96±2.14	16.94±2.15	

NOTES:
[a] Values represent mean ± SD.
[b] Adjusted for effects of maternal prepregnancy weight, type of delivery, and consumption of alcohol and caffeine prior to and during pregnancy and cold remedies during pregnancy.
[c] Adjusted for sex of infant, type of delivery, maternal weight gain during pregnancy, and maternal age.

SOURCE: Fein et al., 1984.

an effect on the nervous system. It was shown, for example, that rats maintained on a diet of PCB-contaminated salmon from Lake Ontario developed behavioral anomalies, compared to controls, in relation to brain concentrations of PCBs. Such a finding can be compared with results obtained from the previously mentioned study of seals exposed to PCBs through their fish diet (Reijnders, 1986). The reproductive process was shown to be "disrupted in the post-ovulation phase." Another study with mink (Reijnders, in press, and 1986) also supported the proposition the PCBs derived from a fish diet have an effect on reproduction at "very low (25 μg per day) levels."

Overall, a limited number of studies have indicated that significant changes in health status may occur in humans consuming contaminated fish; however, obviously many factors impinge on the exact nature of the threat--so many, in fact, that the present findings are best viewed as a stimulus to study the issue in greater detail through carefully controlled field and laboratory investigations.

FILLING THE GAPS IN KNOWLEDGE

Information required for a minimal understanding of the impacts of toxic environmental chemicals on aquatic species and human health substantially exceeds the information available. Important areas for future research include obtaining more knowledge about the nature and extent of exposure on an individual, population, and geographic bases. In addition, biochemical/toxicological data on the scores of chemicals that have the potential to accumulate in edible tissues are also important to obtain, as is information on chronic effects. Moreover, possible human health effects associated with the loss of volatile sediment chemicals to the atmosphere, such as from contaminated subtidal areas, is well worth studying.

The problem of human risk assessment is formidable when one considers that marine life is often exposed to complex mixtures of chemicals in contaminated areas. Moreover, in some cases, assessments conducted thus far with contaminated fish have projected clearly unacceptable human cancer risks (Brown, 1985; Table 5). In addition, the present reliance on PCBs and a small number of other compounds for risk assessment is clearly inadequate. More work needs to be undertaken to make risk assessment more meaningful from a public health point of view, such as by taking into account the fact that complex mixtures of potentially toxic chemicals are likely to be present in edible tissues of fish from polluted areas. Studies in which laboratory animals are fed a diet of contaminated fish tissue or a diet containing extractable, environmentally derived chemicals may well prove to be a useful approach.

Finally, one can only hope that future work will consider a variety of biological end points as indicators of human health effects, rather than focus on the few (e.g., cancer and neurological impairment) that have been studied thus far.

TABLE 5 Contaminant Concentration and Risk Assessment for Consumption of Southern California Fish at Average U.S. Consumption Rate[a]

		Concentration (mg/wet kg)	Risk
White Point			
White Croakers	DDTs	7.6	3.4/10,000
	PCBs	0.38	2.2/20,000
	Total	8.0	5.6/10,000
Rockfish	DDTs	0.44	2.0/100,000
	PCBs	0.057	3.3/100,000
	Total	0.50	5.3/100,000
P. Mackeral	DDTs	0.051	2.3/1,000,000
	PCBs	0.014	8.0/1,000,000
		0.065	1.0/100,000
Santa Monica Bay			
White Croakers	DDTs	0.57	2.6/100,000
	PCBs	0.20	1.1/10,000
	Total	0.77	1.4/10,000
Rockfish	DDTs	0.22	9.9/1,000,000
	PCBs	0.12	6.9/100,000
	Total	0.34	7.9/100,000
P. Mackeral	DDTs	0.057	2.6/1,000,000
	PCBs	0.015	8.6/1,000,000
		0.072	1.1/100,000

NOTE:
[a] 9.3 g/day consumption of domestic estuarine and marine fish.

SOURCE: Brown, 1985.

REFERENCES

Brown, D. 1985. Personal Communication. Based on work conducted while at the Southern California Coastal Water Research Project (S.C.W.R.P.), Long Beach, CA.

Buhler, D. R. and D. E. Williams. 1988. The role of biotransformation toxicity in fish. Aquat. Toxicol. 11:303-311.

Capuzzo, J. M., M. N. Moore, and J. Widdows. 1988. Effects of toxic chemicals in the marine environment: Predictions of impacts from laboratory studies. Aquat. Toxicol. 11:19-28.

Fein, G. G., J. L. Jacobson, S. W. Jacobson, P. W. Schwartz, and J. K. Dowler. 1984. Prenatal exposure to polychlorinated biphenyls: Effects on birth size and gestational age. Pediatr. 105:315-320.

Friberg, L. 1988. The GESAMP evaluation of potentially harmful substances in fish and other seafood with special reference to carcinogenic substances. Aquat. Toxicol. 11:379-393.

Hertzler, D. R. and H. B. Daly. 1985. Ingestion of neurotoxic Lake Ontario salmon influences behaviors of laboratory rats. Meeting of Psychonomic Society, Boston, Mass.

Humphrey, H. E. B. 1983. Population studies of PCBs in Michigan Residents. In PCBs: Human and environmental hazards, F. M. D'itri and M. A. Kamrin, eds. Boston: Butterworth Publishers. pp. 299-310.

Humphrey, H. E. B. 1976. Evaluation of Changes of the Level of Polychlorinated Biphenyls (PCB) in Human Tissue. Final report on W. S. FDA contract. Lansing: Michigan Department of Public Health. p. 86.

Jacobson, J. L., S. W. Jacobson, P. M. Schwartz, G. G. Fein,, and J. K. Dowler. 1984. Prenatal exposure to an environmental toxin: A test of the multiple effects model. Dev. Psychol. 20:523-532.

Jacobson, S. W., J. L. Jacobson, P. M. Schwartz, and G. G. Fein. 1983. Intrauterine exposure of human newborns to PCBs: Measures of exposure. In PCBs: Human and Environmental hazards, F. M. D'itri and M. A. Kamrin, eds. Boston: Butterworth Publishers. pp. 311-343.

MacLeod, W. D., Jr., L. S. Ramos, A. J. Friedman, D. G. Burrows, P. G. Prohaska, D. L. Fisher, and D. W. Brown. 1981. Analysis of residual chlorinated hydrocarbons, aromatic hydrocarbons and related compounds in selected sources, sinks, and biota of the New York Bight. NOAA Technical Memorandum OMPA-6. Seattle, Wash.: National Oceanic and Atmospheric Administration.

Malins, D. C., B. B. McCain, D. W. Brown, S-L. Chan. 1984. Toxic chemicals in marine environments: Food-chain transfers and biological effects. In health and Environmental Research on Complex Organic Mixtures, R. H. Gray, E. K. Chess, P. J. Mellinger, R. G. Riley, and D. L. Springer, eds. Richland, Wash.: Battelle Memorial Institute, Pacific Northwest Laboratory. pp. 591-609.

Malins, D. C., U. Varanasi, D. W. Brown, M. M. Krahn, and S-L. Chan. 1986. Biological transport of contaminants in marine environments: Bioavailability and biotransformations. Rapp. P.-v. Reun. Cons. Int. Explor. Mer. 186:442-448.

Malins, D. C., B. B. McCain, D. W. Brown, S-L. Chan. M. S. Myers, J. T. Landahl, P. G. Prohaska, A. J. Friedman, L. D. Rhodes, D. G. Burrows, W. D. Gronlund, and H. O. Hodgins. 1984. Chemical pollutants in sediments and diseases in bottom-dwelling fish in Puget Sound, Washington. Environ. Sci. Technol. 18:705-713.

Puffer, H. W., M. J. Duda, and S. P. Azen. 1982. Potential health hazards from consumption of fish caught in polluted coastal waters of Los Angeles County. N. Am. J. Fish. Manage. 2:74-79.

Reijnders, P. J. H. 1986. Reproductive failure in common seals feeding on fish from polluted coastal waters. Nature 324:456-457.

Romberg, G. P., S. P. Pavlou, R. F. Stokes, W. Hom, E. A. Crecelius,

P. Hamilton, J. T. Gunn, R. D. Meunch, and J. Vinelli. 1984. Toxicant Pretreatment Planning Study Technical Report CI: Presence, Distribution and Fate of Toxicants in Puget Sound and Lake Washington. Seatle, Wash.: Municipality of Metropolitan Seattle.

Swain, W. R. 1988. Human health consequences of consumption of fish contaminated with organochlorine compounds. Aquat. Toxicol. 11:357-377.

Takeuchi, T. 1972. Distribution of mercury in the environment of Minamata Bay and the inland Ariake Sea. In Environmental Mercury Contamination, R. hartung and B. D. Dinaman, eds. Ann Arbor, Mich.: Ann Arbor Science. pp. 79-81.

U.S. Environmental Protection Agency (U.S. EPA). 1985. Assessment of Human Health Risk from Ingesting Fish and Crabs from Commencement Bay. Final Report. Seattle, Wash.: U.S. Environmental Protection Agency.

MOBILIZATION AND RESUSPENSION

PREDICTING THE DISPERSION AND FATE OF CONTAMINATED MARINE SEDIMENTS

Y. Peter Sheng
University of Florida

ABSTRACT

In order to select the proper remedial action and management strategy to clean up a contaminated marine site, it is essential to be able to predict the dispersion and fate of contaminated marine sediments both under existing conditions and following a variety of proposed remedial actions. This paper reviews our current understanding and predictive ability of the dominant processes controlling the dispersion and fate of contaminated marine sediments. While it is possible to predict the circulation and wave fields and turbulent mixing in marine environments, predictive ability is lacking for the other sediment dispersion processes. In particular, due to the lack of reliable and comprehensive field data, much of our understanding of erosion/resuspension and deposition processes has been obtained from limited laboratory studies which contain many simplifying empirical assumptions. Extrapolation of these empirical process models to field application requires excessive amount of data for calibration. Further field-based research is urgently needed to advance our understanding of erosion, deposition, and flocculation processes.

INTRODUCTION

Marine sediments are the sinks to a variety of contaminants (e.g., heavy metals and toxic chemicals) and nutrients (e.g., phosphorus and nitrogen) from industrial, agricultural, and municipal discharges. These contaminants, both in particulate and dissolved forms, may subsequently reenter the water column due to resuspension of sediments, while the dissolved form may also be diffused into the water column. Once in the water column, contaminants may be transported by the three-dimensional turbulent flow field away from the contaminated site while adsorbed onto the fine sediment particles. Thus, in order to quantitatively assess the long-term fate of contaminants at a contaminated marine site (e.g., New Bedford Harbor, Massachusetts and Puget Sound, Washington), it is essential to be able to perform a mass balance study for the fine sediments within the water body. Basically, this means that we must be able to monitor or predict the long-term dispersion of sediments and contaminants within a large water body. Moreover, in order to assess the feasibility of a proposed remedial action (e.g., dredging or capping), it is necessary to be able to predict the impact of such action on the long-term fate of contaminants.

Due to the complexity of the sediment/contaminant dispersion processes and the tremendous difficulties and costs of comprehensive field measurements, long-term monitoring at contaminated marine sites is usually not done. Hence, it is extremely important to develop predictive capabilities of sediment/contaminant dispersion processes. This paper provides a brief review on our current understanding of the various processes controlling sediment dispersion. The difficulties of comprehensive field studies and the deficiencies of some laboratory studies are discussed. The use of comprehensive models to assist the quantification of sediment dispersion processes is illustrated with a brief discussion of the laboratory studies of erosion and deposition using rotating annuli. Uncertainties in model parameters are discussed throughout the paper. Recommendations for further research are given at the end.

SEDIMENT DISPERSION PROCESSES

As shown in Figure 1, the dominant sediment dispersion processes in marine environments include advection, turbulent mixing, flocculation and settling, erosion/resuspension, and deposition (Sheng, 1986a, 1987).

Mathematically, a mass balance study for sediments within a water body as shown in Figure 1 is equivalent to solving the following mass conservation equation for suspended sediment concentration:

$$\frac{\partial C}{\partial t} + \frac{\partial uC}{\partial x} + \frac{\partial uC}{\partial y} + \frac{\partial (w+w_s)C}{\partial z} = \frac{\partial}{\partial x}(A_H \frac{\partial C}{\partial x}) + \frac{\partial}{\partial y}(A_H \frac{\partial C}{\partial y}) + \frac{\partial}{\partial z}(A_v \frac{\partial C}{\partial z}) \quad (1)$$

where C is the suspended sediment concentration, (u,v,w) is the three-dimensional flow velocities in the (x,y,z) directions, t is the time, w_s is a settling velocity for sediment particles, and A_H and A_v are the horizontal and vertical turbulent eddy diffusivities. Assumptions have been made that sediment particles are of similar shapes and sufficiently small and uniform sizes such that they more or less follow the turbulent eddy motions. Assuming the three-dimensional flow field is known, the above equation can be solved in conjunction with the following boundary conditions:

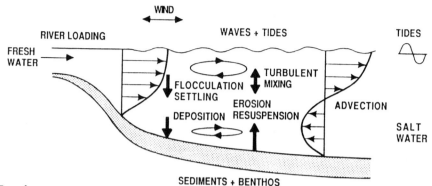

FIGURE 1 Dominant processes controlling sediment dispersion in an estuarine environment.

$$f = -w_s C + A_v \frac{\partial c}{\partial z} = 0 \quad @ \; z = \zeta(x,y,t)$$

$$f = -w_s C + A_v \frac{\partial c}{\partial z} = D - E \quad @ \; z = -h(x,y,t) \quad (2)$$

$$C = C(x,y,t) \quad @ \; \text{Lateral Boundaries}$$

where f represents a net vertical flux of sediments at the free surface, $\zeta(x,y,t)$, or the bottom, $-h(x,y,t)$, D is the deposition, and E is the erosion or resuspension. Both D and E depend on hydrodynamic and sedimentary parameters, and may vary significantly with space and time.

ADVECTION

The advection process is governed by the flow field of (u,v,w), which is generally three-dimensional, time-dependent, and turbulent. In estuarine environments such as Puget Sound, New Bedford Harbor, and Chesapeake Bay, the flow field is driven by tide, wind, and density gradient. Although much research is still needed to understand the hydrodynamic processes, our current understanding on the estuarine circulation and advection process is much better compared to what we know about the other sediment dispersion processes.

Numerous estuarine hydrodynamic and dispersion studies have been performed in the major estuaries in the United States and other countries. The more recent studies have recognized the importance of the three-dimensional aspects of the flow field and their effects on the dispersion (e.g., Sheng, 1983; Byrne et al., 1987; Sheng, 1988a). The formation of turbidity maxima (e.g., Dyer, 1986) due to tidal residual circulation and formation of salt wedge, has also been the topic of numerous studies. Recent reviews of available multidimensional numerical models of estuarine hydrodynamics can be found in Sheng (1986b) and Nihoul and Jamart (1987). It should be noted that generalized curvilinear grid models are presently being developed to simulate the long-term circulation in estuaries with complex bathymetry and geometry.

TURBULENT MIXING

Turbulence in marine environments can be transported across the air-sea interface by wind-induced mixing and breaking of surface waves. It can also be generated due to breaking of internal waves, shearing motion at the bottom, and unstable stratification, etc. In shallow estuarine environments, turbulence may exist over the entire water column, including the bottom boundary layer and the surface boundary layer to enhance the mixing of various materials.

Turbulent transport within the bottom boundary layer plays the important roles in affecting sediment erosion and deposition, while turbulent mixing throughout the water column can affect the flocculation and settling of sediment particles. The transport of turbulence within a water body, however, is very complex and cannot be accurately described in terms of a simple "diffusion" process, which is appropriate for the laminar mixing.

The accurate description of turbulent transport processes (production, advection, damping, diffusion, and dissipation, etc.) is of utmost importance for describing the large-scale circulation as well as the boundary layer dynamics, which can affect the sediment dispersions. A review of the turbulence models suitable for estuarine applications can be found in Sheng (1986b).

It is presently possible to simulate the turbulent transport in bottom boundary layers driven by current, wave, current with wave, and in the presence of vegetation canopy, density stratification, and roughness features. However, turbulent transport in the immediate vicinity of large bottom roughness features and in the near-field of a dredged material disposal plume is fully three-dimensional, highly random, and very complex, and hence still not well understood.

FLOCCULATION AND SETTLING

Marine sediments often contain particles of various sizes ranging from the submicron clay particles to the large flocs or sand particles of hundreds of microns. Flocs are formed as the fine-grained ($d < 60$ μm) sediment particles are brought into frequent collisions by the turbulent shearing motion and differential settling, and if there is sufficient ionic strength in the water to render the suspended sediment particles cohesive. Larger flocs possess weaker strength and may be broken up by increasing turbulent shear. However, little is known quantitatively about the flocculation process and its dependence on numerous physical and chemical properties of the sediment, the fluid, and the flow.

While flocculation can lead to orders of magnitude increase in settling velocity of sediment particles, feeding activities of benthos can also produce larger particles and settling velocity. Furthermore, the settling velocity generally increases with the concentration of suspended sediment particles until the hindered settling starts to reduce the settling velocity (Krone, 1962). At the present time, settling velocity of sediment particles is generally determined in laboratory and there exists no predictive model of settling velocity. Hence, the settling velocity W_s in Eq. (1) is generally treated as a tuning parameter, with values reported in literature varying by more than two orders of magnitude.

EROSION/RESUSPENSION AND DEPOSITION

Erosion or resuspension is the process by which the surficial layer of sediments is removed from the bottom sediments due to increase in hydrodynamic stress and turbulent intensity and/or weakening of sediment resistance. As such, the erosion process depends on the hydrodynamic forces as well as everything that affects the strength of bottom sediments. Deposition, on the other hand, is the process by which the suspended sediment particles arrive at the bottom. Thus, the deposition process depends only on the hydrodynamic process and the properties of suspended sediments. Comprehensive field data are needed in order to derive an erosion model and a deposition model for application with Eq. (1).

Our understanding of the erosion and deposition processes, however, has been rather qualitative and limited due to the complexity of the exchange processes at the sediment-water interface. As shown in Figure 2, a relatively clear water column over bottom sediments with a well-defined sediment-water interface has been regarded by many as a typical

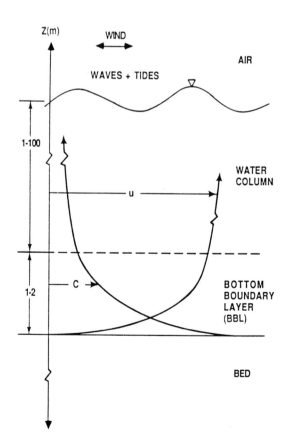

FIGURE 2 Vertical distribution of flow and sediment in an idealized estuarine environment.

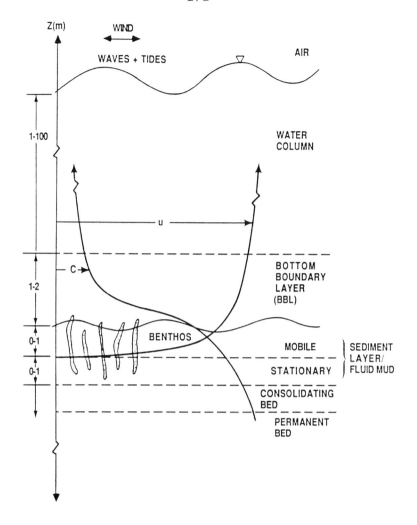

FIGURE 3 Vertical distribution of flow and sediment in a more realistic estuarine environment.

situation for studying erosion and deposition. In realistic estuarine environments, however, relatively high-concentration sediment layers may exist over the consolidating and the permanent beds, with a loosely defined sediment-water interface and the presence of benthos (Figure 3). It is clear that comprehensive field monitoring in such an environment presents tremendous difficulties. Ideally, we would like to measure the vertical profiles of all the following parameters within the bottom boundary layer and the surficial sediments: mean velocity, turbulence, salinity, temperature, suspended sediment concentration, density, size distribution, settling velocity, benthos, pore pressure, and sediment composition.

Even in the absence of any suspended sediments, it is difficult to measure the mean flow and turbulence in the vicinity of the bed where significant vertical variation in flow quantities exists. Field measurement using two-axis submersible LDV failed to capture the wave boundary layer even at 7 cm above the bed (Agrawal et al., 1988). In spite of the fine-scale flow structure, researchers typically rely on mean flow measured at 1 m above the bottom to calculate a turbulent bottom stress using the law of the wall (e.g., Grant et al., 1984). This method may give erroneous results in the presence of large roughness features, strong waves, and significant stratification. While measurement of turbidity and optical scattering at selected locations can be obtained with existing instruments, the accurate determination of suspended sediment concentration and particle size distribution is still unresolved and remains the topic of many researchers.

Due to the difficulties of field measurements, erosion and deposition processes have been studied in laboratory using rotating annuli (Figure 4). A critical review of sediment studies using rotating annuli was recently given by Sheng (1988b). Deficiencies in such laboratory studies include the following:

1. Scaling problem--the small sizes of the rotating annuli (d ~ 1 m) makes it very difficult to generate fully rough turbulent flow, which is often encountered in the field.
2. Secondary flow problem--significant secondary flow (Figure 5) in rotating annuli (with rotating top and/or bottom) alters bottom friction and sediment dispersion patterns.
3. Turbulence damping problem--significant vertical gradient of suspended sediment concentration near the bottom can damp turbulence and reduce the amount of sediment erosion (Sheng and Villaret, 1988).
4. Bioturbation problem--feeding activities of benthos can affect the sediment erosion and deposition, but the extent varies significantly with time and location (McCall and Trevesz, 1982).

Existing laboratory-based models of sediment erosion and deposition generally neglected the above problems and, although capable of reproducing the particular laboatory experiments, may produce large errors when extrapolated to a new field application.

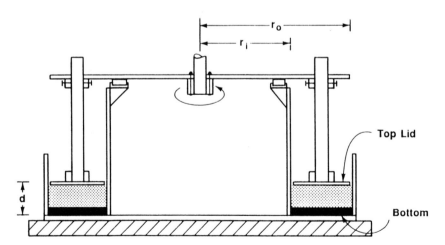

FIGURE 4 Rotating annulus used for sediment erosion and deposition studies.

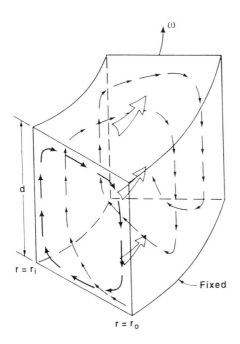 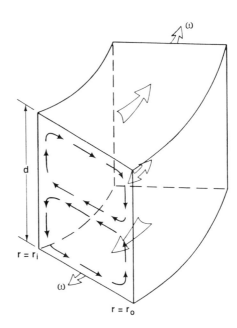

FIGURE 5 Left: flow patterns and particle trajectory in a rotating annulus with a rotating top. Right: flow patterns in the azimuthal and radial planes of a rotating annulus with a rotating top and bottom.

The laboratory-based erosion models are empirical, site-specific, and can vary by several orders of magnitude, as is shown in Figure 6 (Lavelle et al., 1983) which compares 10 erosion models compiled in the form of $E = \alpha |\tau|^\eta$ where τ is the dimensionless bottom stress. Curve is determined from Puget Sound field data by Lavelle et al., curves 3 and 4 are determined from laboratory experiments of Lake Erie sediments (Sheng and Lick, 1979), while curves 9 and 10 are determined from laboratory experiments of San Francisco mud (Partheniades, 1965).

Recently, comprehensive mathematical models have been used to quantify some of the deficiencies of the laboratory sediment experiments. For example, Sheng (1988a) used an integral boundary layer formula to calculate the secondary flow within a rotating annulus (Sheng and Lick, 1979) and found the radial flow to be 20 to 50 percent of the azimuthal flow. In addition, the law of the wall was modified to include the effect of a radial pressure gradient to allow the calculation of the vertical profile of radial and azimuthal velocities. Sheng and Villaret (1988) developed a simplified second-order closure model to investigate the effect of sediment concentration gradient on the flow. They found that erosion models developed without considering such effect can cause very significant error in the prediction of suspended sediment concentration. In some cases, erosion may stop because the turbulent stress becomes significantly reduced by the concentration gradient.

The deposition is treated as a separate process in Sheng and Lick (1979) and Sheng (1986a), which derived a rigorous formula for deposition velocity. Thus, erosion and deposition may be allowed to occur at the same time. Krone (1962) considered the erosion and deposition together, however, and defined a critical stress for net erosion

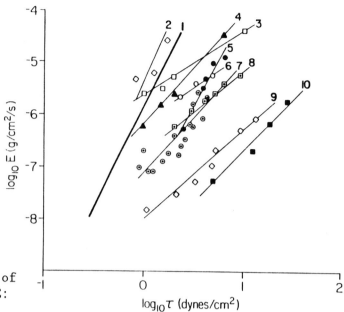

FIGURE 6 Various models of sediment erosion. SOURCE: Lavelle et al., 1983.

(τ_{ce}) and a critical stress for net deposition (τ_{cd}). Although τ_{ce} was assumed to be greater than τ_{cd}, both must be prescribed empirically for each study. Teeter (1988) found the deposition of New Bedford sediments varied by a factor of 256.

CONCLUSIONS AND RECOMMENDATIONS

A brief review of our understanding of the dominant processes of sediment dispersion has been given here. Presently there are sufficient understanding and predictive capabilities of the advection and turbulent mixing processes in marine environments. However, our understanding of the processes of flocculation, settling, erosion, deposition, and bed evolution is rather limited. Existing models of these processes are primarily based on data from laboratory experiments, which often contain limiting simplifying assumptions, and hence are generally site-specific and contain large uncertainties for general application.

Extrapolation of these empirical models to new field application may lead to large errors. It is thus extremely important to develop mechanistic process models (flocculation model, erosion and deposition model, and bed model) using comprehensive field data. These models can then be combined with the circulation model, wave model, and bottom boundary layer model to produce an overall sediment dispersion model (Figure 7), which may then be used for field validation and mass-balance predictions. Unless the process models have been validated by field data, the overall sediment dispersion model cannot be expected to produce reliable "prediction."

Research should be carried out in the following two areas:

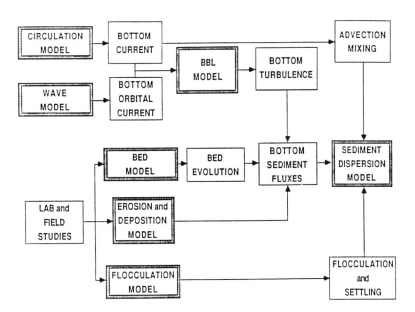

FIGURE 7 Framework of a sediment dispersion model.

1. Re-examination and improvement of existing laboratory-based models of sediment dispersion processes: comprehensive flow and sediment data should be collected in rotating annuli and synthesized to allow the determination of less empirical models. A second-order closure model of turbulent transport can be used to simulate the complete flow field withing the rotating annuli.
2. A comprehensive field program at a contaminated marine site to allow monitoring of the extent of contamination and to allow development of field-based models of sediment dispersion processes: due to the availability of more advanced instrumentation and modeling technique, this study should yield results which are much more useful than previously possible under the DMRP program.

Due to the large uncertainties contained in the various parameters (e.g., size distribution, settling velocity, and bottom sediment distribution, etc.) appearing in a sediment dispersion model, it is important to perform an uncertainty analysis for any sediment mass balance study. Rather than treating the various model parameters as adjustable tuning parameters to achieve a single "best fit" with limited field data, it is more reasonable to attempt to predict the mean value as well as the variance (or uncertainty) of the sediment/contaminant concentration distribution in marine environments.

ACKNOWLEDGMENTS

Support the from U.S. Environmental Protection Agency under Cooperative Agreement AERL-87-01, with Dr. Steve C. McCutcheon as the Scientific Officer, is acknowledged.

REFERENCES

Agrawal, Y. C., D. G. Aubrey, and F. Dias. 1988. Field Observations of the coastal bottom boundary layer under surface gravity waves. Proc. Conf. App. Laser Anemometry to Fluid Dynamics, Lisbon, July 11-14, 1988.

Byrne, R. J., A. Y. Kuo, R. L. Mann, J. M. Brubaker, E. P. Ruzecki, P. V. Hyer, R. J. Diaz, and J.H. Posenau. 1987. Newport Island: An Evaluation of Potential Impacts on Marine Resources of the Lower James River and Hampton Roads. Special Report in Applied Marine Science and Ocean Engineering No. 283. Gloucester Point, Va.: Virginia Institute of Marine Science, College of William and Mary.

Dyer, K. R. 1986. Coastal and Estuarine Sediment Dynamics. New York: John Wiley & Sons. 342 pp.

Grant, W. D., A. J. Williams, and S. Glenn. 1984. Bottom stress estimates and their prediction on the northern California continental shelf during CODE-1. J. Phys. Oceanogr. 14:506.

Krone, R. B. 1962. Flume Studies in the Transport of Sediment in Estuarine Shoaling Processes. Hydraulics Engineering Laboratory Report. Berkeley: University of California.

Lavelle, J. W., H. O. Mofjeld, and E. T. Baker. 1983. An In situ Erosion Rate for a Fine-Grained Marine Sediment. NOAA/ERL PMEL Contribution Number 654. Washington, D.C.: National Oceanic and Atmospheric Administration.

McCall, P. L. and M. Trevesz. 1982. Effects of benthos on physical properties of freshwater sediments. In Animal-Sediment Relations, P. L. McCall and M. J. Trevesz, eds. New York: Plenum Press. Pp. 105-176.

Nihoul, J. C. J. and B. Jamart. 1987. Three-dimensional Models of Marine and Estuarine Dynamics. London: Elsevier.

Partheniades, E. 1965. Erosion and deposition of cohesive soils. J. Hyd. Div. ASCE, 91(HY1):105-138.

Sheng, Y. P. 1988a. Curvilinear-grid model for estuarine and coastal hydrodynamics. Proceedings of the 21st International Conference on Coastal Engineering, Spain, June, 1988. New York: ASCE (in press).

Sheng, Y. P. 1988b. Consideration of flow in rotating annuli for sediment erosion and deposition studies. To be published in J. Coastal Research. In Press.

Sheng, Y. P. 1987. Numerical modeling of estuarine hydrodynamics and dispersion of cohesive sediments. In Sedimendation Control to Reduce Maintenance Dredging of Navigational Facilities in Estuaries. Washington, D.C.: Marine Board, National Research Council. Pp. 94-117.

Sheng, Y. P. 1986a. Modeling bottom boundary layers and cohesive sediment dynamics. In Estuarine Cohesive Sediment Dynamics. New York: Springer-Verlag. Pp. 360-400.

Sheng, Y. P. 1986b. Finite-Difference Models for Hydrodynamics of Lakes and Shallow Seas: Physics-Based Modeling of Lakes, Reservoirs, and Impoundments. New York: American Society of Civil Engineers. Pp. 146-228.

Sheng, Y. P. and W. J. Lick. 1979. The transport and resuspension of sediments in a shallow lake. J. Geophysical Research 84:713-727.

Sheng, Y. P. and C. Villaret. 1988. Second-order closure modeling of sediment-laden turbulent boundary layers. Paper presented at American Geophysical Union Chapman Conference on Sediment Transport Processes in Estuaries, Bahia Blanca, Argentina, June 13-17, 1988. To be published in J. Geophysical Research.

Sheng, Y. P. 1983. Modeling of Three-dimensional Coastal Currents and Sediment Dispersion, Vol. 1, Model Development and Application. Technical Report CERC-83-2. Vicksburg, Miss: U.S. Army Engineer Waterways Experiment Station.

Teeter, A. 1988. Case Study: Physical transport investigations at New Bedford, Mass. Paper presented at the Contaminated Marine Sediments Symposium, Marine Board, National Research Council. Tampa, May 31-June 2, 1988.

COMPUTER SIMULATION OF DDT DISTRIBUTION
IN PALOS VERDES SHELF SEDIMENTS

Bruce E. Logan and Robert G. Arnold
University of Arizona

and

Alex Steele
Los Angeles County Sanitation Districts

ABSTRACT

Prior to imposition of effective source control measures in 1970, large quantities of DDT were discharged to the Los Angeles County municipal sewer system and subsequently to the Pacific Ocean. Much of this material accumulated among sediments of the Palos Verdes shelf. While the bulk of the DDT lies 10 to 40 cm below the sediment surface, its fate may be affected by future wastewater treatment at the Los Angeles County's 385-mgd, Joint Water Pollution Control Plant (JWPCP).

To assess the potential impact of JWPCP secondary treatment requirements on shelf sediment quality, processes that may affect distribution of chemical tracers among those sediments (background sedimentation rate, effluent-related solids contributions, sediment mixing, and diffusive transport through pore waters) were incorporated in a mathematical model. Because physical mixing substantially affects surficial concentrations and the vertical distribution of DDT at the 60-m depth contour, model projections of surface sediment quality at the most heavily contaminated sites are sensitive to projected effluent solids concentrations and thus JWPCP treatment decisions. Dredging and capping do not represent physically or economically feasible remediation strategies at this site. The work described illustrates how empirical models of contaminant fate can be used to (1) simplify or avoid uncertainties associated with some mechanistic models, and (2) serve as a basis for management decisions.

INTRODUCTION

Indicators of biological quality among benthos of the Palos Verdes shelf are sensitive to chemical characteristics of surface sediments (Word, 1978; Sherwood, 1976; Oshida and Wright, 1976; Cross, 1984).

Species composition and primary feeding strategies among benthic invertebrates are functions of both sediment organic content and local surface concentrations (top 5 cm) of hazardous compounds including trace metals, total DDT, and PCBs (Stull et al., 1986). Despite uncertainties caused by statistical correlations among individual chemical parameters, the importance of controlling surface concentrations of DDT in sediments of the Palos Verdes shelf has been accepted. Prior to a ban on its disposal within the Los Angeles County municipal sewage collection system in about 1970 (Norman Ackerman, Supervisor, Oceanographic Monitoring Activities, LACSD, personal communication), DDT was held responsible for endangerment of the California brown pelican (Keith et al., 1970; Risebraugh et al., 1967, 1971).

The Los Angeles County Sanitation Districts (LACSD) provides partial secondary treatment (200-mgd secondary treatment capacity) for approximately 360 mgd (136,300 $m^3 \cdot d^{-1}$) of domestic and industrial wastewater at the Joint Water Pollution Control Plant (JWPCP) in Carson, California. Effluent, including some 150 metric tons of suspended solids per day, is discharged via a system of ocean outfalls to waters of the Palos Verdes shelf. During the last 15 years, JWPCP effluent quality has shown marked improvements, in terms of solids emissions and concentrations of specific contaminants including total DDT (Figure 2), in response to improved treatment and solids handling at JWPCP and implementation of source-control measures on tributary industries.

Much of the DDT that was discharged to the LACSD sewer system prior to 1970 reached the coastal waters and sediments off the Palos Verdes peninsula. Due to its persistence, there are residual links to environmental quality and human health--muscle tissue concentrations of DDT exceed FDA limitations in local populations of white croaker and other bottom-feeding species (Gossett et al., 1982, 1983). Surface sediment DDT concentrations appear to drive fish tissue values, as opposed to the DDT mass emission rate from the Whites Point outfall system (Young et al., 1988).

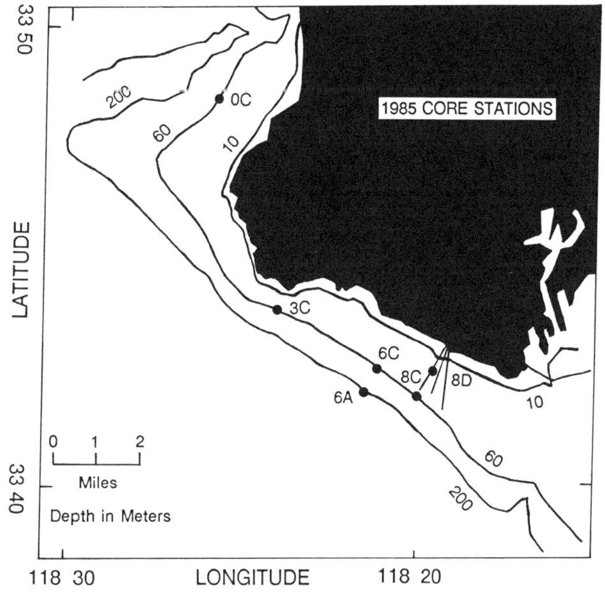

FIGURE 1 The Palos Verdes shelf in relation to the Palos Verdes Peninsula and JWPCP. Positions and depth of the LACSD outfall system and sediment monitoring stations are as indicated.

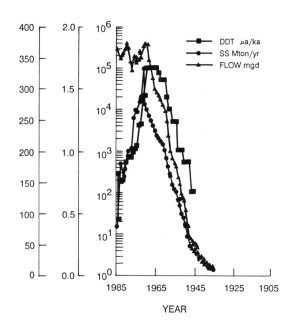

FIGURE 2 Summary of JWPCP mass emissions records to total solids, DDT, and total flow.

Sediment quality on the Palos Verdes shelf is a function of physical and chemical processes that are imperfectly understood. These processes potentially include sedimentation, bioturbation by benthic infauna, periodic resuspension by bottom currents or wave activity, and chemical diffusion. Following implementation of effective source-control measures for DDT in the early 1970s, the bulk of the sediment DDT on the Palos Verdes shelf was buried under less contaminated sediment of both natural and sewage-related origin. Recent sediment profiles (Figure 3, for example) suggest that the bulk of sediment DDT lies buried between 10 and 40 cm below the surface.

The contribution of effluent-related particulates to the overall local sedimentation rate is not well established. The background sedimentation rate (independent of outfall particulates) has been variously estimated at 10 to 200 $mgKcm^{-2}Kyr^{-1}$ (Hendricks, 1984), perhaps in response to variations in local coastline stability, and the fraction of solids discharged from the Whites Point outfall system that is retained on the Palos Verdes shelf has been estimated at 0.01 to 1.0 (Myers, 1974; Hendricks, 1982). There is no consensus regarding the importance of periodic sediment resuspension as a determinant of contaminant profiles.

Hendricks (1978, 1982, 1984, 1988) modeled the fate of chemical tracers in Palos Verdes sediments by combining a solids deposition model designed to yield the discharge- and current-dependent pattern of particulate fluxes to the shelf with a sediment resuspension model that produced time-dependent profiles of select contaminants in the Whites Point sediments. Hendricks' work identified areas of theoretical inadequacy and data gaps, which must be addressed before mixing of marine sediments can be addressed mechanistically. At present, it is impossible to address adequately the following areas of uncertainty:

1. potential interdependence of sediment resuspension parameters (resuspension mass, resuspension frequency, and time between

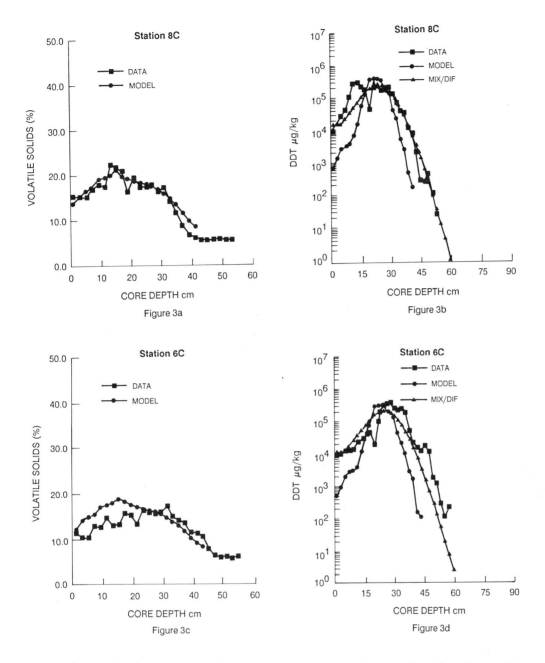

FIGURE 3 Calculated and observed concentrations of volatile solids or DDT in Palos Verdes sediments among monitoring stations along the 61-m depth contour. Volatile solids concentrations were calculated using the basic sedimentation model without considering surface mixing or diffusion. DDT profiles were calculated using the sedimentation/mixing model. Model parameters used in calculations represented here are summarized in Table 1. In all cases, the solid "data" line represents 1985 LACSD monitoring data; the broken "model" line represents best-fit results using the basic sedimentation model; and the broken "mix/dif" illustrates the best-fit model calculations when mixing and diffusion mechanisms are included.

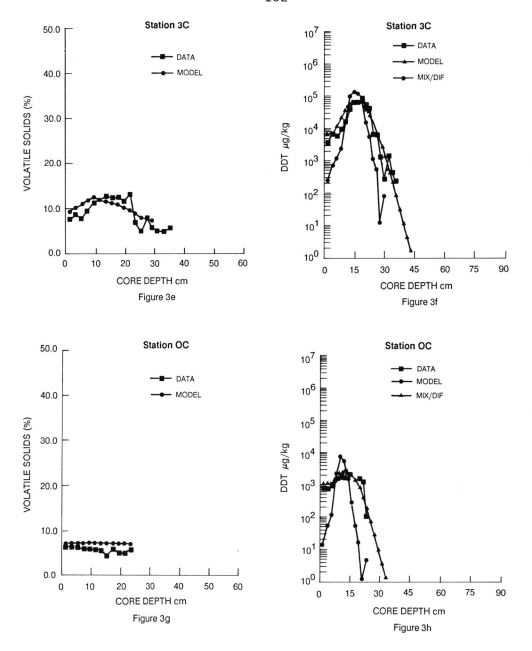

FIGURE 3 Continued.

resuspension and redeposition) and the ratio of effluent-related to natural particulate material among local sediments,
2. biological determinants of sediment resuspension rates,
3. the dynamics of a sediment surface layer approximately 1 mm thick, which may be subject to frequent short-duration resuspension events prior to stable incorporation into bulk sediments, and
4. aggregation processes among biologically active surface sediments.

TABLE 1 Summary of Best-fit Parametric Values Resulting from Calibration of the Sediment Deposition/Mixing Model across the Palos Verdes Shelf

Station[a] (depth)	Background sedimentation rate (mg dry wt cm^2/yr)	% Effluent solids reaching grid location	Local diffusion coefficient (cm^2/sec)	Depth of mixing zone (cm)
8C(61m)	400	0.5	10^{-8}	6
6C(61m)	500	0.5	10^{-8}	6
3C(61m)	500	0.2	10^{-8}	4
oC(61m)	500	0.01	10^{-8}	6
6A	200	0.075	10^{-8}	4

NOTE:
[a]Stations correspond to those indicated in Figure 1.

LACSD and the U.S. Environmental Protection Agency (EPA) are now forced to evaluate the merit of imposing an additional 80 percent reduction in suspended solids emissions by establishing full secondary treatment requirements for JWPCP. It has been suggested that this action could result in gradual reemergence of previously buried DDT and other effluent-related contaminants for which DDT can serve as tracer. Residual uncertainties associated with mechanistic, state-of-the-art sediment models preclude their use for prediction of Palos Verdes surface sediment quality characteristics as a function of the projected treatment level at JWPCP.

Here we apply an empirical approach to sediment quality modeling on the Palos Verdes shelf in order to avoid scientific uncertainties that have frustrated more mechanistic treatments. We have developed a series of sediment models of increasing complexity in order to evaluate the importance of specific processes, or classes of physical processes, as determinants of sediment contaminant profiles. The background sedimentation rate and fraction of discharged solids that reaches the sediment grid are treated as fitted parameters (selected to reproduce local sediment profiles of volatile solids and DDT) as are sediment interstitial dispersion coefficients and mixing parameters in higher order models. Model results include surface sediment concentrations of DDT as a function of projected treatment level (existing level, full secondary, zero discharge, etc). at JWPCP.

METHODS

Deposition Model (Order 1)

The procedure for calculating sediment profiles of specific contaminants is described below and summarized in an appendix to this report.

The Palos Verdes shelf was divided into a two-dimensional (horizontal) grid. The background sedimentation rate and fraction of the JWPCP effluent solids deposited in each grid section were independently estimated. Effluent-related solids were divided into volatile and nonvolatile fractions based on JWPCP monitoring records. The biodegradable fraction of effluent solids was assumed to be completely oxidized in the water column prior to deposition. All solids destroyed in this fashion were assumed to come from the volatile fraction. Residual discharge-related solids were assumed to be refractory, as were solids of natural origin that reached the local sediments.

Necessary measurements and parameter estimates included the record of JWPCP solids emission rates dating to 1935 (Figure 2), the fraction of discharged solids that are volatile, and the percentage of volatile solids that are refractory in nature. With this information, the dry mass of solids that reached the shelf sediments during each year of record was calculated as a function of position on the shelf. The computational procedure also yielded an estimate of the volatile solids fraction in the sediments for comparison with measured concentrations.

The total mass added to the sediment column during each year was dependent upon an empirical relationship between sediment volatile solids and moisture content (Appendix). On the basis of the total mass addition and the calculated (weighted average) sediment density, it was possible to compute the thickness of an incremental layer of sediment that accumulated in response to natural and effluent-related particle deposition during each year. The position-dependent increment of material predicted to have accumulated on the Palos Verdes shelf during the lifetime of the Whites Point discharge (in the absence of resuspension and mixing) was developed by integrating the calculated annual contributions. The computational procedure also yielded depth-dependent estimates of volatile solids and moisture concentrations. These were compared to profile data collected at several grid locations across the shelf to determine goodness-of-fit. By varying input parameter values, it was possible to select the most appropriate local background sedimentation rate and fraction of effluent solids deposited within specific grid boundaries.

In order to predict sediment profiles of specific effluent-related contaminants, results of the foregoing procedure were combined with measurements or estimates of annual mass emission rates (MERs) for the contaminants of interest. In modeling DDT, it was assumed that (1) DDT was uniformly distributed (by mass) among effluent solids, and (2) discharged DDT was entirely refractory in nature. Only total DDT was modeled in this fashion. Sediment computations yielded depth-dependent DDT concentrations in units of mg DDT per kg of dry solids. Annual DDT mass emissions and other input data are summarized in Figure 2.

Fitted parameters in the initial modeling phase included only the background sedimentation rate and the fraction of effluent solids that is deposited in specific sectors of the sediment grid. Results of LACSD work using Hendrick's models were used to suggest appropriate limits for parameter ranges. Grid sampling stations at which the mass contribution of effluent solids is modest (in comparison to background sedimentation), also provided an initial estimate of the background sedimentation rate.

Sediment profiles of volatile solids, moisture content, and DDT concentration were developed in this manner for each sediment grid position at which LACSD measured profile characteristics in 1985. Background sedimentation rate and deposition of effluent-related solids were adjusted to reproduce the volatile solids profile (depth and magnitude of elevated volatile solids concentrations) to the extent possible.

Mixing Model (Order 2)

Perceived shortcomings in the level-one (deposition) model were addressed by adding surface-mixing and DDT-dispersion components to the computation of sediment profile characteristics. While the procedure was limited to calculation of depth-dependent DDT concentrations, it could be adapted to predict profile concentrations of any conservative contaminant for which there is an adequate base of effluent data.

Computational procedures for the mixing model included those of the deposition model mass balance, altered to account for surface mixing and vertical diffusion of DDT throughout the sediment column. Diffusion was incorporated by dividing the sediment profile into 2-cm compartments and permitting concentration-driven transport between adjacent compartments during specific finite time intervals. The 2 cm represents both the depth interval for chemical determinations in LACSD sediment cores and a practical compartment thickness for the diffusion computation. A well-mixed zone, defined to consist of an arbitrary number of the uppermost 2-cm compartments, received the entire input of deposited material during a given time interval; complete mixing was assumed over the time necessary to add 2 cm to the overall profile thickness. At that point, a new compartment was created at the top of the sediment core, and a 2-cm compartment was pushed below the mixing zone. Initial estimates of a molecular diffusion coefficient were developed from the literature of mass transport through porous media and later fitted using 1985 DDT profile data from Figure 1 monitoring stations.

Sediment Quality Projection

Parameter estimates from the mixing model were used to project sediment quality characteristics as a function of the anticipated JWPCP suspended solids MER. Treatment scenarios investigated and corresponding solids emissions follow:

Treatment Level	Suspended solids MER (10^5 MT/yr) (assumed constant from 1987 to 2005)
No change (partial secondary treatment)	0.40
Full secondary	0.10
Zero discharge	0.01
Reevaluation of treatment resulting in 2x solids MER	0.80

RESULTS

Deposition Model

A summary of fitted parameters corresponding to each of the core sampling stations is provided as Table 1. A comparison of best-fit calculations (deposition model) and measured (1985) DDT profiles at 61-m monitoring stations is included within Figure 3b, 3d, 3f, and 3h. A statistical measure of goodness-of-fit has not yet been incorporated into the model structures, so model parameters were visually determined.

Order-one model calculations and sediment measurements of volatile solids and DDT concentrations were within reasonable agreement at all 61-m monitoring stations along the Palos Verdes shelf. However, 1985 surface sediment concentrations of DDT were uniformly high relative to model calculations, suggesting that addition of a mixing mechanism to the deposition model would improve its performance substantially.

Mixing Model

A summary of best-fit parameter estimates (deposition/mixing model) for stations along the 61-m contour is provided in Table 1. Estimated background sedimentation rates are nearly uniform across the shelf (400-500 mg/cm^2-yr) while the local contribution of effluent-related solids varies by a factor of 50. Beyond the shelf, the background sedimentation rate shrinks rapidly. In all cases, the local diffusion coefficient for best fit was 10^{-8}/sec. The depth of the mixing zone was 4-6 cm.

Calculated and measured profiles of volatile solids and DDT concentrations for stations along the 61-m depth contour are provided as Figure 3 (a-h). DDT profiles corresponding to the diffusion (no-mixing) model are included to illustrate the importance of surface-mixing to development of an effective sediment model.

Model Predictions

The calibrated sedimentation/mixing model was utilized to project sediment profiles of DDT concentration along the 61-m depth contour. Results are summarized in Figure 4 (a-d) and Table 2. Individual profiles correspond to JWPCP treatment alternatives ranging from zero discharge to a doubling of the current JWPCP suspended solids, mass emission rate. The projected year-2005 surface concentration of DDT at station dC is independent of JWPCP treatment level. At the other extreme, the model indicates that initiation of full secondary treatment will produce year-2005 surface concentrations of DDT at station 6C that are almost 75 percent higher than those that would result from no additional treatment. In all cases, the predicted year-2005 surface concentrations of DDT are much lower than current levels; there are significant, treatment-dependent differences in those projections.

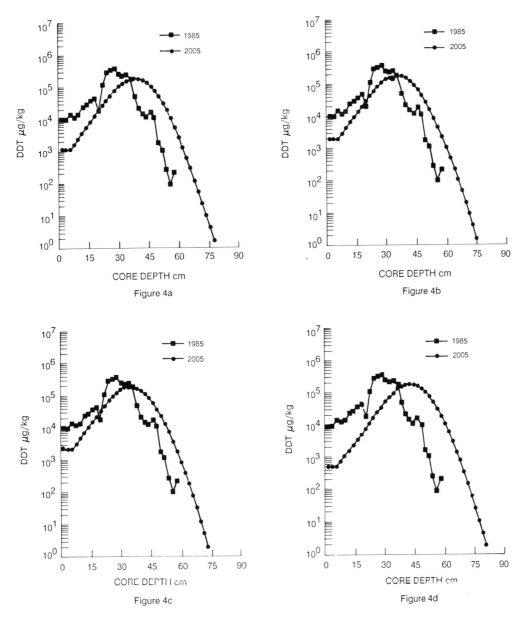

FIGURE 4 Year 2005 projections of sediment DDT concentrations at monitoring station 6C (Figure 1) as a function of depth and treatment level or solids removal efficiency at JWPCP. Corresponding solids MERs are summarized in the text. 4a. No change scenario. Continuation of partial secondary treatment at JWCPC. 4b. Full secondary treatment at JWCPC. 4c. Zero discharge alternative. Hypothetical elimination of ocean discharge via the Whites Point outfall system. 4d. Modification of JWPCP treatment which results in a solids MER 2x the present (1987) level.

Doubling the solids mass emission rate is predicted to lower the year-2005, surface sediment concentration of DDT by a factor of four relative to the secondary treatment alternative.

TABLE 2 Projected[a] Year-2005, Surface Sediment Concentrations of DDT[b] along the 61-M Depth Contour of the Palos Verdes Shelf

Station[c]	Current surface concentration	Zero discharge	Full secondary	No change (Partial secondary) treatment)	2x the current suspended solids MER
8C	11,500	4,000	3,500	2,100	800
6C	9,500	2,200	1,900	1,100	500
3C	3,500	550	500	360	240
dC	780	230	230	230	230

NOTES:
[a] Projections are based on the sedimentation/mixing model.
[b] All figures are in mg/dry kg of sediment.
[c] Station locations are indicated in Figure 1.

DISCUSSION

Sensitivity

In order to demonstrate model response to variation among fitted parameters, results of representative sensitivity analyses are presented in Figure 5 (a-h). Sediment volatile solids concentrations respond to variation in both background sedimentation rate and relative size of the outfall-related solids contribution. DDT profile calculations were compared to measured values at station 6C to assess model sensitivity to variation in the molecular diffusion coefficient and depth of the mixing zone. For each of the four parameters tested, appropriate profiles (volatile solids or DDT) resulting from both artificially high and low parameter estimates are presented.

Results indicate that the sedimentation/mixing model is reasonably sensitive to selection of the background sedimentation rate, although a 50 percent increase in background sedimentation (Figure 5b) did not substantially affect the quality of fit between calculation and measurements. A 50 percent decrease in the assumed natural sedimentation rate (Figure 5a) results in unrealistically high volatile solids concentration.

Volatile solids calculations are very sensitive to change in the relative effluent solids contribution at station 6C. A 40 percent decrease in discharge-related solids (Figure 5c) provides a marked decrease in profile depth; an increase of similar magnitude (Figure 5d) results in overestimation of the sediment organic content.

From the sensitivity of profile depth calculations to the local flux of effluent-related solids, it is apparent that the Whites Point

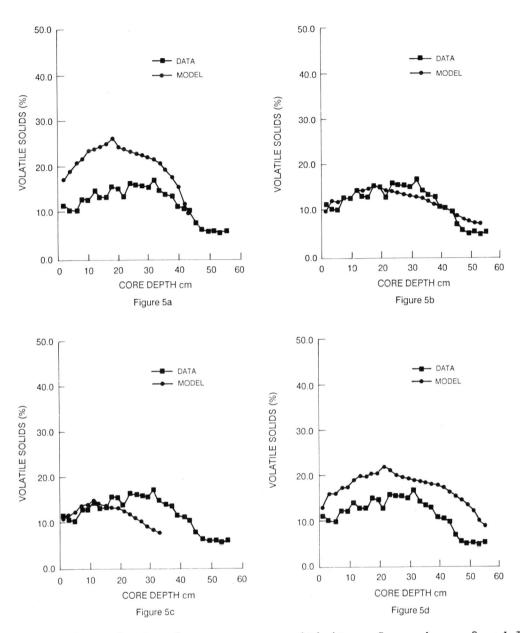

FIGURE 5 Analysis of parameter sensitivity. Comparison of model calculations and 1985 sediment profiles of volatile solids or DDT concentration at monitoring station 6C (Figure 1) in response to systematic variation in model parameters. In all cases, the solid "data" line represents 1985 LACSD monitoring data; the broken "model" line represents results of the sedimentation (no mixing) model calculated using the parametric values summarized below; and the broken "mix/dif" line represents results of the sedimentation/mixing model corresponding to the parameter estimates given.

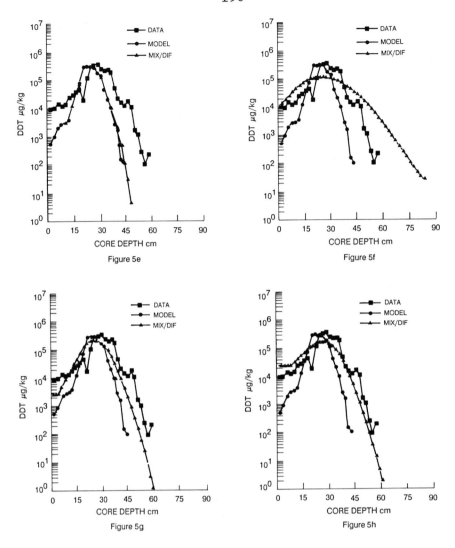

FIGURE 5. Continued

Summary of parameter estimates (by case):

Figure designation	Background sedimentation rate (mg/cm^2/yr)	% Effluent solids deposited in grid sector	Diffusion Coefficient (cm^2/sec)	Mixing depth (cm)
5A	250	0.5	-	-
5B	750	0.5	-	-
5C	500	0.3	-	-
5D	500	0.7	-	-
5E	500	0.5	10^{-9}	-
5F	500	0.5	10^{-7}	-
5G	500	0.5	10^{-8}	4 cm
5H	500	0.5	10^{-8}	8 cm

NOTES:
  The percent effluent solids deposited in the grid sector containing station 6C was uncorrected for degradation of volatile solids.
  Surface mixing was omitted from cases designed to test model sensitivity to variation in the molecular diffusion coefficient.

discharge has contributed significantly to the accumulation of solids on the bottom in the vicinity of the outfall. Based on the current discharge rate and effluent quality, estimates of volatile and refractory solids fractions, etc., outfall-related solids now comprise about 9 percent of the total particulate flux to the bottom at station 6C. From the record of JWPCP solids emissions, it is apparent that this percentage was much higher prior to implementation of partial secondary treatment. Effluent-related contributions at more distant stations are much less significant.

Model simulations indicate that the effect of pore diffusion on the distribution of DDT in Palos Verdes sediment is minimal. At $D = 10^{-8}$ cm^2/sec, molecular diffusion smooths the calculated profile without mitigating the steep (calculated) drop in DDT concentration near the surface. If the diffusion coefficient is reduced by an order of magnitude, its effect is negligible (Figure 5e). Increasing the diffusion coefficient by a factor of 10 (Figure 5f) results in a reasonable representation of near-surface DDT levels at station 6C but broadens the DDT peak and extends elevated concentrations to depths that are not supported by monitoring data. The estimated diffusion coefficient ($D = 10^{-8}$ cm^2/sec) represents an upper limit since values at and below that level are virtually indistinguishable.

DDT profiles are better reproduced by creating a well-mixed zone of arbitrary dimensions among the surface sediments. The physical basis of such a zone lies in either frequent sediment resuspension (unlikely in light of the apparent depth of the mixing zone) or bioturbation. The estimated depth of such a well-mixed or biologically active zone at station 6C is 6 cm. The primary effect of increasing or decreasing the dimension of the mixing zone is evidenced in the estimated surface DDT concentration, which increases nearly an order of magnitude in response to variation in mixing-zone depth between 4 and 8 cm (Figure 5g and 5h).

Sedimentation model (no mixing) results were included in Figure 5(e-h) to illustrate further the importance of mixing to accurate representation of surface sediment DDT concentrations. In Figure 5e and 5f, surface mixing effects were eliminated in order to isolate the effect of order-of-magnitude-scale variation in diffusivity.

Insight and Limitations

Although the sedimentation/mixing model (1) is capable of reproducing sediment profiles of volatile solids and DDT concentrations, and (2) exhibits a satisfactory degree of sensitivity relative to selection of parameter estimates, it provides an incomplete mechanical framework within which to identify the primary determinants of DDT distribution. Nevertheless, we feel that the following mechanistic observations relative to model behavior are justified:

1. Physical rather than chemical processes will determine the long-term fate of DDT among sediments of the Palos Verdes

shelf. Chemical factors of practical importance are that (1) DDT will not chemically or biochemically degrade within the foreseeable future and (2) surface affinity or hydrophobic effects ensure that DDT does not behave as a solute.
2. The primary physical determinants of DDT distribution along the 61-m depth contour are discrete particle sedimentation and sediment mixing due to bioturbation. No attempt was made to rationalize either process--natural and outfall-related sedimentation rates were developed via parameter fitting using sediment profiles of volatile solids; bioturbation was simulated by creating a well-mixed zone of arbitrary depth among the surface sediments.
3. Natural sedimentation rapidly decreases with distance beyond the break of the Palos Verdes shelf.
4. There is little spatial variation in the background sedimentation rate along the 61-m contour suggesting that (1) local variation within the shoreline erosion rate is not a primary determinant of sediment accumulation at that depth and (2) effluent-related solids are not responsible for appreciable flocculation of slow-settling natural particulates.
5. The sedimentation/mixing model could not adequately reproduce the sediment profiles of volatile solids and DDT at the 8D (Figure 1, 30-m depth) sampling location (data not shown). That site is characterized by unusually high surface concentrations of volatile solids and DDT. It is possible that in shallower waters processes not considered here, such as wave-driven sediment resuspension, are responsible for re-exposure of buried materials, including relatively high-level DDT concentrations.

Extensions

Results suggest that there is a great deal to gain from an approach to sediment modeling in which component physical, chemical, and biological processes are added sequentially to the overall model. Such an approach is especially useful when potentially important unit processes are complex--when existing data do not support a mechanistic treatment.

The deposition of effluent-related solids on the Palos Verdes shelf has been successfully modeled by several investigators (Hendricks, 1978, 1982, 1984; Koh, 1982). Comparison of LACSD sediment core measurements with quality characteristics predicted on the basis of effluent quality data and sedimentation pattern predictions *alone* were used to assess the importance of processes (sediment resuspension, biological mixing) that are less amenable to mathematical modeling. The importance of surface-mixing processes as determinants of the vertical distribution of sediment DDT is clear from the exercise. Sediment mixing was then treated empirically to reproduce sediment profiles. From the latter analysis, it is apparent that periodic sediment resuspension is an important determinant of the vertical distribution of sediment contaminants at the shelf 30-m depth contour. A similar approach might be applied to sediment modeling in any situation in which sediment quality data are more reliable or more easily obtained than the physical data necessary for construction of mechanistic models.

In summary, rational treatment of the fate of sediment contaminants awaits development of an adequate base of physical oceanographic data. Shortcomings in this area, even among the most heavily studied coastal sediments in the country, were outlined in the text. Absent reliable theory or supporting data, empirical approaches based in sediment chemistry can identify determinants of sediment quality and contaminant distribution. In cases where current-driven sediment resuspension can be neglected, such models permit prediction of sediment quality changes following perturbations in appropriate forcing functions. Improvements in existing theory and a greater commitment to collection of physical data are necessary to support rational models when sediment resuspension is an important determinant of sediment quality.

In light of Marine Board instructions to consider engineering remedies for contaminated sediments, it is worthwhile considering the magnitude of problems imposed by contaminants in place on the Palos Verdes shelf and the likelihood of finding engineered solutions to these problems. On the bases of specific chemical and biological indicators of sediment quality, Palos Verdes sediments are among the most impacted in the nation. Dr. Zarba's work (this volume) on the severity of national sediment contamination has produced suggested contaminant levels for categorizing the severity of DDT and metals contamination in sediments. A significant portion of the shelf sediments fall among those most heavily contaminated in the nation using either metals or DDT criteria. Dr. Robertson's mussel tissue measurements of metals and pesticides (also in this volume), which were not designed to identify contaminant "hot spots" but to suggest portions of the country likely to contain a high level of general sediment contamination, found that animals exposed to Palos Verdes contaminants contained the highest levels of total DDT and cadmium encountered in his survey. Despite the apparent severity of chemical effects, economic impacts and human health effects of shelf contamination are modest or unproven. Nevertheless, these sediments would be granted high priority for remedial activities should remediation measures prove economically feasible and cost effective. They will not.

The areal extent of severely contaminated sediments (by Dr. Zarba's criteria and others) is on the order of 10 km (alongshore) by 4 km (cross-shore). Contaminant removal would demand dredging to an average depth of about 25 cm over that region. Thus the total volume of sediments to be removed would be on the order of 10^7 m^3, or a cube roughly 2.5 football fields in each dimension. Could these be removed, they would present a remarkable disposal problem. The feasibility of dredging from 30 to 150 m depth is itself uncertain. Dr. Herbich (this volume) indicated that modern dredge capabilities encompass removal of 2,000 m^3/hr^{-1} from a depth of 20 m. Even if this capacity could be extended to the depths of the contaminated sediments, it would take seven months of continuous operation of the best equipment known to remove this material.

To cap the same area with clean material to a depth of even 0.6 m would require a minimum 2×10^7 m^3 of clean sediment, and the environmental problems associated with sediment resuspension during such an activity might be exceptionally severe. Engineered remediation strategies of a nature applicable in shallower waters in which contamination is more contained are inappropriate in this setting.

SUMMARY AND CONCLUSIONS

The primary physical determinants of the distribution of refractile, nondiffusing chemicals among sediments of the Palos Verdes shelf 61-m depth contour include sedimentation of natural and outfall-related particles and bioturbation of surface sediments. Processes of secondary or perhaps negligible importance to the long-term fate of sediment DDT at that depth include sediment resuspension and diffusion through sediment pores. In shallower water, sediment resuspension may be an important factor.

The background or natural sedimentation rate along the 61-m depth contour is approximately 500 dry mg/cm^2/yr. Flocculation of natural particles due to the discharge of effluent-related solids is not apparent. The background sedimentation rate decreases rapidly beyond the shelf break.

No matter what treatment scenario is selected for the JWPCP, the flux of natural particles to the bottom will gradually diminish the surface sediment concentration of DDT along the 61-m depth contour with attendant benefits for fish tissue concentrations of DDT and sediment infaunal ecology. However, the rate at which DDT will decrease at the surface is sensitive to the solids MER from the Whites Point outfall system. Implementation of full secondary treatment at JWPCP results in nearly a 75 percent increase in the predicted, year-2005 surface sediment concentration of DDT at station 6C relative to a partial secondary treatment (or no-change) scenario.

ACKNOWLEDGMENTS

Funding for this research was provided by Los Angeles County Sanitation Districts. We thank Irwin Haydock and Jan Stull for general assistance and advice. The manuscript was prepared by Mrs. Sharon Solomon of the University of Arizona Civil Engineering staff. Portions of this work were presented at the Sediment Dynamics Workshop, October 19-21, California State Polytechnic University, Pomona, California.

REFERENCES

Cross, J. N. 1984. Tumors in fish collected on the Palos Verdes shelf. Southern California Coastal Water Research Project Biennial Report, 1983-1984, W. Bascom, ed. Long Beach, California: SCCWRP.

Gossett, R. W., H. W. Puffer, R. H. Arthur, J. F. Alfafara, and D. R. Young. 1982. Levels of trace organic compounds in sportfish from southern California. Southern California Coastal Water Research Project Biennial Report 1981-1982, W. Bascom, ed. Long Beach, California: SCCWRP.

Gossett, R. W., H. W. Puffer, R. H. Arthur, and D. R. Young. 1983. DDT, PCB and benzo(a)pyrene levels in white croaker (*Genyonemus lineatus*) from Southern California. Mar. Poll. Bull. 14(2):60-65.

Hendricks, T. J. 1988. Development of methods for estimating the changes in marine sediments as a result of the discharge of sewered municipal wastewaters through submarine outfalls: Part II, resuspension processes, draft copy. Final report for U.S. EPA.

Hendricks, T. J. 1984. Predicting sediment quality around outfalls. Southern California Coastal Water Research Project Biennial Report 1983-1984, W. Bascom, ed. Long Beach, California: SCCWRP.

Hendricks, T. J. 1982. An advanced sediment quality model. Southern California Coastal Water Research Project Biennial Report 1981-1983, W. Bascom, ed. Long Beach, California: SCCWRP.

Hendricks, T. J. 1978. Forecasting changes in sediments near outfalls. Southern California Coastal Water Research Project Annual Report 1978, W. Bascom, ed. Long Beach, California: SCCWRP.

Keith, J. O., L. A. Woods, and E. G. Hunt. 1970. Reproductive failure in brown pelicans on the Pacific coast. Trans. North Am. Wildl. Nat. Resour. Conf. pp. 56-63.

Koh, R. C. V. 1982. Initial sedimentation of waste particulates discharged from ocean outfalls. Environ. Sci. Technol. 16:757-763.

Myers, E. P. 1974. The concentration and isotopic composition of carbon in marine sediments affected by a sewage discharge. Ph.D. Thesis. California Institute of Technology, Pasadena, California.

Oshida, P. S. and J. L. Wright. 1976. Acute responses of marine invertebrates to chromium. Southern California Coastal Water Research Project Annual Report 1976. El Segundo, California: SCCWRP.

Risebraugh, R. W., O. B. Mengel, D. J. Martin and H. S. Olcott. 1967. DDT residues in Pacific seabirds: A persistent insecticide in marine food chains. Nature 216:589-591.

Risebraugh, R. W., F. C. Sibley, and M. N. Kirven. 1971. Reproductive failure of the brown pelican on Anacapa Island in 1969. Amer. Birds 25:8-9.

Sherwood, M. 1976. Fin erosion disease induced in the laboratory. Southern California Coastal Water Research Project Annual Report 1976. El Segundo, California: SCCWRP.

Stull, J. K., C. I. Haydock, R. W. Smith, and D. E. Montagne. 1986. Long-term changes in the benthic community on the coastal shelf of Palos Verdes, Southern California. Mar. Biol. 91:539-551.

Word, J. Q. 1978. The infaunal trophic index. Southern California Coastal Water Research Project Annual Report 1978, W. Bascom, ed. Long Beach, California: SCCWRP.

Young, D. R., R. W. Gossett, and T. C. Heesen. 1988. Persistence of chlorinated hydrocarbon contamination in a California marine ecosystem. In Urban Wastes in Coastal Marine Environments, Vol. 5, Oceanic Processes in Marine Pollution Series, D. Wolfe and T. P. O'Connor, eds.

APPENDIX
MODELING APPROACH FOR SEDIMENTATION AND SEDIMENTATION/MIXING MODELS

Computational procedures are summarized below:

1. The uncorrected flux of outfall-related solids to specific sediment locations, S_{uncorr} (mg/cm^2-yr), is estimated as follows:

$$S_{uncorr} = \left\{\begin{array}{l}\text{fraction of effluent}\\ \text{solids which reach}\\ \text{the designated area}\end{array}\right\} (\text{solids MER}) \left\{\frac{1}{\text{area}}\right\}$$

$$= \{\text{solids fraction}\}\ SS \times 10^5\ \frac{MT}{yr} \times \frac{10^3 kg}{MT} \times \frac{10^6 mg}{kg}$$

$$\times \frac{1}{0.25 \times 1\ km^2} \times \left\{\frac{1km}{10^3 m}\right\}^2 \times \left\{\frac{1m}{100cm}\right\}^2$$

$$S_{uncorr}\left(\frac{mg}{cm^2-yr}\right) = \{\text{solids fractions}\}\ (SS) \times 40{,}000$$

Notes:
  1. The fraction of effluent solids reaching specific grid sectors or "solids fraction" was initially estimated from the modeling efforts of LACSD personnel using Hendricks model for sediment deposition on the Palos Verdes shelf. Values were subsequently fitted to reproduce volatile solids profiles in sediment cores.
  2. Grid-sector dimensions are 0.25 km x 1.0 km.
  3. SS values, or suspended solids mass emission rates at JWPCP are taken from Figure 3.

2. Biological degradation of solids reduces the volatile solids content of discharged solids by the fraction fVSd in the water column prior to deposition. Thus, the total solids of sewage origin which reach a core site, S_{sr} (mg/cm^2-yr) is:

$$S_{sr} = S_{uncorr}\ (1 - fVS \times fVSd)$$

where fVS is the ratio of volatile solids to total solids in JWPCP effluent (estimated at 70 percent from LACSD records).

3. The flux of volatile solids of sewage origin to the sediment site is given by:

$$VS_{sr} = (S_{uncorr})\ (fVS)\ (1 - fVSd)$$

4. The total flux of solids to a site is the sum of the sewage-related solids and the background (or natural) sediment flux:

$$S_{flux} = S_{sr} + S_{nat}$$

Note: S_{nat} was originally estimatesd from the depth of elevated DDT concentrations at the ϕC monitoring site. Values were subsequently fitted to reproduce sediment volatile solids profiles.

5. The volatile solids content of natural particulate material (fVS_{nat}) was estimated at 7 percent (Hendricks, 1984). The contribution of background sedimentation to the local flux of volatile solids is equal to S_{nat} times fVS_{nat}. That is:

$$VS_{nat} = S_{nat} \times fVS_{nat}$$

6. The average volatile solids content (VS in percent) of solids which reach the sediments is therefore:

$$VS = \frac{VS_{sr} + VS_{nat}}{S_{flux}} \times 100$$

7. The water content (M, in percent) and the sediment wet density (ρ in g/cm^3) were estimated using the following empirical relationships:

$$M = 1.15 \cdot VS + 26.238$$

$$\rho = -0.01673 \cdot VS + 1.7938$$

Note: Relationships shown were determined using LACSD sediment data and a linear regression algorithm.

8. The depth of sediment (DEP_{tot} in cm) added during time intervals of 1 year is given by:

$$DEP_{tot} = \frac{S_{flux}}{(1 - 0.01M)} \cdot \frac{10^{-3}}{\rho}$$

9. The diffusion of DDT within sediments was calculated using a one-dimensional forward explicit finite difference equation:

$$C^t(x) = C(x) - \frac{D \Delta t}{\Delta x^2} [C(x+1) + C(x-1) - 2 C(x)]$$

where C is the DDT concentration in the sediment [$\mu g/kg$], C^t is the new concentration after a time interval $\Delta t = 1$ day, $\Delta x = 2 cm$ is the node spacing, with 40 nodes per sediment profile. No loss of DDT from sediments was assumed.

10. Mixing was incorporated by equally distributing DDT within the designated number of 2 cm sediment compartments over which mixing would occur. New sediment was incorporated into the mising model only after 2 cm of new material had accumulated.

NOTATION

S_{uncorr}	= solids flux (uncorrected from outfall (mg/cm^2-yr)
MER	= Mass emission rate (metric tons/yr)
fVSd	= fraction of volatile solids that are biologically degraded
fVS	= fraction of total solids that are volatile
VS_{sr}	= volatile solids flux of sewage origin to sediment (mg/cm^2-yr)
S_{sr}	= solids flux of sewage origin to sediment (mg/cm^2-yr)
S_{nat}	= solids flux of natural origin to sediment (mg/cm^2-yr)
fVS_{nat}	= fraction of volatile solids of natural (non-sewage) origin in sedimenting material
S_{flux}	= sum of sewage and natural (non-sewage) sediment solids (mg/cm^2-yr)
VS_{nat}	= volatile solids flux of natural origin to sediment (mg/cm^2-yr)
VS	= percent of sediment solids that is volatile
M	= moisture, or percent water content of sediment
ρ	= sediment wet density (g/cm^3)
DEP_{tot}	= depth of sediment accumulation each year (cm)
C	= concentration of DDT in sediment ($\mu g/kg$)
D	= diffusivity (cm^2/s)
Δx	= distance between nodes in finite difference model (cm)
Δt	= time interval in finite difference model (d)

ASSESSMENT AND SELECTION OF REMEDIAL TECHNOLOGIES

MANAGEMENT STRATEGIES FOR DISPOSAL OF CONTAMINATED SEDIMENTS

M. R. Palermo, C. R. Lee, and N. R. Francingues
U.S. Army Engineer Waterways Experiment Station

ABSTRACT

A comprehensive and consistent strategy for selecting the most appropriate disposal alternative from an environmental standpoint is essential when the disposal of contaminated or potentially contaminated dredged material is required. The U.S. Army Corps of Engineers (COE) has recently developed a management strategy for use in selecting disposal alternatives for materials ranging from clean sand to highly contaminated sediments. A decision-making framework has also been developed to supplement the management strategy and provide a logical basis for comparison of test results with standards or reference information to determine if contaminant control measures are required in a given instance. This approach been adopted as official COE policy for studies involving disposal of contaminated sediments.

BACKGROUND

Beginning in the early 1970s, considerable attention was focused on the potential environmental effects of dredged material disposal. The U.S. Army Corps of Engineers (COE) has since devoted major research efforts toward development of testing protocols and contaminant control measures for both open-water and confined disposal alternatives. In 1984, efforts were initiated to develop an overall management strategy based on these efforts. The management strategy presented here has been adopted by COE as an environmentally sound framework for selecting alternatives for the disposal of dredged material with any level of contamination.

Over 95 percent of the total volume of material dredged in the United States is considered noncontaminated. However, the potential presence of contamination has generated concern that dredged material disposal may adversely affect water quality and aquatic or terrestrial organisms. Since many of the waterways are located in industrial and urban areas, sediments may be contaminated with wastes from these sources. In addition, sediments may be contaminated with chemicals from agricultural practices.

Since the nature and level of contamination in sediment vary greatly on a project-to-project basis, the appropriate method of disposal may involve any of several available disposal alternatives. Further, control measures to manage specific problems associated with the presence or mobility of contaminants may be required as a part of any given disposal alternative. An overall management strategy for disposal of dredged material is therefore required. Such a strategy must provide a framework for decision making to select the best possible disposal alternative and to identify appropriate control measures to offset problems associated with the presence of contaminants.

The lead responsibility for the development of specific ecological criteria and guideline procedures regulating the transport and disposal of dredged and fill material was legislatively assigned to the U.S. Environmental Protection Agency (EPA) in consultation or conjunction with the COE. The enactment of various U.S. laws concerned with the transport and disposal of dredged and fill material, required the COE to participate in developing guidelines and criteria for regulating dredged and fill material disposal. The focal point of research for these procedures is the Dredged Material Research Program (DMRP), which was completed in 1978; the ongoing Dredging Operations Technical Support (DOTS) Program and the Long-term Effects of Dredging Operations (LEDO) Program; and the COE/EPA Field Verification Program (FVP).

Scope

The management strategy presented here is based on findings of research conducted by the COE, EPA, and others, and experience in actively managing dredged material disposal. Approaches for evaluating potential for contaminant-related problems, testing protocols, and the applicability of various disposal alternatives are discussed. Procedures for conducting tests or for design and implementation of management strategies are not presented but are appropriately referenced. A more detailed presentation of the management strategy is available from the COE Waterways Experiment Station (Francingues et al., 1985).

MANAGEMENT STRATEGY

The selection of an appropriate strategy is partially dependent on the nature of the dredged material, nature and level of contamination, the physicochemical nature of the disposal site environment, available dredging alternatives, project size, and site-specific physical and chemical conditions, all of which influence the potential for environmental impacts. Technical feasibility, economics, and other socioeconomic factors must also be considered in the decision-making process. The technical management strategy presented here mainly considers the nature and degree of contamination, physicochemical conditions at disposal sites, potential environmental impacts, and related technical factors. A flow chart illustrating the strategy is shown in Figure 1.

The steps for managing dredged material disposal consist of the following:

1. evaluate contamination potential,
2. consider potential disposal alternatives,
3. identify potential problems,
4. assess the need for disposal restrictions,
5. select an implementation plan,
6. identify available control options,
7. evaluate design considerations, and
8. select appropriate control measures.

The initial screening consists of examining available historical data and information on pollutant discharges and spills at the dredging site to determine whether there is a reason to suspect the presence of significant concentrations of contaminants.

If the dredged material is clean and/or environmental impacts are within acceptable limits, conventional open-water or confined disposal methods may be used. If impacts resulting from conventional disposal techniques would not be within acceptable limits, contaminated material may be disposed by either open-water or confined methods with appropriate restrictions. Each disposal alternative may pose problems for managing contaminated dredged material. Based on the initial evaluation, site-specific conditions, dredging methods, and anticipated site use, the potential contaminant problems can be identified. For open-water disposal, contaminant problems may be either water column or benthic related. Confined disposal contaminant problems may be related to either water quality (effluent, surface runoff, or leachate) or contaminant uptake (plants or animals).

The magnitude and potential impacts of specific contaminants must be evaluated using appropriate testing protocols. Such protocols, designed for evaluation of dredged material, consider the unique nature of dredged material and the physicochemical environment of each disposal alternative. The results of all testing are compiled and evaluated to determine the potential for environmental harm from contamination, to examine the interrelationships of the problems and potential solutions, and to determine what restrictions on open-water or confined disposal are appropriate. If impacts as evaluated using the testing protocols are acceptable, conventional open-water or confined disposal may again be considered.

Specific environmental problems identified using the testing protocols must be addressed by implementation plans appropriate for the level of potential contamination. Restrictions may also be required for open-water or confined disposal that could eliminate certain options from consideration. Several options may be available for the selected implementation strategy. Options for controlling water column and benthic impacts include bottom discharge via submerged diffusers, treatment, contained aquatic disposal, and subaqueous capping using clean sediments. Options for controlling confined disposal impacts include containment, treatment, long-term storage, and reuse. The degree of contaminant control finally selected may range anywhere between disposal in open water with no special restrictions to a completely controlled confinement. Many of the technologies identified

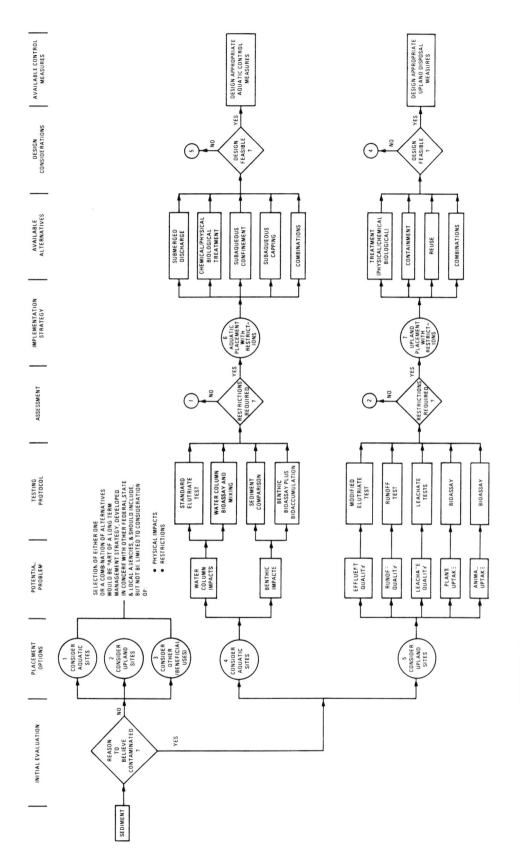

FIGURE 1 Management strategy flow chart.

are either commonly used in COE dredging activities or are presently being evaluated as part of COE's ongoing research and operations.

POTENTIAL PROBLEMS AND TESTING PROTOCOLS

The properties of a dredged material affect the fate of any contaminants present, and the short- and long-term physical and chemical environment of the dredged material at the disposal site influences the environmental consequences of contaminants (Gambrell et al., 1978). These factors should be considered in evaluating the environmental risk of a proposed disposal method for contaminated sediment. Where the physical and chemical environment of a contaminated sediment is altered by disposal, chemical and biological processes important in determining environmental consequences of potentially toxic materials may be affected.

The major disposal alternatives are open water (subaqueous environment) and confined (subaqueous, intertidal, or upland environment). A number of variations exist for each of the major alternatives, each having some influence on the fate of contaminants at disposal sites. Environmentally sound disposal of dredged material can be achieved using any of the major alternatives if appropriate management practices are employed.

Water Column

Although the vast majority of heavy metals, nutrients, and petroleum and chlorinated hydrocarbons are usually associated with the fine-grained and organic components of the sediment (Burks and Engler, 1978), there has been little evidence of biologically significant release of these constituents from typical dredged material to the water column during or after dredging or disposal operations. Turbidity due to fine particulates suspension is only of limited short-term impact.

Water column impacts can best be evaluated by chemical analyses of dissolved contaminants for which water quality criteria exist. The standard elutriate test is used for this purpose (U.S. EPA/COE, 1977). Results must be considered in light of mixing and dilution. If the criteria are exceeded after consideration of mixing, a bioassay can be used to determine the potential consequences of exceeding the criteria for a short time.

Benthic

The DMRP results conclusively indicated that most subaqueous disposal in low-energy aquatic environments where stable mounding will occur will favor containment of contaminated materials. Dredging and disposal do not introduce new contaminants to the aquatic environment, but simply redistribute the sediments, which are the natural depository of contaminants introduced from other sources. The potential for accumulation of a contaminant in the tissues of an organism (bioaccumulation) may be affected by several factors, such as duration of exposure, salinity, water hardness, exposure concentration, temperature, chemical form of the contaminant, and the particular organism under study. The relative importance of these factors varies. Elevated concentrations of contaminants in the ambient medium or associated sediments are not always indicative of high levels of contaminants in tissues of benthic invertebrates. Bulk analysis of sediments for contaminant content alone cannot be used as a reliable index of availability and potential ecological impact of dredged material, but only as an indicator of the presence of contaminants and total contaminant content. Bioaccumulation of most contaminants from sediments is generally minor.

Potential benthic impacts can be evaluated by comparing contaminant concentrations in the sediments of both the dredging and disposal sites. If the concentrations are higher in the dredged material than in the disposal site sediment, a bioassay/bioaccumulation test can be used to determine the environmental consequences of the contaminant levels.

Effluent Quality

Dredged material placed in a confined disposal area undergoes sedimentation, while clarified supernatant waters are discharged from the site as effluent during active dredging operations. The effluent may contain levels of both dissolved and particulate-associated contaminants. A large portion of the total contaminant level is particulate associated.

A modified elutriate test procedure, developed under the LEDO program (Palermo, 1986), can be used to predict both the dissolved and particulate-associated contaminant concentrations in confined disposal area effluents (water discharged during active disposal operations). The laboratory test simulates contaminant release under confined disposal conditions and reflects sedimentation behavior of dredged material, retention time of the containment, and chemical environment in ponded water during active disposal. The acceptability of the proposed confined disposal operation can be evaluated by comparing the predicted contaminant concentrations with applicable water quality standards while considering an appropriate mixing zone. In some cases appropriate water column bioassays would be required if water quality criteria are exceeded.

Surface Runoff Quality

After dredged material has been placed in a confined disposal site and the dewatering process has been initiated, contaminant mobility in rainfall-induced runoff is considered in the overall environmental impact of the dredged material being placed in a confined disposal site.

The quality of the runoff water can vary depending on the physicochemical processes that occur during drying and the contaminants present in the dredged material.

An appropriate test for evaluating surface runoff water quality must consider the effects of the drying process to adequately estimate and predict runoff water quality. At present there is no single simplified laboratory test to predict runoff water quality. A laboratory test using a rainfall simulator has been developed and is being used to predict surface runoff water quality from dredged material as part of the FVP (Lee and Skogerboe, 1983).

Leachate Quality

Subsurface drainage from confined disposal sites in an upland environment may reach adjacent aquifers. Fine-grained dredged material tends to form its own disposal area liner as particles settle with percolation drainage water, but the settlement process may require some time for self-sealing to develop. Since most contaminants potentially present in dredged material are closely adsorbed to particles, only the dissolved fraction will be present in leachates. A potential for leachate impacts exists when a dredged material from a saltwater environment is placed in a confined site adjacent to freshwater aquifers. The site-specific nature of subsurface conditions is the major factor in determining possible impact (Chen et al., 1978).

An appropriate leachate quality testing protocol must predict which contaminants may be released in leachate and the relative degree of release. Laboratory testing protocols to predict leachate quality from dredged material disposal sites have been developed and applied, however additional evaluations of available leaching procedures are needed before a leaching test protocol for confined dredged material can be recommended. These evaluations are now an ongoing COE research effort.

Plant Uptake

After dredged material has been placed in either an intertidal, wetland, or upland environment, plants can invade and colonize the site. There is potential for movement of contaminants from the dredged material into plants and then eventually into the food chain.

A test protocol for plant uptake was developed under the LEDO program based on the results of the DMRP. This procedure has been applied to testing a number of contaminated dredged materials and has given appropriate results and information to predict the potential for plant uptake of contaminants from dredged material (Folsom and Lee, 1981, 1983).

Animal Uptake

Animals have also been known to invade and colonize confined dredged material disposal sites. In some cases, prolific wildlife habitats have become established on these sites. Concern has developed recently on the potential for animals inhabiting either wetland or upland, terrestrial, confined disposal sites to become contaminated and contribute to the contamination of food chains associated with the site.

A test protocol is being tested under the FVP that utilizes an earthworm as an index species to indicate toxicity and bioaccumulation of contaminants from dredged material (Simmers et al., 1983).

Other Impacts

Potential impacts could arise from flammable or noxious emissions released from the dredged material during dredging and disposal operations. Standard safety precautions will eliminate adverse human health effects and are normally required under contract specifications.

SELECTION OF A DISPOSAL ALTERNATIVE

Disposal alternatives are divided into general classes: open water, confined, open water with restrictions, and confined disposal with restrictions. Disposal alternatives with restrictions are used whenever results of the testing protocols indicate they are needed. Conventional disposal alternatives are well documented in DMRP reports (Herner and Co., 1978) and are described only briefly in this section. The preference of open-water disposal over confined disposal, or vice versa, is dependent on many factors other than contaminants, as discussed earlier.

Open-Water Disposal

This disposal alternative involves conventional open-water disposal techniques. This alternative would be selected if the initial evaluation and testing protocols as discussed earlier indicated that water column and benthic effects are acceptable.

Dredged material can be placed in open-water sites by direct pipeline discharge, hopper dredge discharge, or dumping from scows. For conventional open-water disposal, no special placement techniques are used and the material is normally discharged at a selected point within a designated disposal site.

Ocean open-water disposal sites are designated using a set procedure (EPA, 1977). Criteria for site designation include storage capacity requirements and chemical/biological considerations. Procedures for site selection are under review with the objective of improving the efficiency of the overall site designation process.

The capacity of open-water disposal sites is determined by the volume of accumulated material that can be placed without exceeding the designated site boundaries or exceeding water-depth constraints.

Capacity also may be determined by the assimilative ability of the waters within the designated site boundaries, i.e., their ability to reduce concentrations of suspended material and associated contaminants to an acceptable level. Procedures for evaluation of open-water disposal site capacity to include descent and spread of discharges, dispersion, erosion and resuspension from mounds, and consolidation of mounds is currently under study by the COE.

The open-water environment is physically dynamic, and materials placed in open water will be dispersed, mixed, and diluted to some degree. Therefore, all evaluative procedures must be interpreted in light of the mixing expected at the disposal site. Any of several methods or models (Holliday et al., 1978) may be used to estimate the maximum concentration of the liquid and suspended particulate phases found at the disposal site after initial mixing.

Confined Disposal

Conventional confined disposal consists of placing or pumping the dredged material into a diked containment area where the material settles and consolidates. The area should be designed to provide good sedimentation and sufficient volume for storage (Palermo et al., 1978). The supernatant water is discharged over a weir, which is designed to maintain good effluent quality by minimizing resuspension of settled material. If the turbidity of the effluent exceeds applicable water quality standards, a chemical clarification system may be used for additional solids removal (Schroeder 1983). Following completion of the disposal operation, the site should be managed to promote consolidation and drying (Haliburton, 1978). The containment area can then be used for additional disposal, mined for productive use of the material, or returned to the sponsor for other uses (Montgomery et al., 1978).

Open-Water Disposal with Restrictions

In cases where testing protocols indicate that water column or benthic effects will be unacceptable when conventional open-water disposal techniques are used, open-water disposal with restrictions may be considered. This alternative involves the use of dredging or disposal techniques that will reduce water column and benthic effects. Such techniques include use of subaqueous discharge points, diffusers, subaqueous confinement of material, or capping of contaminated material with clean material. The same basic considerations for conventional open-water disposal site designation, site capacity, and dispersion and mixing also apply to open-water disposal with restrictions.

Submerged Discharge

The use of a submerged point of discharge reduces the area of exposure in the water column and the amount of material suspended in the water column and susceptible to dispersion. The use of submerged diffusers also reduces the exit velocities for hydraulic placement, allowing more precise placement and reducing both resuspension and spread of the discharged material. Considerations in evaluating feasibility of a submerged discharge and/or use of a diffuser include water depth, bottom topography, currents, type of dredge, and site capacity. Diffusers have been successfully demonstrated in the Netherlands and in the United States (Hayes et al., 1988).

Subaqueous Confinement

The use of subaqueous depressions or borrow pits or the construction of subaqueous dikes can provide confinement of material reaching the bottom during open-water disposal. Such techniques reduce the areal extent of a given disposal operation, thereby reducing both physical benthic effects and the potential for release of contaminants. Considerations in evaluating feasibility of subaqueous confinement include type of dredge, water depth, bottom topography, bottom sediment type, and site capacity. Subaqueous confinement has been utilized in Europe and to a limited extent by the COE New York District. Precise placement of material and use of submerged points of discharge increase the effectiveness of subaqueous confinement.

Capping

Capping is the placement of a clean material over material considered contaminated. Considerations in evaluation of the feasibility of capping include water depth, bottom topography, currents, dredged material and capping material characteristics, and site capacity. Both the Europeans and the Japanese have successfully used capping techniques to isolate contaminated material in the open-water disposal environment. Capping is also currently used by the COE's New York District and New England Division as a means of offsetting the potential harm of open-water disposal of contaminated or otherwise unacceptable sediments. The London Dumping Convention has accepted capping, subject to careful monitoring and research, as a physical means of rapidly rendering harmless contaminated material dumped in the ocean. The physical means are essentially to seal or sequester the unacceptable material from the aquatic environment by a covering of acceptable material.

The efficiency of capping in preventing the movement of contaminants through this seal and the degradation of the biological community by leakage, erosion of the cover (cap), or bioturbation are being addressed by research under the LEDO program. The engineering aspects of cap design and placement are being addressed under the COE's Dredging Research Program (DRP). It is possible that techniques and equipment can be developed that will provide a capped dredged material disposal area as secure from potential environmental harm as upland confined disposal areas. The capping technique for disposal of dredged material has potential for relieving some pressure on acquiring sites for confined disposal areas in localities where land is rapidly becoming unavailable.

Chemical/Physical/Biological Treatment

Treatment of discharges into open water may be considered to reduce certain impacts. For example, the Japanese have used an effective in-line dredged material treatment scheme for highly contaminated harbor sediments (Barnard and Hand, 1978). However, this strategy has not been widely applied and its effectiveness has not been demonstrated for solution of the problem of contaminant release during open-water disposal.

Confined Disposal with Restrictions

Site Selection and Design

Conventional confined disposal methods, described previously, can be modified to accommodate disposal of contaminated sediments in new, existing, and reusable disposal areas. The design or modification of these areas must consider the problems associated with contaminants and their effects on conventional design.

Site location is an important consideration since it can mitigate many contaminant mobilization problems. Proper site selection may reduce surface runon and therefore contaminated runoff and contaminant release by flooding. Groundwater contamination problems can be offset through selection of a site with natural clay foundation instead of a sandy area and through avoidance of aquifer recharge areas (Gambrell et al., 1978).

Careful attention to basic site design as discussed previously will aid in implementing many of the controls outlined. Retention time can be increased to improve suspended solids removal and, therefore, contaminant removal. Additional ponding depth can also improve sedimentation. Decreasing the weir loading rate and improving the weir design to reduce leakage and control the discharge rate can also reduce the suspended solids and contaminant concentration of the effluent. Dewatering should be examined carefully before selecting a method, since it promotes oxidation of the material and thereby increases the mobility of certain contaminants (Gambrell et al. 1978). Care must also be taken to reduce loss of contaminated sediment by erosion during drainage and storm events.

Four options are considered available for confined disposal with restrictions. These options include

1. containment--dredged material and associated contaminants are contained within the disposal site;
2. treatment--dredged material is modified physically, chemically, or biologically to reduce toxicity, mobility, etc.;
3. storage and rehandling--dredged material is held for a temporary

period at the site and later removed to another site for ultimate disposal; and
4. reuse--dredged material is classified and beneficial uses are made of reclaimed materials;

Obviously, combinations of the above options are available for a particular dredging operation.

Effluent Controls

Effluent controls at conventional confined disposal areas are generally limited to chemical clarification. The clarification system is designed to provide additional removal of suspended solids and associated adsorbed contaminants as described in Schroeder (1983). Additional controls can be used to remove fine particulates that will not settle or to remove soluble contaminants from the effluent. Examples of these technologies are filtration, adsorption, selective ion exchange, chemical oxidation, and biological treatment processes. Beyond chemical clarification, only limited data exists for treatment of dredged material (Gambrell et al., 1978).

Runoff Controls

Runoff controls at conventional sites consist of measures to prevent erosion of contaminated dredged material and dissolution and discharge of oxidized contaminants from the surface. Control options include maintaining ponded conditions, planting vegetation to stabilize the surface, liming the surface to prevent acidification and to reduce dissolution, covering the surface with synthetic geomembranes, and/or placing a lift of clean material to cover the contaminated dredged material (Gambrell et al., 1978).

Leachate Controls

Leachate controls consist of measures to minimize groundwater pollution by preventing mobilization of soluble contaminants. Control measures include proper site selection as described earlier, dewatering to minimize leachate production, chemical admixing to prevent or retard leaching, lining the bottom to prevent leakage and seepage, capping the surface to minimize infiltration and thereby leachate production, vegetation to stabilize contaminants and to increase drying, and leachate collection, treatment, or recycling (Gambrell et al., 1978).

Control of Contaminant Uptake

Plant and animal contaminant uptake controls are measures to prevent mobilization of contaminants into the food chain. Control measures include selective vegetation to minimize contaminant uptake, liming or chemical treatment to minimize or prevent release of contaminants from the material to the plants, and capping with clean sediment or excavated material (Gambrell et al., 1978).

Other Controls

The control of gaseous emissions that might present human health hazards can consist of physical measures such as covers, vertical barriers, control trench vents, pipe vents, and gas-collection systems. Wind-erosion control of contaminated surface materials is another type of management or operating control to minimize transport of contaminants off site. Techniques for limiting wind erosion are generally similar to those employed in dust control and include physical, chemical, or vegetative stabilization of surface soils (U.S. COE, 1983).

Many of the contaminant controls described in the preceding paragraphs are directly applicable to the control of highly contaminated sediments. These controls will be extremely site specific. Special considerations that are based on the physical nature and chemical composition of the dredged material will be required to effectively design a confined disposal facility. For example, some contaminated dredged material may require in-pipeline treatment prior to discharging the material into the containment facility. Similarly, if the facility requires a bottom liner system, the liner materials (synthetic membrane or clay) must be chemically compatible (resistant) with the dredged material to be placed on them. Special compatibility testing will be needed for selection of appropriate liner materials. Other requirements such as leachate detection and monitoring are likely due to the potentially adverse environmental effects of the liner leaking.

DECISION-MAKING FRAMEWORK

A decision-making framework has been developed that utilizes the management strategy described above and incorporates the results from the suite of test protocols (Lee et al., 1986). Reference information and data from the test protocols are used to make the decisions called for in the framework. Detailed procedures for using the framework and example applications using data from reference sites and testing protocols are found in Lee et al. (1986).

Responsibility for Local Authority Decisions

There are certain decisions that must be made initially and then periodically within the decision-making framework that are the sole responsibility of the local authorities. These local authority decisions (LADs) are required to initially set specific goals to be achieved. For example, a LAD must establish the environmental quality ultimately desired at the site and the rate at which this goal is to be achieved. A LAD must determine whether or not to consider mixing zones when test results exceed reference site values or water quality criteria. A LAD must determine the appropriate reference site(s) for test result comparisons in the decision-making framework in order to achieve the ultimate and intermediate goals. The selection of reference sites can vary from the actual disposal site to a pristine background site. This selection is dependent on the goal established for the area such as a goal of nondegradation (reference site is disposal site) or cleaner-than-present condition (reference site is pristine background site) or some other goal. The clear identification of the ultimate and intermediate goals and selection of appropriate references to achieve them is a crucial responsibility of the local authorities and will influence the outcome of all test result interpretations.

Evaluation of Respective Contaminant Pathways

Evaluation of respective contaminant pathways under the framework is illustrated by flowcharts. Examples of the flow charts for the open water contaminant pathways (water column and benthic) are shown in Figures 2 and 3. Similar flowcharts are available for the confined disposal pathways of effluent discharge, surface runoff, leachate, and direct uptake by plants and animals.

Test results are compared to established numerical values where these are available and appropriate for test interpretation. When such values do not exist, the framework provides guidance on interpreting test results in comparison to results of the same test performed on a reference sediment. For each test, guidance is provided on these bases for determining whether or not restrictions on the discharge are required to protect against contaminant impacts or whether further evaluation is required to determine the need for restrictions. In some cases, there is inadequate scientific knowledge to reach a decision solely on the basis of test results, and LADs that incorporate both scientific and administrative judgments are required to reach a decision. In such cases, guidance is given on evaluating the scientific considerations involved.

In this manner, guidance is provided for systematically interpreting the results of each test required to evaluate potential impacts of aquatic disposal and upland disposal. Applying the systematic detailed guidance will lead to a decision that restrictions are or are not required for aquatic disposal and/or upland disposal.

IMPLEMENTATION

COE Policy

The Management Strategy/Decision-Making Framework approach was adopted as official COE policy in 1985 for studies involving disposal of contaminated sediments (Kelly, 1985). Additional guidance on use of these approaches under COE's regulatory program was provided in 1987 (U.S. COE, 1987). The recently adopted Dredging Regulation (33 Code of

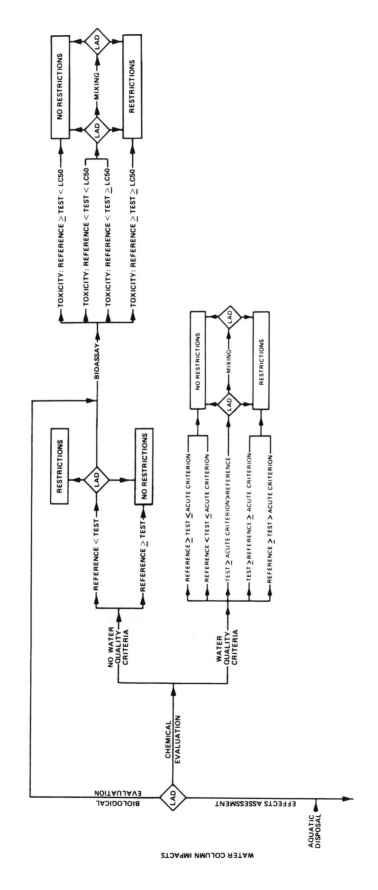

FIGURE 2 Flow chart for decision making for aquatic disposal, water column impacts.

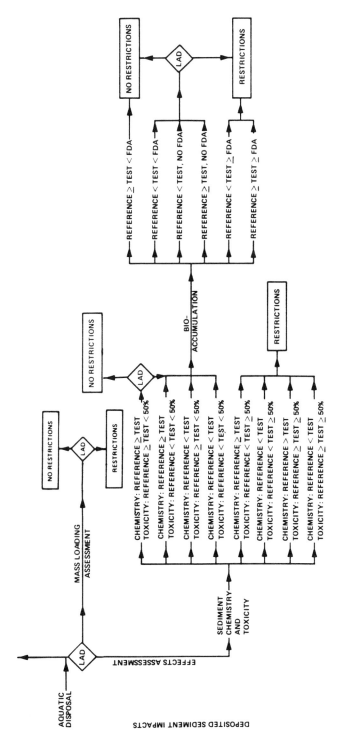

FIGURE 3 Flow chart for decision making for aquatic disposal, benthic impacts.

Federal Regulations 320) incorporated the strategy approaches by reference. The regulation also describes the approach as the basis of a "federal standard," intended to meet environmental requirements at least cost within a consistent national framework.

The technical approaches used in the management strategy have received widespread acceptance by federal and state agencies. In fact, the initial development of the decision-making framework was funded by the Washington State Department of Ecology as a part of its implementation of the EPA Superfund program. Environment Canada has adopted a technical approach to disposal alternative evaluation closely patterned after the management strategy. This approach, illustrated in Figure 4, was proposed for use in evaluation of dredging projects along the St. Lawrence River (Rochon, 1985).

Applications

Strategy is now being applied routinely by the COE and the private sector. Recently, three studies of disposal alternatives incorporated the strategy in a comprehensive manner, utilizing testing approaches for both open-water and confined disposal:

1. Indiana Harbor, Indiana, a project in COE's Chicago District involving PCB-contaminated sediment (U.S. COE Environmental Laboratory, 1987);
2. Everett Harbor, Washington, a Navy homeport project involving approximately 1 million yd^3 of contaminated sediment (Palermo et al., 1986); and
3. New Bedford Harbor, Massachusetts, a Superfund project involving sediments highly contaminated with PCB's and metals (Francinques and Averett, 1988).

These projects are example applications of the Management Strategy/ Decision-Making Framework.

Refinement

Refinement of the technical approaches used in the management strategy is an ongoing effort under COE research programs concerned with the environmental effects of dredged material disposal. The objectives of these efforts are to develop appropriate tests and procedures, improve accuracy of predictions, and reduce the costs of testing and evaluations.

REFERENCES

Averett, D. E., Palermo, M. R., Otis, M. J., and Rubinoff, P. B. 1988. Evaluation of disposal alternatives for New Bedford Harbor Superfund project. Report in preparation, U.S. Army Engineer Waterways Experiment Station, Vicksburg, Miss.

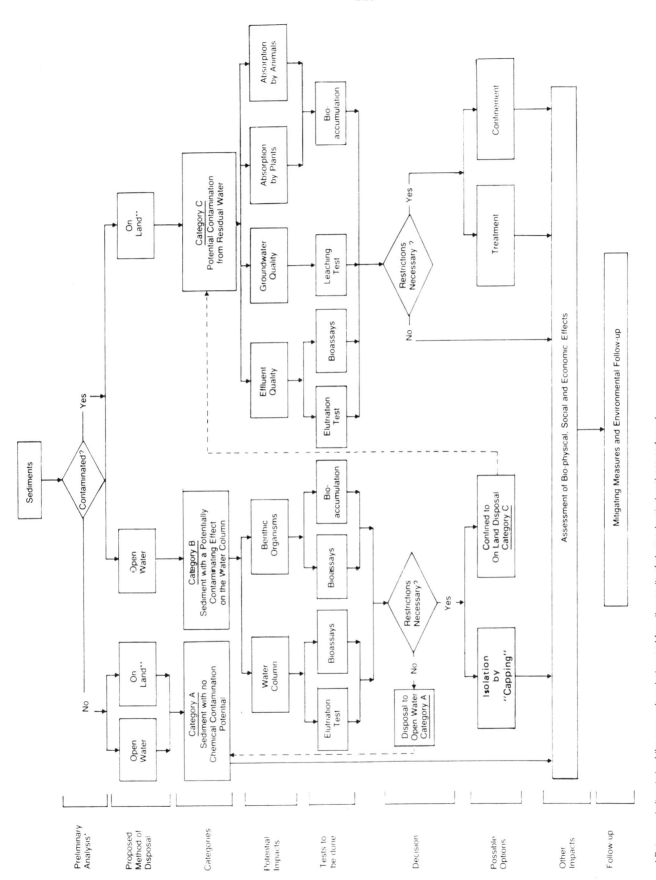

FIGURE 4 Analysis requirements for sediment categories and disposal methods adopted by Environment Canada. SOURCE: After Rochon, 1985.

Barnard, W. D. 1978. Prediction and Control of Dredged Material Dispersion Around Dredging and Open-Water Pipeline Disposal Operations. Technical Report DS-78-6. Vicksburg, Miss.: U.S. Army Engineer Waterways Experiment Station.

Barnard, W. D. and T. D. Hand. 1978. Treatment of Contaminated Dredged Material. Technical Report DS-78-14. Vicksburg, Miss.: U.S. Army Engineer Waterways Experiment Station.

Burks, S. A. and R. M. Engler. 1978. Water Quality Impacts of Aquatic Dredged Material Disposal (Laboratory Investigations). Technical Report DS-78-4. Vicksburg, Miss.: U.S. Army Engineer Waterways Experiment Station.

Chen, K. Y., D. Eichenberger, J. L. Mang, and R. E. Hoeppel. 1978. Confined Disposal Area Effluent and Leachate Control (Laboratory and Field Investigations). Technical Report DS-78-7. Vicksburg, Miss.: U.S. Army Engineer Waterways Experiment Station.

Cullinane, M. J., D. E. Averett, R. A. Shafer, J. W. Male, C. L. Truitt, and M. R. Bradbury. 1986. Guidelines for Selecting Control and Treatment Options for Contaminated Dredged Material Requiring Restrictions. Final Report, Puget Sound Dredged Disposal Analysis (PSDDA) Reports. Seattle, Washington: U.S. COE Seattle District, Washington State Department of Ecology and U.S. EPA Region 10.

Folsom, B. L., Jr., and C. R. Lee. 1981. Zinc and cadmium uptake by the freshwater marsh plant *Cyperus esculentus* grown in contaminated sediments under reduced (flooded) and oxidized (upland) disposal conditions. J. Plant Nutrition 3:233-244.

Folsom, B. L., Jr., and C. R. Lee. 1983. Contaminant uptake by *Spartina alterniflora* from an upland dredged material disposal site--application of a saltwater plant bioassay. Proc. Internat. Conf. on Heavy Metals in the Environment, Heidelberg, West Germany. Edinburgh, Scotland: CEP Consultants, Ltd. Pp. 646-648.

Francingues, N., M. R. Palermo, R. Peddicord, and C. R. Lee. 1985. Management strategy for the disposal of dredged material: Contaminant testing and controls. Miscellaneous Paper D-85-1. U.S. Army Engineer Waterways Experiment Station, Vicksburg, Miss.

Gambrell, R. P., R. A. Khalid, and W. H. Patrick. 1978. Disposal Alternatives for Contaminated Dredged Material as a Management Tool to Minimize Adverse Environmental Effects. Technical Report DS-78-8. Vicksburg, Miss.: U.S. Army Engineer Waterways Experiment Station.

Haliburton, T. A. 1978. Guidelines for Dewatering/Densifying Confined Dredged Material. Technical Report DS-78-11. Vicksburg, Miss.: U.S. Army Engineer Waterways Experiment Station.

Hayes et al. 1988. Demonstration of innovative and conventional dredging equipment at Calumet Harbor, Illinois. Miscellaneous Paper EL-88-1, U.S. Army Engineer Waterways Experiment Station, Vicksburg, Miss.

Herner and Company. 1978. Dredged Material Research Program Publication Index and Retrieval System. Technical Report DS-78-23. Vicksburg, Miss.: U.S. Army Engineer Waterways Experiment Station.

Holliday, B. W., B. H. Johnson, and W. A. Thomas. 1978. Predicting and Monitoring Dredged Material Movement. Technical Report DS-78-3.

Vicksburg, Miss.: U.S. Army Engineer Waterways Experiment Station.

Kelly, BG P. J. 1985 (17 Dec). Policy guidance regarding management and disposal of contaminated dredged material. Water Resources Support Center, Fort Belvoir, Va.

Lee, C. R. and J. G. Skogerboe. 1983. Prediction of surface runoff water quality from an upland dredged material disposal site. Proc. Internat. Conf. on Heavy Metals in the Environment, Heidelberg, West Germany.

Lee, C. R., Peddicord, R. K., Palermo, M. R., and N. R. Francinques. 1986. General decision-making framework for dredged material, example application of Commencement Bay Washington. Miscellaneous Paper D-86-X. U.S. Army Engineer Waterways Experiment Station, Vicksburg, Miss.

Montgomery, R. L., A. W. Ford, M. E. Poindexter, and M. J. Bartos. 1978. Guidelines for Dredged Material Disposal Area Reuse Management. Technical Report DS-78-12. Vicksburg, Miss.: U.S. Army Engineer Waterways Experiment Station.

Palermo, M. R., R. L. Montgomery, and M. E. Poindexter. 1978. Guidelines for Designing, Operating and Managing Dredged Material Containment Areas. Technical Report DS-78-10. Vicksburg, Miss.: U.S. Army Engineer Waterways Experiment Station.

Palermo, M. R. 1986. Interim guidance for conducting modified elutriate tests for use in evaluating discharges from confined dredged material disposal sites. Miscellaneous Paper D-86-1. U.S. Army Engineer Waterways Experiment Station, Vicksburg, Miss.

Rochon, R. 1985. Problems Associated with Dredging Operations on the St. Lawrence, Situation, Methods and Priority Areas for Research. Technical Report EPA 4/MA/1. Ottawa: Environmental Protection Service, Environment Canada.

Schroeder, P. R. 1983. Chemical Clarification Methods for Confined Dredged Material Disposal. Technical Report D-83-2. Vicksburg, Miss.: U.S. Army Engineer Waterways Experiment Station.

Simmers, J. W., R. G. Rhett, and C. R. Lee. 1983. Application of a terrestrial animal bioassay for determining toxic metal uptake from dredged material. Proc. Internat. Conf. on Heavy Metals in the Environment, Heidelberg, West Germany.

U.S. Army Corps of Engineers (COE). 1983. Preliminary Guidelines for Selection and Design of Remedial Systems for Uncontrolled Hazardous Waste Sites. Draft Engineer Manual 1110-2-600. Washington, D.C.: COE.

U.S. Army Corps of Engineers. 1987. Testing requirements for dredged material evaluation. Regulatory Guidance Letter RGL-87-8, COE, Washington, D.C.

U.S. Army Corps of Engineers Environmental Laboratory. 1987. Disposal Alternatives for PCB-contaminated sediments from Indiana Harbor, Indiana. Miscellaneous Paper EL-87-9, Vols. I and II. U.S. Army Engineer Waterways Experiment Station, Vicksburg, Miss.

U.S. Environmental Protection Agency (EPA). 1977. Ocean dumping, final revision of regulations and criteria. Federal Register 42(7).

U.S. Environmental Protection Agency. 1980. Guidelines for specification of disposal sites for dredged or filled material.

Federal Register 45(249):85336-85358.

U.S. Environmental Protection Agency/Corps of Engineers (EPA/COE). 1977. Ecological Evaluation of Proposed Discharge of Dredged Material into Ocean Waters, Implementation Manual for Section 103 of Public Law 92-532 (Marine Protection, Research, and Sanctuaries Act of 1972). Vicksburg, Miss.: U.S. Army Engineer Waterways Experiment Station.

ALTERNATIVES FOR CONTROL/TREATMENT OF CONTAMINATED DREDGED MATERIAL

M. John Cullinane, Jr., Daniel E. Averett, Richard A. Shafer
Clifford L. Truitt, and Mark R. Bradbury
U.S. Army Engineer Waterways Experiment Station

and

James W. Male, University of Massachusetts

ABSTRACT

As concern over dredging and disposal of contaminated sediments increases, unconfined open-water disposal of dredged material from harbors and navigation channels is closely scrutinized by state and local governments and numerous federal agencies. Although control of potential contaminant release from dredging and disposal of contaminated sediments is a relatively new concern, practical methods are available for handling such materials in an environmentally sound manner. Alternative technologies for dredging, transport, and disposal of contaminated dredged material are reviewed in this paper. Contaminant control/treatment during three basic operations are discussed: dredging, material transport, and disposal operations.

BACKGROUND

Because many contaminants become attached to sediment particles, concentrations in sediment are generally much greater than in water. As the concern over dredging and disposal of contaminated sediments increases, unconfined open-water disposal of dredged material from harbors and navigation channels is being closely scrutinized by state and local governments as well as numerous federal agencies. This paper presents recent concepts and technologies for handling contaminated dredged material.

CONTAMINANT CONTROL DURING DREDGING OPERATIONS

Dredge Selection

During dredging operations all dredge plants disturb bottom sediment, creating a plume of suspended solids around the dredging operation. Limitations may be placed on levels of suspended solids even during normal dredging operations (Lunz et al., 1984). Contaminated sediment may release contaminants into the water column through resuspension of the sediment solids, dispersal of interstitial water, or desorption from the resuspended solids. Control of sediment resuspension during dredging reduces the potential for release of contaminants and/or their spread to previously uncontaminated areas.

Selection of dredging equipment and method in general depends on the following factors: physical characteristics of material to be dredged, quantities of material, depth, distance to disposal area, physical environment of and between the dredging and disposal areas, contamination level of sediment, mobility of contaminants, method of disposal, production required, and type of dredges available. Dredging of contaminated sediments requires the additional consideration of contaminant loss during the extraction process and meeting of applicable criteria pertaining to removal efficiencies and/or environmental protection.

Different dredging methods appear more appropriate for certain contaminant classes. For volatile contaminants, mechanical dredges are likely to produce less loss than hydraulic dredges. Soluble contaminants can be removed more efficiently by a hydraulic dredge, but are difficult to control at the disposal site and treatment of the effluent water may be required.

Equipment and Operational Controls

Hopper Dredge Operation

The rate of solids loss in the overflow (which may determine if overflow is acceptable) will vary with amount of water in the hopper, hopper capacity and drainage characteristics, material characteristics (settleability), pumping rate, and elapsed time of overflow. Reduction of sediment resuspension can be accomplished by reducing the flow rate of the slurry being pumped into the hopper during the latter phases of the hopper-filling operation, reducing the solids concentration in the plume by reducing the sediment concentration in the overflow. By using this technique, the solids content of the overflow can be reduced by as much as 50 percent while the loading efficiency of the dredge is simultaneously increased. In extreme cases, pumping past overflow may be prohibited. Another approach is a submerged discharge system for hopper dredge overflow, called an antiturbidity overflow system (ATOS) (Ofuji and Naoshi, 1976).

Cutterhead and Suction Dredge Operation

Concentrations of suspended sediments from a cutterhead dredging operation range from 200 to 300 mg/liter near the cutterhead to a few mg/liter at 1,000-2,000 ft from the dredge. Resuspension of sediment during cutterhead excavation is dependent on the operating techniques used. The sediment resuspended by a cutterhead dredge depends on thickness of cut, rate of swing, and cutter rotation rate (Barnard, 1978). Proper balance of these operational parameters can decrease sediment resuspension while having little or no effect on production (Hayes et al., 1984). Modifications to cutterhead and suction dredges have improved their production capabilities and reduced dredged sediment resuspension. Greater production rates are achieved by pumping a higher solids concentration, reducing the quantity of return water that may be contaminated and require treatment. Recent modifications include matchbox heads, walking spuds, ladder pumps, flow and density instrumentation, underwater video and sensor equipment, shape of the cutterhead, and rake angle.

Dust Pan Dredges

Dust pan dredges are not well suited for dredging contaminated materials. However, when used in this application, the angle of the water jets on the head and the water pressure from these jets should be adjusted to achieve the minimum amount of sediment resuspension.

Special Purpose Dredges

Special-purpose dredging systems have been developing during the last few years in the United States and overseas to pump dredged material slurry with a high solids content and/or to minimize the resuspension of sediment. Most of these systems are not intended for use on typical maintenance operations; however, they may provide alternative methods for unusual dredging projects, such as contaminated sediments.

Clamshell Bucket Dredge Operations

Resuspension of sediments during clamshell dredging operations can be reduced by implementing operational controls and/or altering the bucket design. Operational controls can be applied to hoist speed, placement of the dredged material in the hopper barge, loading the hopper past overflow and dragging the bucket along the bottom. Equipment design includes the fit of the bucket and the use of enclosed clamshell buckets. Watertight buckets have been developed in which the top is enclosed so that the dredged material is contained within the bucket (Barnard, 1978). Comparisons between standard open clamshell bucket and a watertight clamshell bucket indicates that watertight buckets generate 30 to 70 percent less resuspension in the water column than the open buckets. The enclosed bucket did, however, produce increased resuspension near the bottom, due to a shock wave that precedes the watertight bucket.

Additional Control Techniques

Several additional techniques and/or considerations have been suggested to assist in controlling resuspension and contaminant release during dredging. Typical control techniques that are commonly evaluated include silt curtains; barriers, such as dikes, weirs, and sheet pile enclosures; and operational controls, such as dredging only during a specific time in the tidal cycle. Success with these has been varied and their application is very site specific.

CONTAMINANT CONTROL DURING TRANSPORT

Primary transportation methods used to move dredged material include pipelines, barges, scows, trucks, and rail. The primary emphasis during this phase of the overall dredging process is toward spill/leak prevention. Accidental release of contaminated materials into a previously uncontaminated environment has extremely costly consequences in monetary and public relations aspects. Thus, each step in the transport system must be carefully evaluated.

Controls for Pipeline Transport

Pipelines are commonly used to transport bulk materials over relatively short distances. During the design stage, planners should carefully consider pipeline routes, climatic conditions expected, corrosion resistance of the material, redundancy of safety devices (i.e., additional shutoff valves, loop/by-passes, pressure relief valves), coupling methods and systems to detect leaks. Souder et al. (1978) outlines specific pump and pipeline design procedures.

Controls for Scow/Barge Transport

Barge/scow transport of dredged material has historically been one of the most used methods to move large quantities over long distance. Controls to prevent spread of contaminated materials when utilizing barge transport are primarily concerned with loading/unloading procedures, fugitive emissions, route and navigation hazards, and decontamination of equipment.

Loading and unloading operations present the greatest potential for uncontrolled release of contaminated materials. Use of clamshell and dragline attachments at the dredging site will release substantially more dredged material into the water column than vacuum/suction systems. However, when planning for pumping dredged material into barges, planners should consider how the material will be transferred from the dredge onto a barge. Overflow during such operations can cause a significant return of contaminants to the water column. Flexible connections from dredge to barge will reduce the possibility of pipe damage due to wave action. If the dredged material is tremied into the barge, then movement of the boom between barges or dredge and barges must be carefully controlled to prevent material from falling directly into the waterway.

Controls for Truck/Rail Transport

Trucks are used for dredged material when the distance from the dredging site is beyond the range normally used for overland pipelines and less than the distance for rail car transport (> 50 to 100 mi). Controls associated with transporting dredged material by truck/rail parallel those for barge/scow transport. Primary concerns include weight restrictions, routing, and loss in transport, loading, and unloading operations. Loading and unloading operations present the greatest potential risk of contaminating nearby clean areas. Controls suggested for consideration are drainage of water from loading and unloading area into central sump for periodic removal, daily removal of spilled material, specially designed loading ramps to collect spilled material, use of watertight clamshells for transferring materials from barges into truck. Decontamination of truck/rail under carriages may be necessary to control contaminated materials from falling onto public roadways when leaving loading/unloading areas.

CONTAMINANT CONTROL DURING UPLAND/NEARSHORE DISPOSAL OPERATIONS

Six categories of contaminated media may be associated with the disposal of contaminated sediment. These include dredged material slurry, dredged material solids, site effluent, site runoff, site leachate (including flow-through dikes), and residual solids.

Upland disposal of contaminated dredged material must be planned to contain the dredged material within the site and restrict contaminant mobility out of the site in order to control or minimize potential environmental impacts. Francingues et al. (1985) identified and described five possible mechanisms for transport of contaminants from upland disposal sites:

1. release of contaminants in the effluent during dredging operations,
2. surface runoff of contaminants in either dissolved or suspended particulate form following disposal,
3. leaching into ground water and surface waters,
4. plant uptake directly from sediments, followed by indirect animal uptake from feeding on vegetation, and
5. animal uptake directly from the sediments.

TABLE 1 Site Characteristics Affecting the Need for Control/Treatment Technologies

Site area	Depth to aquicludes
Site configuration	Direction and rate of groundwater flow
Dredging method	
Climate (precipitation, temperature, wind, evaporation)	Existing land use
	Depth of groundwater
Soil texture and permeability	Ecological areas
Soil moisture	Drinking water wells
Topography	Receiving streams (lakes, rivers, etc.)
Drainage	
Vegetation	Level of existing contamination
	Nearest receptors

Site Control Strategies

Site Selection

Site location is an important, if not the most important, consideration in minimizing the cost of required restrictions. Selection of a technically sound site can reduce or eliminate the need for applying contaminant control/treatment technologies. Site characteristics that may affect the need for, or type of, treatment/control are listed in Table 1.

Covers

Covers are control measures designed to seal or isolate the surface of contaminated dredged material from physical, chemical, or biological processes that could release contaminants from a confined upland or nearshore disposal site. Surface covers can be as simple as a 1- to 3-ft thick layer of clean dredged material or as complex as a multilayer cap that includes impermeable membranes, filters, gas channels, biobarriers, and top soil. Functions of a cover could include one or more of the following:

- prevent or minimize surface water infiltration,
- promote aesthetics,
- reduce water erosion and dissolution of contaminants in surface water runoff,
- reduce wind erosion and fugitive dust emissions,
- contain and control gases and odors,
- provide a surface for vegetation and/or site reclamation, and
- prevent direct bioturbation (human and animal).

Since these functions address all of the migration pathways (i.e.,

TABLE 2 Normal Duration of Surface Water Diversion and Collection Measures

Technology	Duration of normal use
Dikes and berms	Temporary
Channels (earthen and CMP)	Temporary
Waterways	Permanent
Terraces and benches	Temporary and Permanent
Chutes	Permanent
Downpipes	Temporary
Seepage ditches and basins	Temporary
Sedimentation basins	Temporary
Levees	Temporary
Floodwalls	Permanent

SOURCE: U.S. Environmental Protection Agency (U.S. EPA, 1985).

surface water, groundwater, air, and direct contact), some type of surface cover will likely be a component of any upland or nearshore disposal system.

Surface-Water Controls

The overall objective of surface water controls is to minimize the volume of water that becomes contaminated via contact with the contaminated sediment. Surface-water controls accomplish this objective by preventing surface water runon from areas adjacent to the disposal site, by draining the disposal site efficiently to reduce infiltration and leachate generation, and by preventing erosion and sediment loss from the cover of the site. Surface-water controls also aid in collecting and transferring water that may be contaminated to treatment or disposal systems. Surface-water control methods are well established and are familiar to the engineering and construction industry. Lee et al. (1985a) provides a detailed discussion of management practices of U.S. Army Corps of Engineers (COE) construction sites. Table 2 lists surface-water control measures and their duration of use at disposal sites (U.S. EPA, 1985).

Groundwater Controls

Liners

Lining a site is a technique designed to contain leachate within the site and minimize groundwater contamination. A variety of liner materials are available for use in confined disposal operations. Soil liners are suitable for use as the only liner in most dredged material upland and nearshore sites. However, in certain upland applications, a combination of synthetic membrane and soil liner may be required to achieve maximum containment of contaminants. To ensure continued effectiveness of the liners whether soil or flexible membrane, they must be compatible with the dredged material and leachate they are to contain and be properly installed (Phillips et al., 1985).

Groundwater Recovery

Groundwater recovery technologies are usually considered as remedial actions where sites containing hazardous materials have released contaminants to the groundwater. Control of groundwater contamination involves one of four options:

1. containment of a plume;
2. removal of a plume after measures have been taken to halt the source of contamination;
3. diversion of groundwater to prevent clean groundwater from flowing through a source of contamination or to prevent contaminated groundwater from contacting a drinking water supply; or
4. prevention of leachate formation by lowering the water table beneath a source of contamination.

Ideally, adequate site investigation and installation of appropriate controls at a newly selected disposal site will prevent groundwater contamination and hence the need for groundwater controls. The reader is referred to other documents such as U.S. EPA 1982a and 1985 for more detailed information.

Leachate Collection

Disposal sites for dredged material must accommodate the interstitial water associated with the sediment, dilution water that may be mixed with the sediment by the dredging operation, and precipitation or other sources of water added to the disposal area surface. A leachate collection system is usually a network of perforated pipes placed under and around the perimeter of the site. The pipes drain to a sump or series of sumps from which the leachate may be withdrawn either by gravity, if topography allows, or by pumping. Spacing and sizing of the pipes depends on the allowable leachate head in the site and the rate at which water must be removed. Detail design of a collection system for leachate control is described in U.S. EPA, 1985.

Dewatering

Two mechanisms exist for dewatering and densifying fine-grained dredged material using pervious underdrainage layers: gravity underdrainage or vacuum-assisted underdrainage. The gravity underdrainage technique consists of providing free drainage at the base of the dredged material. Downward flow of water from the dredged material into the underdrainage layer takes place by gravity. Vacuum-assisted underdrainage is similar to gravity underdrainage, but a partial vacuum is maintained in the underdrainage layer by vacuum pumping.

Site Security

Any time contaminated sediment is being dredged, transported, or disposed, site security for the protection of safety and health of the public and of workers must be addressed. In addition to the time when the site is being filled, site security must be considered for the time after disposal is completed. The extent of security measures will depend on the nature and concentration of contaminants, the migration pathways affected by the contaminants, the risk to humans and wildlife, and future use of the site. For unusual conditions, where justified by the risk presented by the nature and location of the site, a site-specific safety plan may be developed in accordance with guidance presented in EM 1110-1-505 (U.S. Army COE, 1986).

Treatment of Dredged Material Slurries

Solids Separation and Classification Processes

The objective of separating solids from slurries is to attain two distinct waste streams: a substantially liquid waste stream that can be subsequently treated for removal of dissolved and fine suspended contaminants, and a concentrated slurry of solids and minimal liquid that can be dewatered and treated. The most appropriate solids separation method for a given site depends upon several factors, including the following:

- volume of contaminated solids;
- composition of sediment, including gradation, percent clay, and percent total solids;
- types of dredging or excavation equipment used, which determines the feed rate to solids separation and, in the case of slurries, the percent solids; and
- site location and surroundings.

Types of available solids separation equipment includes settling basins, clarifiers, impoundment basins, screens, and cyclones.

Solidification/Stabilization

Solidification and stabilization are terms which are used to describe treatment that accomplishes one or more of the following objectives (U.S. EPA, 1982b): improves waste handling or other physical characteristics of the waste, decreases the surface area across which transfer or loss of contained pollutants can occur, limits the solubility or toxicity of hazardous waste constituents. Methods involving combinations of solidification and stabilization techniques are often used (U.S. EPA, 1982b; Cullinane and Jones, 1985).

Thermal Destruction Processes

Thermal destruction is a treatment method that uses high temperature oxidation under controlled conditions to degrade a substance into products that generally include gases, vapors, and ash. The most common incineration technologies applicable to the treatment of dredged material slurries include rotary kiln, fluidized bed, and multiple hearth. Because of the cost of incineration and the extremely low fuel value of most dredged material slurries, it is doubtful that thermal destruction technologies would ever be an economically viable option for treating dredged material slurries. However, projects involving small volumes of highly contaminated material may be candidates for application of thermal destruction technologies.

Treatment of Dredged Material Solids

Dredged material solids are those solid materials remaining after initial or final dewatering of the dredged material slurry. Treatment of the dredged material solids can be accomplished before or after placement in a disposal area. Conceptually, dredged material solids can be treated with a variety of technologies. Among these are incineration, solidification/stabilization, extraction, immobilization, degradation, attenuation, and reduction of volatilization. Incineration, although a demonstrated technology for organics destruction is believed to be far too costly for the treatment of contaminated dredged material. In addition, the technology has limited application for treating dredged material solids contaminated with heavy metals. Solidification/stabilization technologies have been demonstrated at the field scale for hazardous wastes and at the laboratory scale for dredged material. However, this technology has not been proven for the containment of organics or in the marine environment. The remaining technologies are in various stages of development for application to hazardous waste sites and, although they may have some potential for application to dredged material solids, are many years away from being demonstrated technologies.

Treatment of Site Waters

A variety of physical, chemical, and biological processes have been developed for municipal and industrial water and waste treatment requirements. Many of these processes have potential in treating site waters generated by the disposal of contaminated dredged material at confined nearshore and upland disposal sites. However, few processes have actually been required or applied to dredged material disposal. Among the processes widely applied in confined disposal operations are plain sedimentation for solids and sediment-bound contaminant removal, and chemical clarification and filtration for enhanced removal of particulate (suspended solids) and sorbed metals and organics. Use of activated carbon for removal of soluble organics has received some limited application to dredged material. Other processes not previously applied to dredged material include organics oxidation, dissolved solids removal methods (e.g., distillation), and volatiles stripping. A comparison of the relative efficiencies of the treatment levels is given in Table 3.

TABLE 3 Contaminant Removal Efficiency of Water Treatment Levels[a]

Level	Class of contaminant	Percent removal	Water concentration remaining
I	Solids	99.9+	mg/liter range
	Metals	80 to 99+	ppb to ppm range[b]
	Organics	50 to 90+	ppb to ppm range[b]
II	Metals	99+	ppb range[b]
	Organics	50 to 90	ppb to ppm range[b]
III	Metals	99+	ppb range
	Organics	95+	ppb range
IV	Nutrients	90 to 98+	mg/liter range
V	Metals	99+	highest quality attainable
	Organics	99+	highest quality attainable
VI	Pathogens	90 to 99+	

NOTES:
[a] Assumes influent strength defined by dredged sediment that are not classifiable as "extremely hazardous waste" under RCRA (i.e., low saturation influents).
[b] Concentrations based on capability of best-available treatment technology.

Reuse of Contaminated Dredged Material

Reuse has been proposed as a potential alternative for long-term management of contaminated dredged material. Reuse of contaminated dredged material serves at least two beneficial functions: continued use of confined sites located close to dredging areas and creation of a potential construction material resource. The concept of a reuse alternative may also incorporate beneficial uses of materials such as sand and gravel reclaimed by classification/separation processes. The development and evaluation of reuse alternatives is extremely site specific and will depend on several factors: physical and chemical characteristics of the material to be dredged, availability of temporary storage and/or treatment sites, and identification of long-term disposal sites or suitable beneficial uses.

CONTAMINANT CONTROL/TREATMENT FOR RESTRICTED OPEN-WATER DISPOSAL

Restricted open-water disposal as used here simply suggests that one or more controls beyond those normally applied in conventional projects are required to address either known risks or uncertainties associated with disposal of contaminated sediments. Most positive control measures are based on the concept of isolating the contaminants from the water column or benthic environment. Recently, concepts based on either the separation of contaminants from the dredged material slurry or chemically stabilizing the contaminants in the dredged material have also been proposed.

Site Characteristics as a Control Technology

A level of increased control or restriction can be achieved during disposal simply by taking advantage of the best features of the site, by considering natural mixing processes, and by using conventional techniques and equipment to their best potential. At least six considerations can be identified that are important in evaluating the engineering acceptability of a proposed open-water disposal site: currents (velocity and structure), average water depths, salinity/temperature stratifications, bathymetry (bottom contours), dispersion and mixing, and navigation and positioning (location/distance, surface sea state, etc.).

Engineered Control Technologies

Submerged Discharge

The use of a submerged discharge or closed conduit of some type to place dredged material is a second level of restriction or control available. In general, a conduit is used primarily to ensure more accurate placement of the material and to reduce the exit velocity during formation of the surge phase. A conduit extending from the surface to the bottom will isolate the material from the water column during descent, reduce entrainment, and negate the effects of currents or stratifications. Submerged diffusers have been successfully field tested in the Netherlands at Rotterdam Harbor and as part of an equipment demonstration project at Calumet Harbor, Illinois (Hayes et al., 1984). The diffuser minimizes upper water column impacts, and especially improves placement accuracy, and controls sediment spreading, reducing benthic impacts.

Some hopper dredges have pump-out capability by which material from the hoppers can be discharged like a conventional hydraulic pipeline dredge. In addition, some have further modifications that allow pumps to be reversed so that material can be pumped down through the dredge's extended dragarms. Because of the expansion at the draghead, the result is similar to use of a diffuser section.

Lateral Confinement at the Site

An increased degree of positive control over the movement of the material placed at a site can be achieved by using lateral barriers to confine the disposed material. Such confinement can be accomplished by using depressions or contour irregularities existing at a site, by excavating such depressions, or by construction of subaqueous dikes. Lateral confinement addresses the short-term benthic impact by ensuring accurate initial placement and attenuation of the spreading dredged material. It also addresses long-term benthic and water column impacts by providing an inherent degree of isolation from the aquatic environment, reducing the effects of convective currents, and increasing the ease and effectiveness of capping when used.

Capping

Capping is simply the addition of a layer of some type of material over the mass of dredged sediment at the disposal site to effect isolation from the environment. The long-term impacts associated with soluble diffusion, convective transport, and bioturbation are reduced when a capping control measure is used. Physical stability of the disposal mass over time is also increased by capping, although short-term instability may be a concern if capping material is applied too rapidly over weak underlying dredged material.

Phillips et al. (1985), using the technologies discussed previously, described five conceptual designs for restricted open-water disposal sites: deep-water mound, deep-water confined, shallow-water mound, shallow-water confined, and waterway confined. The general features of these concepts are shown in Figure 1.

Dredged Material Treatment and Open-Water Disposal

Restricted open-water disposal is necessitated by the presence of contaminants associated with the sediment. On a mass basis, these

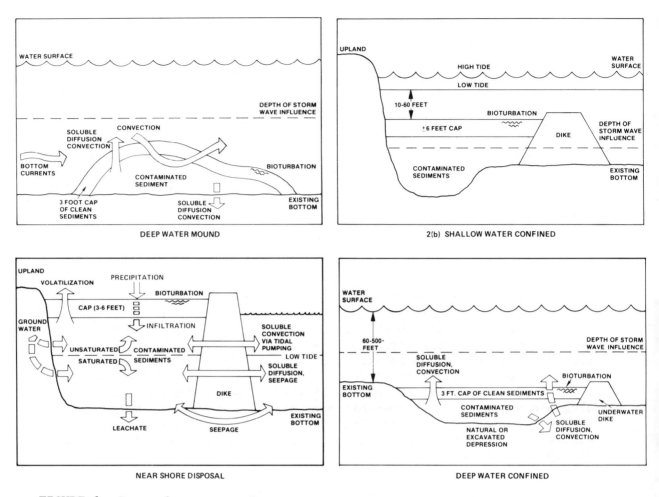

FIGURE 1 General concepts for restricted open-water disposal.

contaminants are a very small fraction of the total amount of dredged material. Recently, concepts based on the treatment of the dredged material followed by either unrestricted open-water disposal or open-water disposal with less stringent restrictions than would be applied to the untreated dredged material have been proposed. These concepts generally fall into three categories: separation of the contaminants from the dredged material, immobilization of the contaminants in the dredged material, or contaminant destruction.

Although treatment of the contaminated dredged material followed by unrestricted open-water disposal is an attractive concept, there have been no field-scale demonstrations. Floating and shore-based equipment is not readily available and the cost is uncertain.

SUMMARY

Although the control of potential contaminant release from the dredging and disposal of contaminated sediments is a relatively new concern, practical methods are available for handling such materials in an environmentally sound manner. The current state of the art for handling contaminated dredged materials is summarized below.

1. The short and long-term release of contaminants via various migration pathways from dredged material disposal sites cannot be ignored. Several techniques for predicting releases through specific pathways have been developed; however, the development of additional techniques and more information is needed to assess environmental effects and the need for implementing control/ treatment design features.
2. Control/treatment technologies are available and have been proposed for use at dredged material disposal sites. Beyond removal of suspended sediment from disposal area overflow, few technologies have been demonstrated for control/treatment of contaminated dredged material.
3. Design procedures for site water treatment technologies at upland and nearshore disposal sites are available and proven. Nearshore sites that involve saline waters present unusual, but not insurmountable, design problems.
4. A variety of site-control measures such as lining and capping have been developed for control of hazardous waste materials. Such control measures are not easily adaptable to conditions at a confined disposal site for dredged material. Placement of liners, particularly at nearshore sites, has not been sufficiently demonstrated. Dewatering of confined contaminated dredged material will require special equipment, treatment of site water, and a management plan for controlling contaminant release.
5. Procedures for designing restricted open-water disposal sites are not well developed. In particular, designs for submerged diffusers and downpipes for deep open-water sites have not been thoroughly developed and their implementation has not been documented. To date, the feasibility of implementing lateral confinement and capping in deep water has not been demonstrated. Projects are presently under design that will be used to demonstrate these technologies.
6. The selection of an appropriate control/treatment alternative depends on both site (dredging and disposal) and sediment characteristics. Because of site specificity and lack of experience in applying the available control/treatment alternatives to dredged material, no single alternative will emerge as the best alternative.
7. With the assurance of major cost increases, selection of control/ treatment alternatives for very highly contaminated dredged material could rely on technologies developed and being implemented for control of hazardous wastes, i.e., Resource Conservation and Recovery Act (RCRA) and Comprehensive Environmental Response, Compensation, and Liability Act (CERCLA) programs. A variety of proven and demonstrated technologies for disposal of low-level contaminated dredged material is also readily available.
8. A recurring limitation is the evaluation of alternative technical

feasibility, environmental effectiveness, costs interactions. Technical feasibility can only be addressed through the continued development and demonstration of new control/treatment technologies. The evaluation of environmental effectiveness will require analysis of the results obtained applying the proposed control/treatment technologies combined with the continued development of criteria against which the effectiveness of a control/treatment alternative can be evaluated. Procedures must be developed that enable planner or engineers to perform site-specific contaminant migration analysis. The costs of both the control/treatment alternatives and testing protocols are inadequately documented and are highly variable. Additional effort must be expended to refine the costs associated with controlling contaminant migration from contaminated dredged material disposal sites, evaluate the potential for contaminant migration, and assess the environmental impacts associated with contaminant migration.

ACKNOWLEDGMENTS

Funding for preparation of this paper was provided under the Dredging Operations Technical Support Program of the U.S. Army Corps of Engineers. Information and data presented in this paper were developed as part of the Puget Sound Dredged Disposal Analysis (PSDDA) segment of the Puget Sound Estuary Program sponsored by the U.S. Environmental Protection Agency, the Washington State Department of Ecology, and the Washington Department of Natural Resources. The lead agency for PSDDA was assigned to the U.S. Army Engineer District, Seattle. Permission to publish this information was granted by the Chief of Engineers.

REFERENCES

Barnard, W. D. 1978. Prediction and Control of Dredge Material Dispersion Around Dredging and Open-Water Pipeline Disposal Operation. Technical Report DS-78-13. Vicksburg, Miss.: U.S. Army Engineer Waterways Experiment Station.

Cullinane, M. J. and L. W. Jones. 1985. Technical Handbook for Stabilization/Solidification of Hazardous Waste. Cincinnati, Oh.: EPA Hazardous Waste Engineering Research Laboratory.

Cullinane, M. J., D. E. Averett, R. A. Shafer, J. W. Male, C. L. Truitt, and M. R. Bradbury. 1986. Guidelines for selecting control and treatment options for contaminated dredged materials requiring restrictions. Prepared for U.S. Army Engineer District, Seattle. U.S. Army Engineer Waterways Experiment Station, Vicksburg, Miss.

Francingues, N. R., Jr., M. R. Palermo, C. R. Lee, and R. K. Peddicord. 1985. Management strategy for disposal of dredged material: Contaminant testing and controls. Miscellaneous Paper D-85-1. U.S. Army Engineer Waterways Experiment Station, Vicksburg, Miss.

Hayes, D. F., G. L. Raymond, and T. N. McLellan. 1984. Sediment resuspension from dredging activities. In Dredging and Dredged Material

Disposal. New York: American Society of Civil Engineers. Pp 73-82.

JBF Scientific Corporation. 1978. An Analysis of the Functional Capabilities and Performance of Silt Curtains. Technical Report D-78-39. Vicksburg, Miss.: U.S. Army Engineer Waterways Experiment Station.

Keitz, E. L. and C. C. Lee. 1983. A profile of existing hazardous water incineration facilities. Proc. Ninth Ann. Research Symp. Incineration and Treatment of Hazardous Waste. EPA-600/9-83-003. Cincinnati, Oh.: EPA Industrial Environmental Research Laboratory.

Lee, C. C. 1983. A comparison of innovative technology for thermal destruction of hazardous waste. Proc. 1st Ann. Hazardous Materials Management Conference, July 12-14, 1983, Philadelphia, Pa.

Lee, C. R., J. G. Skogerboe, K. Eskew, R. A. Price, N. R. Page, M. Clar, R. Kort, and H. Hopkins. 1985a. Restoration of Problem Soil Materials at Corps of Engineers Construction Gates. Instruction Report EL-85-2. Vicksburg, Miss: U.S. Army Engineer Waterways Experiment Station.

Lee, C. R., R. K. Peddicord, M. R. Palermo, and N. R. Francingues. 1985b. Decision-making framework for management of dredged material: Application to Commencement Bay, Washington. Miscellaneous Paper D-85. (Draft). U.S. Army Engineer Waterways Experiment Station, Vicksburg, Miss.

Lunz, J. D., D. G. Clarke, and T. S. Fredette. 1984. Seasonal restrictions on bucket dredging operation. In Dredging and Dredged Material Disposal. New York: American Society of Civil Engineers. Pp. 371-383.

Monsanto Research Corporation. 1981. Engineering Handbook in Hazardous Waste Incineration. NTIS-PB81-248163. Springfield, Va.: National Technical Insformation Service.

Ofugi, I. and I. Naoshi. 1976. Antiturbidity Overflow System for Hopper Dredge. Proceedings, World Dredging Conference WODCON VII. Pp. 207-234.

Peddicord, R. K., C. R. Lee, S. Kay, M. R.. Palermo, and N. R. Francinques. 1985. General Decision-making Framework for Management of Dredged Material: Example Application to Commencement Bay, Washington. Miscellaneous Paper D-85. (Final Draft Report), prepared for State of Washington Department of Ecology, U.S. Army Engineer Waterways Experiment Station, Vicksburg, Miss.

Phillips, K. E., J. F. Malek, and W. B. Hammer. 1985. Commencement Bay Nearshore/Tide Flats Superfund Site, Tacoma, Washington: Remedial Investigations, Evaluation of Alternative Dredging Methods and Equipment Disposal Methods and Sites, and Sites and Treatment Practices for Contaminated Sediments. Seattle, Wash.: COE Seattle District.

Souder, P. S. Jr., L. Tobias, J. F. Imperial, and F. C. Mushal. 1978. Productive Use Concepts. Technical Report D-78-28. Vicksburg, Miss.: U.S. Army Engineer Waterways Experiment Station.

U.S. Army Corps of Engineers (COE), Office of Chief of Engineers. 1986. Guidelines for Preliminary Selection of Remedial Action for Hazardous Waste Sites. Engineer Manual EM 1110-2-505. Washington, D.C.: Headquarters, Department of the Army.

U.S. Environmental Protection Agency. 1985. Handbook: Remedial Action at Waste Disposal Sites (Revised). EPA/625/6-85/006. Cincinnati, Oh.: EPA

Hazardous Waste Engineering Laboratory and Office of Emergency and Remedial Response, Washington, D.C.

U.S. Environmental Protection Agency. 1984. Slurry Trench Construction for Pollution Migration Control. EPA-540-/2-84-001. Washington, D.C.: EPA Office of Emergency and Remedial Response and Municipal Environmental Research Laboratory, Incinnati, Oh.

U.S. Environmental Protection Agency (EPA). 1982a. Handbook for Remedial Action at Waste Disposal Sites. EPA-625/6-82-006. Cincinnati, Oh.: EPA Municipal Environmental Research Laboratory.

U.S. Environmental Protection Agency. 1982b. Guide to Disposal of Chemically Stabilized and Solidified Waste. SW-872. Washington, D.C.: EPA Office of Solid Waste and Emergency Response.

DEVELOPMENTS IN EQUIPMENT DESIGNED FOR HANDLING CONTAMINATED SEDIMENTS

John B. Herbich
Texas A&M University

ABSTRACT

Conventional dredging equipment typically handles large volumes of material in maintaining or deepening navigational channels. Such equipment may be operated in a modified procedure to handle relatively small volumes of contaminated material. However, in some cases, it may be more appropriate to use special purpose dredges (either specially developed, or adapted), which are more suitable for handling contaminated sediments. Several special purpose dredges are described and their capabilities discussed.

INTRODUCTION

The selection of proper dredging equipment for any project is important to achieve an efficient operation (Andrassy and Herbich, 1988). In the case of contaminated sediments, it is even more important since any additional contamination generated during dredging must be avoided. Selection depends on a number of factors:

1. characteristics of sediments,
2. quantity of sediments to be removed,
3. degree of contamination,
4. toxicity of contaminants,
5. location,
6. environmental conditions at the site (waves, currents, tides, etc.),
7. distance to the disposal site,
8. type of disposal, and
9. availability of particular equipment.

There are several types of dredges for conventional operations designed principally for moving large volumes of material efficiently. Conventional equipment operated in a modified procedure can be effective, as reported by Hayes et al. (1988); however, in some cases it may be more appropriate to use special purpose dredges (either developed or adapted) suitable for handling contaminated sediments.

There are several dredges that may be placed in a special purpose category:

1. mechanical--enclosed clamshell;
2. mechanical-hydraulic--Mud Cat, remotely controlled Mud Cat, and Clean-up system;
3. hydraulic--Refresher, waterless, matchbox, and wide sweeper, cutterless dredge; and
4. pneumatic--Pneuma and oozer.

MECHANICAL DREDGES

Enclosed Clamshell

The Japanese have developed a watertight clamshell for use with grab bucket dredges. An evaluation of the watertight bucket was made by the U.S. Army Engineer Waterways Experiment Station in 1982 (Figure 1). Experiments conducted at the Jacksonville District indicated that the watertight bucket significantly reduced water column turbidity and did not reduce production.

Figure 2 shows the benefit of using an enclosed bucket. Operation of the dredge can be modified slightly to reduce sediment resuspension by slowing the raising and lowering of the bucket through the water column. It must be noted that this operation modification reduces the production rate of the dredge, and generally high unit costs are associated with this type of mechanical dredging.

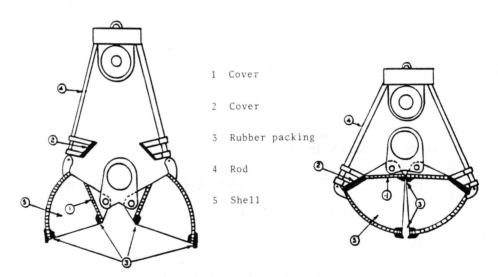

FIGURE 1 Open and closed positions of the watertight clamshell bucket. SOURCE: Hayes et al., 1984.

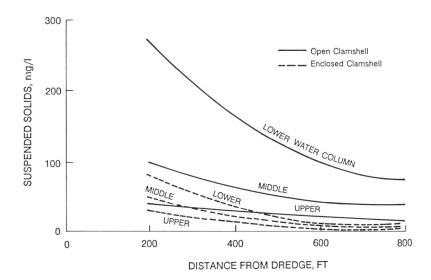

FIGURE 2 Resuspended sediment levels from open and enclosed clamshell dredge operations in the St. John's River. SOURCE: Hayes, 1986.

MECHANICAL HYDRAULIC DREDGES

Mud Cat

The Mud Cat has a horizontal cutterhead equipped with knives and spiral augers that cut the material and move it laterally toward the center of the augers where it is picked up by the suction (Figure 3). The dredge can remove sediments in a 2.6-m width and in water depths up to 4.9 m. The dredge operates on anchor cables, and the manufacturer claims that it leaves the bottom of the dredged area flat and free of windrows characteristic of typical cutterhead and hopper dredge operations.

FIGURE 3 The Mud Cat--notice that the cover is lifted to show two augers.

By covering the cutter-auger combination with a retractable mud shield the amount of turbidity generated by Mud Cat's operation can be minimized.

Remotely Controlled Mud Cat

A remotely controlled unit has been developed in which the control cab is located on land and is remotely connected to the Mud Cat by an umbilical cord. This allows dredging of hazardous or toxic materials. The remote control provides the shore-based dredge master with a variable traversing winch control, a variable auger control, a variable dredge pump speed control, and a manually controlled emergency shut down. The usual instrumentation is also displayed in the control cab and visual alarms are provided.

Clean-up System

To reduce or minimize resuspension of the sediment, Toa Harbor Works, Japan has developed a unique Clean-up system for dredging highly contaminated sediment (Herbich and Brahme, 1988; Sato, 1984). The Clean-up head consists of a shielded auger that collects sediment as the dredge swings back and forth and guides it toward the suction of a submerged centrifugal pump (Figure 4). To minimize sediment resuspension, the auger is shielded and a moveable wing covers the sediment as it is being collected by the auger. Sonar devices indicate the topography of the bottom. An underwater television camera also indicates the amount of material being resuspended during a particular operation. Figure 5 shows details of a shielded auger (Sato, 1984). Fairly large volumes (2.2 million m^3 up to 1981) have been excavated by Clean-up dredges in soft muds and sand containing various contaminants such as mercury, cadmium, PCBs, oily and organic substances. Table 1 summarizes the specifications of Clean-up dredges.

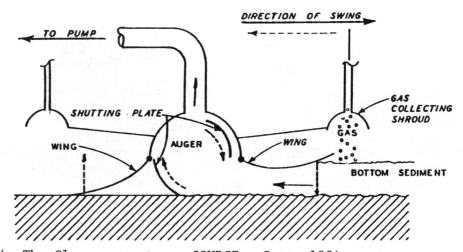

FIGURE 4 The Clean-up system. SOURCE: Sato, 1984.

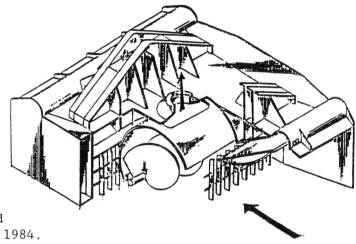

FIGURE 5 Clean-up shielded auger head. SOURCE: Sato, 1984.

Performance of various Clean-up dredges between 1973 and 1981 is summarized in Appendix A.

HYDRAULIC SUCTION DREDGES

Refresher

The Refresher dredge was developed purposely for removal of contaminated materials by a Penta-Ocean Construction Company, Ltd. (Shinsha, 1988). The dredge material is confined by a specially designed flexible enclosure that completely covers the cutter, preventing escape of sediments to the outside of the immediate dredging area (Figure 6). The working open section is always on the swing side of the cutterhead. A gas removal system is also installed and can be activated as needed to prevent gas moving up the section pipe. The flexible enclosure of the cutterhead is automatically adjusted to bottom contours.

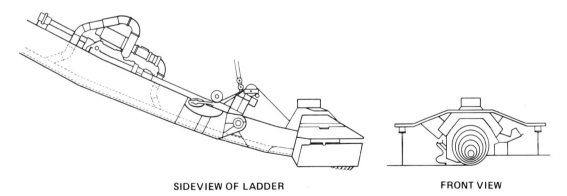

FIGURE 6 Refresher dredge. SOURCE: After Shinsha, 1988.

General Specifications

The Refresher dredge is equipped with the main pump and an additional pump on the ladder to provide a high level of production. Automatic valves in the suction pipe prevent sediment-water mixture from flowing back in case of power failure (Figure 7).

Refresher No. 3 and Mini-Refresher *Tokyo Maru* specifications are given in Table 2. Pump specifications for both the ladder pump and the main pump are given in Table 3.

Waterless Dredge

Waterless Dredging Company developed a dredging system in which the cutter and centrifugal pump are enclosed within a half-cylindrical shroud. By forcing the cutterhead into the material, the cutting blades remove the sediment near the front of the cutterhead with little entrainment of water. According to the manufacturer, this waterless system is capable of pumping slurry with a solids content of 30 to 50 percent by weight with little generation of turbidity. The dredge pipeline sizes range from 15 to 30 cm. The waterless dredge development is relatively new and and experience with it is quite limited.

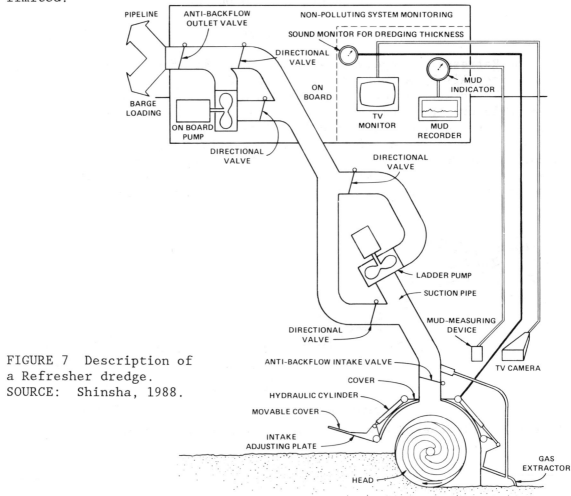

FIGURE 7 Description of a Refresher dredge.
SOURCE: Shinsha, 1988.

TABLE 1 Specifications of Clean-up Dredge

Dredge	Type	Dimensions (m) L x B x D x d	Main pump	Booster pump	Pump suction capacity (m/hr)	Maximum discharge distance (m)	Maximum dredging depth (m)	Minimum dredging depth (m)	Remarks
Clean-up No. 1	DE[a]	26.9x10.1x1.2x 0.4	Centrifugal pump 100 PS	—	500	1,000	Under surface of water 6.2	1.5	—
Clean-up No. 2	DE	36.0x11.0x3.2x 2.0	Centrifugal oil pump 147 PS	Centrifugal electric pump 253 PS	0–1,500	1,500	Under surface of water 23	2.5	Discharge variable pump with barge loading equipment
Clean-up No. 3	DE	42.5x13.4x3.3x 1.9	Centrifugal oil pump 147 PS	Centrifugal oil pump 147 PS	0–2,000	1,500	Under surface of water 23	3.0	Discharge variable pump with barge loading equipment
Clean-up No. 5	DE	21.5x8.0x1.5x 0.7	Centrifugal oil pump 50 PS	—	0– 500	500	Under surface of water 11	1.5	Discharge variable pump pump
Clean-up SIRSI	D[b]	35.0x9.7x2.4x 1.6	PNEUMA PUMP 300/60 Type	—	300	1,200	Under surface of water 15	3.5	Equipped with barg loading system

NOTES:
[a] DE = Diesel electric
[b] D = Diesel.

SOURCE: Sato, 1984.

TABLE 2 General Specifications for the Refresher Dredge.

GENERAL SPECIFICATIONS		
	REFRESHER NO. 3	MINI REFRESHER TOKYO MARU
SHIP DIMENSIONS (LxBxD)	45.60x13.50x3.30 m	17.00x6.40x1.35 m
DISPLACED TONNAGE	1,200 t	80 t
DISPLACED DEPTH	2.20 m	0.80 m
DISCHARGE DISTANCE	BARGE LOADING ~3.000m	BARGE LOADING ~500 m
DREDGING CAPACITY	150 ~ 400 m^3/h	50 ~ 120 m^3/h
MAXIMUM DREDGING DEPTH	20 m	7.5 m

SOURCE: After Shinsha, 1988.

TABLE 3 Pump Specifications for Refresher No. 3 and Mini-Refresher *Tokyo Maru* Dredges.

PUMP SPECIFICATIONS			
		REFRESHER NO. 3	MINI REFRESHER TOKYO MARU
LADDER PUMP	MODEL	CENTRIFUGAL TYPE	CENTRIFUGAL TYPE
	RATED HORSEPOWER	150PS	100PS
	PUMPING CAPACITY	800 ~ 2,000 m^3/h	250 ~ 600 m^3/h
	PUMPING HEIGHT	10 ~ 20m	8 ~ 15 m
ON-BOARD PUMP	MODEL	CENTRIFUGAL TYPE	CENTRIFUGAL TYPE
	RATED HORSEPOWER	1,800PS	150PS
	PUMPING CAPACITY	800 ~ 2,000 m^3/h	250 ~ 600 m^3/h
	PUMPING HEIGHT	20 ~ 50	10 ~ 35m

SOURCE: After Shinsha, 1988.

Matchbox Dredge

A special suction head was developed by a dredging contractor in the Netherlands to replace the traditional cutterhead (d'Angremond et al., 1984). The main design points are as follows (Figure 8):

1. A large plate covers the top of the dredge head to avoid inflow of water and escape of gas bubbles.
2. Adjustable angle between the drag head and the ladder to create an optimum position of the drag head independent of the dredging depth.
3. There are openings on both sides of the drag head to improve dredging efficiency. During swinging action the leeward side is closed to prevent water inflow.
4. Dimensions of the head must be carefully designed for the average flow rate and swing rate (Figure 9).

A diffuser may be installed at the submerged end of the discharge pipe to reduce the dispersion of fine sediment in the water column (d'Angremond et al., 1984; Neal et al., 1978). By its gradually widening cross section, the flow could decelerate to an acceptable velocity to reduce turbulence. Outflow velocities are designed to be between 0.2 to 0.3 m/sec (Figure 10); however, it is unlikely that contaminated material would be discharged in open water. A possible application may be to employ such a diffuser in a containment area. A degassing system is also installed to prevent or reduce the amount of gas moving up the suction pipe.

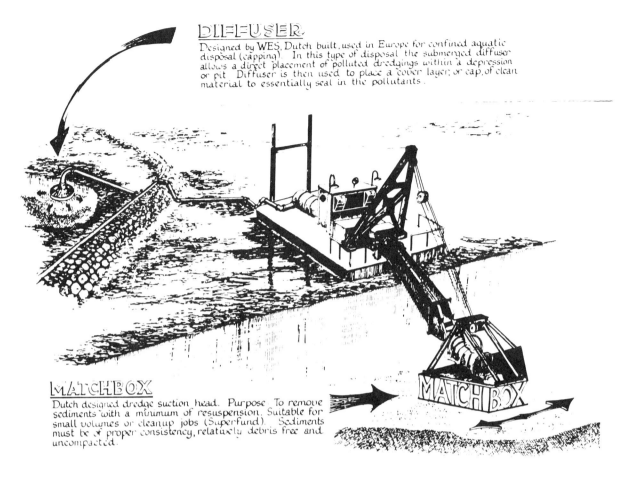

FIGURE 8 Dutch Matchbox dredge. SOURCE: After U.S. Army Engineer District, Chicago.

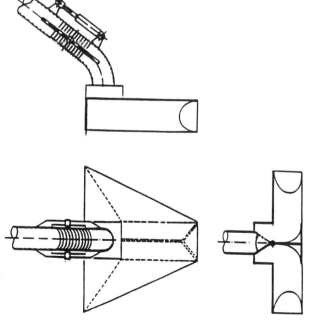

FIGURE 9 Schematic of Matchbox suction head. SOURCE: After d'Angremond et al., 1984.

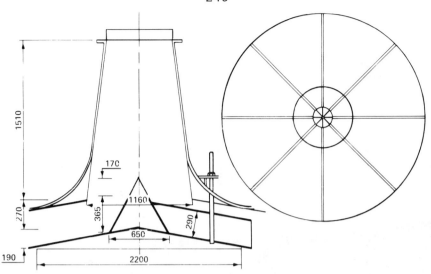

FIGURE 10 Schematic of diffuser (all dimensions are in millimeters).
SOURCE: d'Angremond et al., 1984.

TABLE 4 Plume Area for 10-mg/liter Contour for
the Cutterhead, Clamshell, and Matchbox Dredges.

Depth percent	Cutterhead acres	Clamshell acres	Matchbox acres
5	0	1.7	0
50	0	1.8	0
80	0	---	0.4
95	1.2	3.5	2.95

SOURCE: Hayes et al., 1988.

A direct comparison between a Matchbox suction head and a conventional cutterhead was made by the Waterways Experiment Station in Calumet Harbor (Hayes et al., 1988). The Matchbox was specifically designed to be fitted on the ladder of the U.S. Army Corps of Engineers' dredge *Dubuque*. The Calumet Harbor demonstration indicated that the clamshell dredge generated the largest suspended sediment plume affecting the entire water column. The cutterhead slightly outperformed the Matchbox dredge as shown in Table 4.

Wide Sweeper Cutterless Dredge

Wide Sweeper No. 6 hydraulic suction dredge does not have a cutter and is principally employed for removal of contaminated materials without resuspending the sediment particles (Shinsha, 1988). The main features of Wide Sweeper are that

1. bottom sediments can be removed essentially without resuspension of particles,
2. acoustic sensors determine the characteristics of sediment to be dredged,
3. the suction head can follow seabed configuration to some extent, or can be kept in a horizontal position,
4. turbidity generated is monitored by a television camera, and
5. the dredge is equipped with a ladder pump and a main pump.

General specifications are shown in Table 5.
Figure 11 shows the general arrangement of the suction head.

TABLE 5 General Specifications of Wide Sweeper No. 6

Name of Ship	Type	Hull Dimensions	Main Pump	Ladder Pump	Including
"Wide Sweeper No. 6"	Diesel Electric	Length: 58.2m Breadth: 14m Depth: 3.7m Draft: 2.3m	3,200 PS, Single-stage, Single-suction. Centrifuge type	950 PS, Single-stage. Single-Suction. Centrifuge type	Sludge observation system Operation control system

SOURCE: Shinsha, 1988.

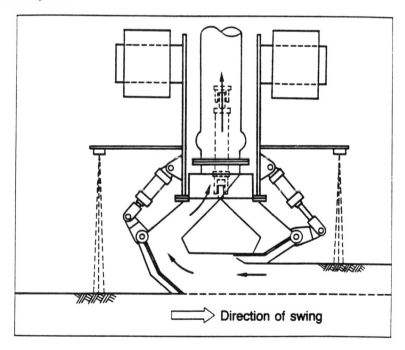

FIGURE 11 General arrangement of the suction head on dredge Wide Sweeper No. 6. SOURCE: Shinsha, 1988.

PNEUMATIC DREDGES

Pneuma Pump

The Pneuma pump is a compressed-air-driven, displacement-type pump with several major components (Herbich, 1975). The pump body (Figure 12), the largest of these components in dimensions and weight, incorporates three large cylindrical pressure vessels, each having a material intake on the bottom and an air port and discharge outlet on top. Each intake and discharge outlet is fitted with a check valve, allowing flow in one direction only. Pipes leading from the three discharge outlets join in a single discharge directly above the pressure vessels. Different types of attachments may be fitted on the intakes for removal of varying types of bottom material. The Pneuma system was the first dredging system to use compressed air instead of centrifugal motion to pump slurry through a pipeline. It has been used extensively in Europe and Japan. According to the literature published by the manufacturer, this system can pump slurry with a relatively high solids content with little generation of turbidity. Pneuma pump was evaluated by the U.S. Army Engineer Waterways Experiment Station (Richardson et al., 1982). The operation principle of the pump body is illustrated in Figure 12.

The system is based on the principle of employing static water head and compressed air. During the dredging process the pump is submerged and sediment and water are forced into one of the empty cylinders through an inlet valve. After the cylinder is filled, compressed air is forced into the cylinder closing the inlet valve and simultaneously forcing the material out of an outlet valve and into a discharge line. When the cylinder is empty, the air pressure is reduced to atmospheric pressure, the outlet valve closes and the inlet valve opens. The two-stroke cycle is then repeated. The distribution system controls the cycling phases of all three cylinders so there is always one cylinder operating in the discharge mode. The system has been used in water depths up to 50 m.

Oozer Dredge

The oozer pump was developed by Toyo Construction Company, Japan. The pump operates in a manner similar to that of the Pneuma pump system; however, there are two cylinders (instead of three) and a vacuum

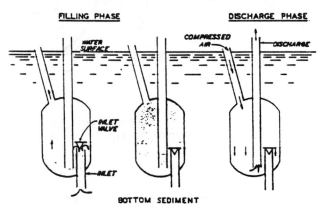

FIGURE 12 The pneumatic pump system. SOURCE: Herbich and Brahme, 1988.

is applied during the cylinder-filling stage when the hydrostatic pressure is not sufficient to rapidly fill the cylinders. The pump is usually mounted at the end of a ladder and equipped with special suction heads and cutter units depending on the type of material being dredged. The conditions around the dredging system, such as thickness, bottom elevation after dredging, and amount of resuspension, are monitored by high-frequency acoustic sensors and an underwater television camera. A large oozer pump has a dredging capacity ranging from 300 to 500 m^3/hr. During one dredging operation, suspended solids levels within 3 m of the dredging head were all within background concentrations of less than 6 mg/liter. Figure 13 is a sketch of the oozer dredge; Figure 14 describes the ooze dredging system DREX, consisting of a suction mouth and a device that permits a back and forth movement of the suction mouth. This modified system is said to increase solids concentration up to 60 percent.

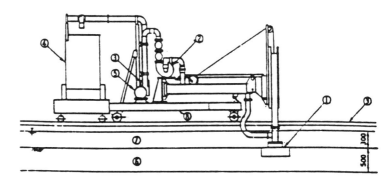

FIGURE 13 Outline of oozer dredge (dimensions are in millimeters).
SOURCE: Herbich and Brahme, 1988.

1 suction mouth   2 pump   3 magnetic flow-meter
4 ooze collecting tank   5 driving DC motor   6 test soil
7 clear   8 carriage   9 rail

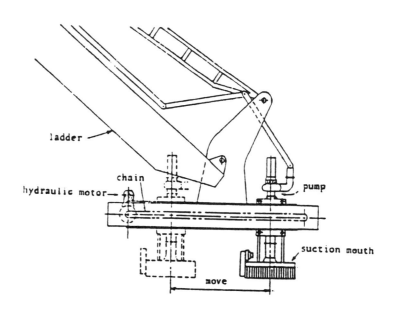

FIGURE 14 Sketch of oozer dredging system.
SOURCE: Herbich and Brahme, 1988.

The main features of the oozer dredge are as follows:

1. The dredge can effectively remove contaminated sediments from a maximum depth of 18 m.
2. Since the swing speed can be adjusted from 0 to 20 m/min, the dredge can be effective in removing suspended sediments.
3. Five acoustic sediment sensors can measure the bearing pressure of sediment to be removed and the thickness of various sediment layers.
4. Underwater television cameras monitor the presence of turbidity near the suction intake.
5. Toxic gases released during dredging pass through gas scrubbers to remove toxic content before gases are released to the atmosphere.
6. A screen is located at the suction mouth to prevent large objects from entering. Double-suction valves and electrically controlled check valves provide secondary protection.
7. Oozer dredges can, under ideal conditions, pump sediments at in situ density.
8. Different cutters and suction heads are available for dredging sediments ranging from clay to sand.

Specifications for oozer dredge *Taian Maru* are given in Table 6 and her performance between 1974 and 1980 is listed in Table 7.

RESUSPENSION LEVELS OF SEDIMENT FOR SPECIAL PURPOSE DREDGES

The special purpose dredges that appear to have the most potential in limiting resuspension are shown in Table 8.

SUMMARY

Conventional dredging equipment may be operated in a modified procedure to handle contaminated material. A decision about whether to use modified equipment should be made on economic grounds. Several special purpose dredges were developed, principally overseas, and have been successfully employed in removal of contaminated sediments. Capabilities of mechanical, mechanical-hydraulic, and hydraulic suction dredges are summarized in Tables 9, 10, and 11. Capabilities of pneumatic dredges are shown in Table 12.

ACKNOWLEDGMENTS

This symposium paper was reviewed by Mr. Charles C. Calhoun, Jr., Assistant Chief, Coastal Engineering Research Center (CERC), U.S. Army Engineer Waterways Experiment Station, and by Dr. Cliff L. Truitt, CERC. Their comments were appreciated.

TABLE 6 Specifications of Oozer Dredge *Taian Maru*

Hull:

    Overall length .... 37 m
    Breadth .... 12 m
    Depth .... 3 m
    Draft .... 2.2 m
    Dredging depth .... 17 m

Engine:

    Oozer pump 1

        Type: Cylindrical twin-barrel, negative pressure suction and positive pressure discharge.

    Dredging capacities (Pressure intensity: $7 kg/cm^2$)

Density (%)	Discharge distance (m)	Pumping production (m^3/h)	Dredging production (m^3/h)
60	100	580	350
60	500	500	300
60	1,000	420	250

    Discharge pipe: 450 A

    Air compressors 3

        Type: Screw rotary system
        Capacity: $34.2 m^3/min$ x $7 kg/cm^2$ x 1,770 r.p.m.
        Driving generator: 190 KW x AC 440 V x 60 Hz x 4P

    Vacuum pump 1

        Type: Roots system
        Capacity: -400 mmHg x $44.8 m^3/min$
        Driving generator: 110 KW x AC 440 V x 60 Hz x 6P

    Main generator

        Type: Horizontal drip-proof rest self-excite
        Capacity: 450 KVA x 3 0 x 60 Hz x 445 V x 8P
        Continuous output: 530 ps x 900 r.p.m.

    Winch

        Ladder winch: 12T x 24 m/min x 75 KW x 6P..... 1
        Swing winch: 15T x 0-12 m/min x 70 KW......... 1
        Spud winch: 12T x 17 m/min x 50 KW x 6P....... 1

TABLE 7 Performance of Oozer Dredge *Taian Maru*

Place	Construction Period	Soil Condition	Undisturbed Moisture Content (%)	Deleterious Material	Volume of Dredging Soil (m^3)	Treatment
Iyo-Mishima Ehime Pref.	April-May 1974	Sandy silt	144	Pulpwood	30,000	Direct discharge length 356 m; Natural sedimentation; No treatment of return flow water
-Do-	July 1974	-Do-	400 - 800	-Do-	27,000	Direct discharge length 356 m 1000 m; Natural sedimentation; No treatment of return flow water
Chiba Port Chiba Pref.	January-June 1975	Silt	300 - 400	Hg, Pb	40,800	Transport with barge; Secondary pumping with centrifugal pump; Natural sedimentation
Takasago Port Hyogo Pref.	August-December 1975	Sandy silt	150 - 200	PCB	224,816	Direct discharge length 300-1,200 m; Solidity; Removal of deleterious material
Sakaide Port Kagawa Pref.	July-September 1975	Silt	90 - 100	Hg	14,000	Direct discharge length 460 m; Natural sedimentation; Removal of deleterious material
Iwakuni Port Yamaguchi Pref.	January-February 1976	Sandy silt	100 - 200	Pulpwood	31,000	Direct discharge length 300-800 m; Natural sedimentation; Removal of deleterious material
Tokuyama Port Yamaguchi Pref.	July-December 1976	Sandy silt	50 - 300	Hg	128,160	Transport with barge; Natural sedimentation; Removal of deleterious material
-Do-	December 1976-March 1977	Sandy silt	50 - 300	Hg, C_6H_{14}	82,000	Transport with barge; Natural sedimentation; Removal of deleterious material
Yokkaichi Port Mie Pref.	April 1977-February 1978	Sandy silt	280 - 500	Hg	655,000	Discharge with booster pump; Natural sedimentation; Removal of deleterious material
Mizushima Port Okayama Pref.	July-September 1978	Silt	40 - 100	C_6H_{14}	92,800	Transport with barge; Natural sedimentation; Removal of deleterious material
Osaka Bay Hyogo Pref.	March 1980	Silt	150 - 250	Organics	6,000	Pilot works
Yokkichi Port	June 1980	Silt	150	-Do-	44,000	Excavation of sea bed

TABLE 8 Resuspended Sediments by Special Purpose Dredges

Dredge	Reported suspended sediment concentrations[a]
Pneuma pump	48 mg/liter, 3 ft above bottom
	4 mg/liter, 23 ft above bottom (16 ft in front of pump)
Clean-up system	1.1 to 7.0 mg/liter above suction
	1.7 to 3.5 mg/liter at surface
Oozer pump	Background level (6 mg/liter), 10 ft from head
Refresher system	4 to 23 mg/liter, 10 ft from head

NOTES:
[a]Suspended solids concentrations were adjusted for background concentrations.

SOURCE: Herbich and Brahme, 1988.

TABLE 9 Summary Table of Mechanical Dredges

Type	Production	Depth limitation	Resuspension of sediment	Comments
Open clam-shell bucket	Low	30-40 ft	High	
Watertight clamshell bucket	Low	30-40 ft	Low	Experiments conducted in the St. John's River

TABLE 10 Summary Table of Mechanical-Hydraulic Dredges

Type	Production	Depth limitation	Resuspension of sediment	Comments
Mud Cat	Moderate	15 ft	Low to moderate	Extensively used
Remotely controlled Mud Cat	Low	15 ft	Low to moderate	New development
Clean-up system	Moderate	70 ft	Low to moderate	Extensively used in Japan

TABLE 11 Summary Table of Hydraulic Suction Dredges

Type	Production	Depth limitation	Resuspension of sediment	Comments
Refresher	Moderate to high	60-115 ft	Low	Extensively used in Japan
Waterless	Moderate		Low	Limited experience
Matchbox	Moderate to high	85 ft	Low	Experiments conducted at Calumet Harbor
Wide Sweeper	Moderate	100 ft	Low	Used in Japan

TABLE 12 Summary Table of Pneumatic Pumps (Dredges)

Type	Production	Depth limitation	Resuspension of sediment	Comments
Pneumatic	Low to moderate	+100 ft	Low	Evaluated by COE Waterways Experiment Station
Oozer	Moderate to high	59 ft	Low	Used extensively in Japan

REFERENCES

Andrassy, C., and J. B. Herbich. 1988. Generation of suspended sediment at the cutterhead. The Dock and Harbour Authority 68(797):207-216.

d'Angremond, K., A. J. de Jong, C. P. de Waard. 1984. Dredging of polluted sediment in the first petroleum harbor, Rotterdam. Proc. 3rd U.S.-the Netherlands meeting on Dredging and Related Technology. Fort Belvoir, Va.: U.S. Army Engineer Water Resources Support Center.

Hayes, D. F., G. L. Raymond, and T. N. McLelland. 1984. Sediment resuspension from dredging activities. Dredging '84. Clearwater, Fla.: American Society of Civil Engineers.

Hayes, D. F. 1986. Guide to selecting a dredge for minimizing resuspension of sediment. Environmental Effects of Dredging, Technical Notes, EEDP-09-1. U.S. Army Engineer Waterways Experiment Station, Vicksburg, Miss.

Hayes, D. F., T. N. McLelland, and C. L. Truitt. 1988. Demonstration of innovative and conventional dredging equipment at Calumet Harbor, Illinois. MP EL-88-1. U.S. Army Engineer Waterways Experiment Station, Vicksburg, Miss.

Herbich, J. B. 1975. Coastal and Deep Ocean Dredging. Houston, Tex.: Gulf Publishing Company.

Herbich, J. B. and S. B. Brahme. 1988. A Literature Review and Technical Evaluation of Sediment Resuspension During Dredging. TR-88 (in press). Vicksburg, Miss.: U.S. Army Engineer Waterways Experiment Station.

McLellan, T. N., R. N. Davis, and D. F. Hayes. 1988. Field studies of sediment resuspension characteristics of selected dredges. TR HL-88-(in press). Vicksburg, Miss.: U.S. Army Engineer Waterways Experiment Station.

Neal, R. W., G. Henry, and S. H. Greene. 1978. Evaluation of the Submerged Discharge of Dredged Material Slurry During Pipeline Dredge Operations. TR D-78-44. Vicksburg, Miss.: U.S. Army Engineer Waterways Experiment Station.

Richardson, T. W., J. E. Hite, R. A. Shafer, and J. D. Ethridge. 1982. Pumping Performance and Turbidity Generation of Model 600/100 Pneuma Pump. TR HL-82-8. Vicksburg, Miss.: U.S. Army Engineer Waterways Experiment Station.

Sato, E. 1984. Bottom sediment dredge CLEAN UP. Principles and results, management of bottom sediments containing toxic substances. Proc. 8th U.S./Japan Experts Meeting, T. R. Patin, ed. Vicksburg, Miss.: U.S. Army Engineer Waterways Experiment Station. Pp. 403-418.

Shinsha, H. 1988. Personal Communication. Refresher Dredge, Technical and Research Institute, Penta-Ocean Construction Company, Ltd., Japan.

Yamaguchi, A. 1988. Personal Communication. Kumamoto Prefectural Government, Kumamoto City, Japan.

APPENDIX A TABLE A-1 Performance of Clean-up Dredges, 1973-1981[1]

List of Dredging Work and Removal of Bottom Sediments

No.	Client	Description of Works	Contract Period From	Contract Period Until	Dredge Employed	Area (m^2)	Volume (m^3)	Depth (m)	Thickness (m)	Soil Characteristics	Discharge Distance (m)	Pumping Volume (m^3/hr)	Density (%)	Dredged Volume (m^3/hr)	Ooze Treatment
1	Ministry of Transport Port Construction Div. 4	Experiment to invent a special self-propelled dredging barge	Feb. 1973	-	Cleanup SIRSI	1,000	1,000	9.0	1.0	Silt	80	266	41.6	111	Dredging with cleanup SIRSI; transportation with box barge; hydro extraction treatment
2	Kanegafuchi Chemical Industry	Dredging at Takasago East Port	Sep/27 1973	Oct/20 1973	Cleanup SIRSI Cleanup by night only 18:00-6:00(next day)	19,000	11,240	-4.5 -5.0	0.5 1.0	Upper layer: silt lower layer: clay or sand	1,500	206	25.7	53	Dredging with cleanup SIRSI; settling basin; wastewater treatment; transportation with discharge pipe; solidification treatment
3	Japan Aquatic Resources Protection Association	Experiment for feasibility study on protection works against red water for the year 1973	Feb/17 1974	Feb/25 1974	Cleanup SIRSI	2,400	473	-10.0 -11.0	0.22	Organic dirty soils	Offshore discharge with box barge	197	19.3	38	Dredging with cleanup SIRSI; transportation with box barge; thickener; pumping
4	Mitsubishi Mining & Smelting	Reclamation at Nishiyama port	May/1 1974	Oct/31 1974	Cleanup SIRSI	46,200	58,000	-4.0 -5.0	1.25	Fine sand with silt	300	260	8.8	23	Dredging with cleanup SIRSI; reclaimed area; disposal through buffer tank method; transportation with discharge pipe
5	Ohtsu Paperboard Mfg.	Dredging & removal of ooze for Ohtsu Paperboard Mfg.	May 1974	Oct 1974	Cleanup No. 1	140,00	23,700	1.5 4.5	0.17	Pulp & sludge	887	370	20	74	Cleanup No. 1; hydro-extraction treatment; disposal area; transportation with discharge pipe & solidification treatment
6	Japan Aquatic Resource Protection Association	Experiment for feasibility study on protection works against red water for the year 1974	Oct/18 1974	Nov/24 1974	Cleanup No. 2	12,390	3,717	-3.0	0.30	Organic dirty soils	250	159	22	35	Cleanup No. 2; filtration treatment; disposal pond; transportation with discharge pipe; hydro-extraction treatment; sand carrier with grab bucket
7	Yokka-Ichi Port Administration	Experiment for anti-polution measures at Yokka-Ichi Port	Mar/5 1975	Mar/11 1975	Cleanup No. 2	3,432	2,712	-12.0	0.80	Oily dirty soils	80	600	47.7	286	Cleanup No. 2; transportation with sand carrier with grab bucket; dirty soils reservoir
8	Sumitomo Metal Co., Ltd.	Removal of sludge near No. 8 drainage outlet	Mar/8 1975	Mar/24 1975	Cleanup No. 2	10,500	5,854	-12.0	0.56	Dirty soils with tar	Offshore Discharge with box barge	500	20	100	Cleanup No. 2; box barge; disposal area within the premises
9	Okayama Pref.	Dredging (-5.5m) in front of slipway at Tamashima Port	Mar 1975	Apr 1975	Cleanup No. 2	12,125	18,000	-5.5	1.50	Silty sand	1,500	500	10	50	Cleanup No. 2; transportation with discharge pipe; reclaimed area
10	Japan Aquatic Resources Protection Association	Experiment for feasibility study on protection works against red water for the year 1975	Aug/25 1975	Sep/30 1975	Cleanup No. 2	5,650	3,140	-2.5	0.70	Organic dirty soils	450	194	29.9	58	Cleanup No. 2; filtration treatment; disposal area hydro-extraction treatment; sand carrier with grab bucket

No.	Client	Description of Works	Contract Period From	Contract Period Until	Dredge Employed	Area (m²)	Volume (m³)	Depth (m)	Thickness (m)	Characteristics	Distance (m)	Pumping Volume (m³/hr)	Density (%)	Dredged Volume (m³/hr)	Ooze Treatment
11	Yamagata Pref.	Pollution prevention dredging at Sakata Port (Area No.2)	Sep/16 1975	Jul/31 1976	Cleanup No. 1	13,675	22,200	-6.0	1.50	Dirty soils with mercury (silt)	870	330	14.8	49	Cleanup No. 1; disposal pond; wastewater treatment through natural sedimentation method
12	The Tokyo Metropolis	Dredging of dirty soils at Takahama Canal for 1976	Dec/22 1975	Mar/31 1976	Cleanup No. 5	2,793	2,100	-3.5	0.75	Silty organic dirty soils	130	300	32.5	98	Cleanup No. 5; tank barge; disposal area
13	Nippon Mining	Dredging & removal of dirty soils at Sagano Seki Port for 1976	May/10 1976	Oct/9 1976	Cleanup SIRSI	58,500	30,700	-5.0	0.50	Silt or sand	485	383	40	153	Cleanup SIRSI; sedimentation pit; wastewater treatment
14	Yamagata Pref.	Pollution prevention dredging at Sakata Port (Area No. 2)	Sep/16 1975	Jul/20 1976	Cleanup No. 1	9,118	14,800	-6.0	1.50	Silty dirty soils with mercury	870	330	14.8	49	Cleanup No. 1; disposal area; wastewater treatment through natural sedimentation method; transportation with discharge pipe
15	Yamagata Pref.	Pollution prevention dredging at Sakata Port (Area No. 2)	Sep/16 1975	Jul/20 1976	Cleanup No. 1	7,207	11,700	-6.0	1.50	Silty dirty soils with mercury	870	330	14.8	49	Cleanup No. 1; disposal area; wastewater treatment through natural sedimentation method; transportation with discharge pipe
16	Yokkaichi Port Administration Association	Dredging & removal of accumulative dirty soils at Yokka-ichi Port (Phase 1 & 2)	Oct/2 1976	Mar/25 1977	Cleanup No. 1 No.2, No.3	226,500	510,900	-6.5 -13.0	0.5 3.3	Oily dirty soils	280 1,300	441	74.3	328	Cleanup No. 1, 2, 3; station barge; dirty soils reservoir; transportation with discharge; wastewater treatment
17	The Tokyo Metropolis	Dredging at Takahama Canal and preparation of disposal area for dredged material	Nov/28 1976	Mar/21 1977	Cleanup No. 5	4,000	7,296	-2.7 -3.5	0.7 0.7	Organic dirty soils	50	180	28.8	52	Cleanup No 5; tank barge; disposal area
18	Japan Aquatic Resources Protection Association	Experiment for feasibility study on protection works against red water for the year 1976	Jan/15 1977	Feb/5 1977	Cleanup No. 2	34,000	9,370	-16 -21	0.3	Organic dirty soils	400	356	37.0	132	Cleanup No. 2; sedimentation barge; filtration treatment; transportation and disposal; hydro-extraction treatment
19	Nippon Mining	Dredging & removal of dirty soils at Sagano Seki Port for 1977	Apr/1 1977	Oct/31 1977	Cleanup SIRSI	32,500	15,800	-5.0	0.50	Silt with fine sand	635	180	16.3	29.3	Cleanup SIRSI; sedimentation pit; wastewater treatment
20	Miyagi Pref.	Sludge dredging & removing for improvement of fishery environment at Kesen-numa Bay for 1976	Aug/26 1976	Nov/25 1976	Cleanup No. 2	57,000	28,500	-4 -7	0.5	Organic dirty soils	350	190	18.6	74.2	Cleanup No. 5; soil carrier with grab bucket; offshore disposal
21	Miyagi Pref.	Sludge dredging & removing for improvement of fishery environment at Kesen-numa Bay for 1977	Jul/12 1977	Oct/15 1977	Cleanup No. 5	26,860	20,820	-5 -7	0.7 1.0	Organic dirty soils	90	190	18.6	74.2	Cleanup No. 5; soil carrier with grab bucket; offshore disposal

No.	Client	Description of Works	Contract Period From	Contract Period Until	Dredge Employed	Specifications for Dredging Work Area (m²)	Thickness Volume (m³)	Soil Depth (m)	Discharge (m)	Characteristics	Distance	Efficiency Pumping Volume (m³/hr)	Density (%)	Dredged Volume (m³/hr)	Ooze Treatment
22	Japan Aquatic Resources Protection Association	Pilot works for prevention of red water for 1977	Aug/25 1977	Sep/18 1977	Cleanup No. 2	64,500	24,405	-6 -7.5	0.3 0.6	Organic dirty soils	250	180	17.8	32.0	Cleanup No. 2; sedimentation barge; filtration treatment; transportation and disposal; hydro-extraction treatment
23	Yokkaichi Port Administration Association	Dredging & removal of accumulative dirty soils at Yokka-ichi Port (Phase 3)	Aug/10 1977	Feb/14 1978	Cleanup No. 3	210,860	408,300	-9.8 -13.5	0.3 2.5	Oily dirty soils	300	238	73.5	175	Transportation with discharge pipe; transportation with discharge pipe; Cleanup No. 3; station barge; dirty soils reservoir; wastewater treatment
24	Yokkaichi Port Administration Association	Dredging & removal of accumulative dirty soils at Yokka-ichi Port (Phase 5)	Nov/20 1977	Mar/25 1978	Cleanup No. 3	97,600	202,000	-7.5 -8	2.1	Oily dirty soils	837	316	76.9	243	Transportation with discharge pipe; transportation with discharge pipe; cleanup No. 3; station barge; dirty soils reservoir; wastewater treatment
25	Yokkaichi Port Administration Association	Dredging & removal of accumulative dirty soils at Yokka-ichi Port (Phase 6)	Jun/18 1977	Mar/31 1978	Cleanup No. 3	15,500	15,500	-11	1.0	Oily dirty soils	800	268	67.3	181	Transportation with discharge pipe; transportation with discharge pipe; Cleanup No. 3; station barge; dirty soils reservoir; wastewater treatment
26	The Tokyo Metropolis	Dredging & removal of dirty soils in front of Hamarikyu Works for 1978	Oct/26 1977	Mar/31 1978	Cleanup No. 5	22,800	48,300	-4 -4.6	0.5 0.8	Organic dirty soils	56	200	26.0	53.0	Cleanup No. 5; transportation with box barge; disposal area
27	Okayama Pref.	Pollution prevention	Jun/15 1978	Jan/15 1978	Cleanup No. 2	204,500	90,000	-12 -16	0.4 0.8	Oily dirty soils	Offshore discharge with box barge	550	9.0	50.0	Clean No. 2; transportation with box barge; disposal area
28	Miyagi Pref.	Sludge dredging & removing for improvement of fishery environment at Kesen-numa Bay for 1976	Jul/26 1978	Jan/15 1979	Cleanup No. 3	65,000	51,700	-4 -7	0.8 1.0	Organic dirty soils	Offshore discharge with box barge	350	48.5	169.7	Cleanup No. 3; transportation with box barge; disposal area
29	The Tokyo Metropolis	Dredging of dirty soils at Shibaura-Nishi Canal & preparation of disposal area for dredged material	Sep/15 1978	Jan/31 1979	Cleanup No. 1 No. 5	11,006	16,860	-2.9	1.5	Organic dirty soils	Offshore discharge with box barge	350	16.0	56.0	Cleanup No. 1 & 5; transportation with box barge; disposal area
30	The Tokyo Metropolis	Dredging of dirty soils at Shiohama Hirahima Shiomi Canal for 1978	Feb/2 1979	Feb/23 1979	Cleanup No. 5	2,000	2,700	-2.3 -3.0	1.4	Organic dirty soils	Offshore discharge with box barge	320	17.2	55.0	Cleanup No. 5; transportation with box barge; disposal area
31	Okayama Pref.	Pollution prevention work for 1978	Feb/16 1979	Mar/20 1979	Cleanup No. 2	72,900	37,200	-12 -16	0.4 0.8	Oily dirty soils	Offshore discharge with box barge	850	15.0	165.0	Cleanup No. 2; transportation with box barge; disposal area
32	The Tokyo Metropolis	Dredging of dirty soils at Akebonokita Canal and Tsukishima/Shin-Tsukishima River	Mar/1 1979	Mar/30 1979	Cleanup No. 5	88,005	6,942	-2.4	0.8	Organic dirty soils	Offshore discharge with box barge	300	20.0	60.0	Cleanup No. 5; transportation with box barge; disposal area

No.	Client	Description of Works	Contract Period From	Contract Period Until	Dredge Employed	Area (m²)	Thickness Volume (m³)	Soil Depth (m)	Discharge ness (m)	Characteristics	Distance (m)	Pumping Volume (m³/hr)	Density (%)	Dredged Volume (m³/hr)	Ooze Treatment
33	Okayama Pref.	Pollution prevention work for 1979	Jun/12 1979	Sep/30 1979	Cleanup No. 2	455,200	215,800	-12 -16	0.4 0.8	Oily dirty soils	Offshore discharge with box barge	860	19.8	170.0	Cleanup No. 1; transportation with box barge; disposal area
34	Kobe Steel Co., Ltd.	Removal of soft mud in front of Kakogawa raw material handling berth	Sep/1 1979	Sep/15 1979	Cleanup No. 3	3,100	1,700	-17 -17.6	0.6	Silt	370	500	22.0	110	Cleanup No. 3; transportation with discharge pipe; reclaimed area
35	The Tokyo Metropolis	Dredging work at Tatsumi Canal for 1979	Aug/20 1979	Jan/15 1980	Cleanup No. 5	51,010	40,430	-4.0 -4.5	0.5	Silt	Offshore discharge with box barge	350	17.1	60.0	Cleanup No. 5; transportation with box barge; disposal area
36	The Tokyo Metropolis	Dredging of dirty soils at Shinshibaura, Shibaura & Nishi Canals for 1979	Jan/17 1980	Apr/20 1980	Cleanup No. 5	11,640	18,235	-2 -3.2	0.7 1.2	Silt	Offshore discharge with box barge	300	21.7	65.0	Cleanup No. 5; transportation with box barge; disposal area
37	Niigata Pref.	Toyanogata River purification work	Feb/24 1979, Feb/7 1980, Mar/21 1980	Mar/13 1979, Mar/5 1980, Apr/6 1980	Cleanup No. 1	39,440	7,500	-1.7	0.5	Silt	500 2,000	56	69.6	39	Cleanup No. 1; booster pump; sedimentation area; transportation with discharge pipe; transportation with discharge pipe
38	The Tokyo Metropolis	Dredging of dirty soils at Takahama/ Shibaura-Nishi Canal	Jun/12 1980	Mar/31 1981	Cleanup No. 5	64,266	44,986	-2.7 -3.0	0.7	Silt	130	300	31.2	95	Cleanup No. 5; box barge; disposal area
39	Port Construction Div. 3	Investigation for dredging work at Osaka Bay	Aug/4 1980	Oct/15 1980	Cleanup No. 3	4,800	3,840	-13.5 -14.6	0.4 1.2	Silt	Unloading with box barge	285	75.8	216	Cleanup No. 3; transportation with box barge; unloading with barge unloader
40	Miyagi Pref.	Sludge dredging & removing for improvement of fishery environment at Kesennuma Bay for 1980	Aug/15 1980	Oct/31 1980	Cleanup No. 3	27,600	13,800	-4.0 -7.0	0.5	Organic dirty soils	Offshore disposal with sand carrier with grab bucket	350	60.0	210	Cleanup No. 3; sand carrier with grab bucket; offshore disposal
41	Okayama Pref.	Dredging of basin at Yobimatsu Channel	Feb/23 1981	Mar/26 1981	Cleanup No. 2	57,750	23,100	-4.0	0.4	Silt	250	800	25.0	200	Cleanup No. 2; box barge; unloading with barge unloader
42	Mitsubishi chemical industries & other private companies	Maintenance dredging at Yobimatsu Port & other works	May/9 1981	Sep/10 1981	Cleanup No. 2	76,800	79,070	-4.5 -17.0	0.45 1.80	Silt with coarse sand	250	800	13.8	110	Cleanup No. 2; box barge; unloading with barge unloader

MONITORING THE EFFECTIVENESS OF CAPPING
FOR ISOLATING CONTAMINATED SEDIMENTS

Robert W. Morton
Science Applications International Corporation

ABSTRACT

Disposal of contaminated sediments in the marine environment through capping with cleaner materials is a management option that has been used extensively during recent years, particularly in New England. Most capping projects have been restricted to quiescent, shallow waters (20-30 m); however, as a result of monitoring programs associated with these projects, a body of knowledge concerning the creation of capped disposal mounds has been developed to predict the consequences of extending such procedures to other waters. In particular, the application of capping technology to deeper water is extremely important, because disposal site designation programs currently underway throughout the United States are predominantly aimed at water depths of 100 m or greater.

Since many capping projects are considered experimental or controversial, the monitoring procedures to assess them must be carefully devised to answer specific questions regarding the overall ability of the cap to isolate contaminated materials. Therefore, the results of previous monitoring efforts, such as those conducted under the Disposal Area Monitoring System (DAMOS), can provide a baseline approach for future monitoring of capping operations. This paper presents an overview of the results of capping projects conducted under DAMOS and the rationale for the existing monitoring program developed from those efforts.

INTRODUCTION

Capping of contaminated dredged material with sediment relatively free of contaminants has developed into a commonly used management technique for reducinge potential environmental impact of open-water disposal. Capping was first employed in 1977 by the New England Division of the U.S. Army Corps of Engineers (COE) at the New London Disposal Site (NLON). This project, which took place in 20 m of water in the eastern end of Long Island Sound, has led to continued application and field observations of the technique, including major capping operations at the Central Long Island Sound Disposal Site (CLIS) and the New York Mud Dump Site (EMD) in open ocean waters.

Additional studies stressing laboratory observations on the effectiveness of capping have been conducted at the COE Waterways Experiment Station.

As a result of these studies, a great deal has been learned regarding the effectiveness of capping in the marine environment. However, during each capping operation, the fact that contaminated sediments are involved means that major issues must be addressed and fully understood to ensure that minimal adverse impacts occur as a result of the operation. These issues include

- thickness of the cap--related to the effectiveness of the capping material in isolating the contaminants, particularly the potential for leaching of contaminants and effects of bioturbation;
- placement of the cap--related to navigation control during disposal to ensure coverage of contaminated sediments and to the mixing and displacement of contaminated sediment by the capping material; and
- stability of the cap--related to the support of the cap by typical high-water content contaminated material and resistance to erosion and transport of capping material.

Previous studies of capping have indicated that with careful management the operation can be successful in relatively quiescent, shallow waters. However, designation or use of new disposal sites in water depths greater than 100 m, where capping will be a management option, are currently underway in the New England region, at the Foul Area Disposal site (FADS) in Massachusetts Bay, in the Seattle area at the Everett Homeport Disposal Site, and potentially in the New York region at a site to be designated offshore of the mud dump site.

Permits for capping of contaminated sediments at these and other disposal sites will certainly require monitoring of the disposal operation and the resulting deposit. Therefore, development of a rational, practical, and meaningful monitoring approach is critical to the future application of capping technology to the management of contaminated sediments.

HISTORY OF CAPPING OPERATIONS

The first use of capping as a disposal management strategy occurred at the New London Disposal site (NLON) in 1977 when contaminated sediments from the vicinity of dock areas were dredged first and then covered with cleaner sediments as dredging proceeded from the head of the estuary to the mouth. Capping of the contaminated sediments was assured because the mass of material used for capping was more than 30 times greater than that of the contaminated material. However, such an abundance of capping material is not always available and, for capping to become a truly feasible management strategy, procedures for capping with much less material had to be developed.

The first field study of controlled capping of contaminated

material with more reasonable amounts of capping material took place at the Central Long Island Sound Disposal site (CLIS) in 1979 (Figure 1). During this project, two disposal mounds were formed, each with approximately 30,000 m^3 of contaminated sediments from Stamford, Connecticut. These deposits were then capped, one with approximately 76,000 m^3 of silt (STNH-S), and the other with 33,000 m^3 of sand dredged from New Haven Harbor (STNH-N). This study produced several important conclusions, which were applied to future capping projects.

- Disposal of contaminated sediments must be tightly controlled. This is necessary reduce the spatial distribution of material to be capped and can be accomplished through use of taut-wire disposal buoys and/or precision navigation control.
- Disposal of capping material must be spread over a larger area. Dispersal of cap material is necessary to ensure adequate capping of the margins of the contaminated deposit and is particularly important for silt capping material, which does not spread as evenly as sand.
- Silt develops a thicker cap than sand and therefore requires more cap material. Silt caps do not spread as readily during disposal; however, the greater thickness is needed because the depth of bioturbation is deeper in silt than in sand.
- Silt caps recolonize with fauna similar to the surrounding silt environment, sand caps with completely different species. Recolonization of both mounds occurred as expected and impacts to surrounding environment were negligible.
- Caps are resistant to erosion. Once stabilized, both the silt and sand caps have remained essentially unchanged for more than eight years (including two hurricanes).

A second study, utilizing similar sediments was conducted two years later at Cap Sites 1 and 2 (CS-1 and CS-2) with comparable results. In this study, the placement of capping material was the most significant factor affecting the isolation of contaminated sediments. In spite of efforts to distribute the cap evenly, additional disposal of silt material was required to achieve adequate coverage.

Other capping operations have been successfully accomplished since 1979.

- Disposal and capping in borrow pits: this approach has been suggested as an alternative for New York Harbor, but is currently on hold pending studies of the environmental significance of the borrow pits.
- Dredging of a depression, filling with contaminated sediments and capping with displaced material: this procedure was used successfully in Norwalk, Connecticut, but is restricted to shallow-water environments. This approach has also been proposed for disposal of PCB-contaminated sediments at the New Bedford Superfund site.
- Open-water capping at the New York Mud Dump Site: approximately 522,000 m^3 of contaminated material have been successfully

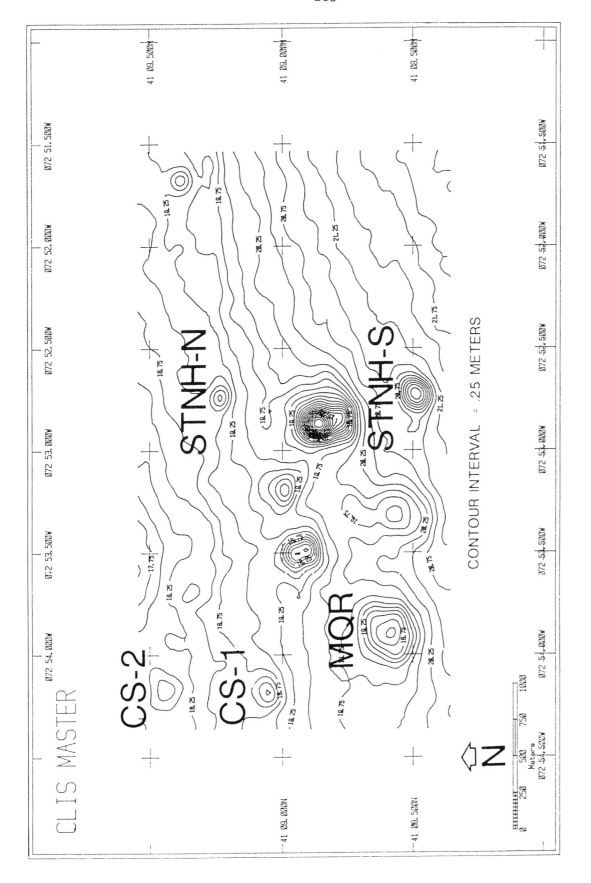

FIGURE 1 Contoured bathymetry chart of the Central Long Island Sound disposal site.

covered by 1.2 million m³ of clean sand in a mound that has persisted on the open shelf for seven years. Excellent management of continuous capping operations within the site requires identification of cap material prior to issuing permit for disposal of contaminated sediments.

As a result of these studies, the factors affecting capping can be predicted with some accuracy, particularly the amount of material needed to create an effective cap and the controls necessary to dispose of the contaminated sediment and capping material in a stable deposit.

DAMOS MONITORING PROGRAM

As stated above, future capping projects are certain to require extensive monitoring programs, which will be required to evaluate the effectiveness of the capping operation and address site-specific issues. A starting point for design of such programs could be the Disposal Area Monitoring System (DAMOS). DAMOS has been in existence for more than ten years and has developed into a multidisciplined program that provides the COE New England Division (NED) with the necessary information to manage open-water disposal of dredged material in a scientific, rational manner. An overview of the DAMOS monitoring approach, which begins with site designation and extends through the disposal operation to post-disposal monitoring is presented in Figure 2. DAMOS has always been a flexible program designed to respond to changes in technology that result in a better understanding of the effects of dredged material disposal on the marine environment. As a result, it has been closely involved with the development, execution, and monitoring of the capping projects described above.

	PHYSICAL	BIOLOGICAL	CHEMICAL
SITE DESIGNATION (CHARACTERIZATION)	BATHYMETRY/SSCAN REMOTS CURRENTS/WAVES SEDIMENT GRAIN SIZE	REMOTS – HABITAT BENTHIC – TYPE PRESENT BRAT – FISH HABITAT FISH – TYPE PRESENT	BULK SEDIMENT ANALYSIS
PRE-DISPOSAL (BASELINE)	BATHYMETRY/SSCAN REMOTS HARBOR CHARACTERIZATION (DENSITY, GS, GEOTECH) DISPOSAL CONTROL	BENTHIC BODY BURDEN COMPOUNDS SELECTED BASED ON WASTE CHARACTERIZATION IF > ONE YEAR – REMOTS	WASTE CHARACTERIZATION BULK SEDIMENT ANALYSIS BIOASSAYS ETC.
DURING DISPOSAL	BATHYMETRY/REMOTS PLUME STUDIES MUSSELS/DAISY		
POST DISPOSAL	BATHYMETRY/SSCAN REMOTS MUSSELS/DAISY	REMOTS (WITHIN 2 WEEKS)	
MONITORING	BATHYMETRY/REMOTS (NEXT SEASON, THEN ANNUALLY, AUG/SEPT) MUSSELS	REMOTS (NEXT SEASON, THEN ANNUALLY, AUG/SEPT) IF RECOLONIZED: BENTHIC, BRAT, BODY BURDEN	IF NOT RECOLONIZED; BULK SEDIMENT ANALYSIS

FIGURE 2 Proposed integrated DAMOS monitoring/management program for dredged material disposal.

Through years of intense field observations, the DAMOS program been able to develop a comprehensive data base that confirms the viability of several important parameters required for capping operations:

- Operational feasibility: it has been shown (Figure 3) that navigation control and disposal operating procedures are adequate to create mounds of contaminated material and to spread sufficient cap material to effectively cover those mounds.
- Minimal dispersion during disposal: extensive plume tracking studies have demonstrated that most dredged material is transported to the bottom during the convective descent phase of disposal (Figure 4). This is required to ensure that, with proper navigation, contaminated materials will be contained within a reasonable area for capping and not dispersed throughout the water column or spread over a broad area of the bottom.
- Long-term stability of disposal mounds: repeated measurements over the past 10 years have shown that following initial reworking and consolidation, capped disposal mounds remain unchanged for extended periods of time. This means that disposal sites do exist where currents and wave activity have insufficient energy to erode and transport dredged material and that those areas can be considered as containment sites. These conditions are necessary for initial control of contaminated material prior to capping and to ensure that once caps have been deposited, the capping material will remain in place over the long term.
- Sand or silt cap material: all studies to date have indicated that either sand or silt are adequate for capping of contaminated material, although silt caps require more material and must be spread by controlling disposal over a larger area. This conclusion is extremely important because the economic feasibility of capping depends to a large extent on the availability of clean sediment and frequently sand is not common in regions where contaminated silts occur.
- Isolation of contaminated material: both chemical and biological monitoring have demonstrated that, given sufficient cap thickness and stability, neither bioturbation nor chemical leaching will expose contaminated sediments to the surrounding environment. Therefore, use of uncontaminated sediments as a cap material is a viable method for isolation.

As a result of monitoring both capped and uncapped dredged material deposits for a number of years, the DAMOS program has developed and utilized an extensive array of instrumentation and procedures for evaluating the parameters described above. Examples of the instrumentation required for execution and monitoring of capping projects follow.

- Precision navigation: navigation control on the order of 5 m is necessary to ensure correct placement of contaminated and capping sediments and to accurately sample the steep gradients associated with the dredged material deposit during post-disposal monitoring.

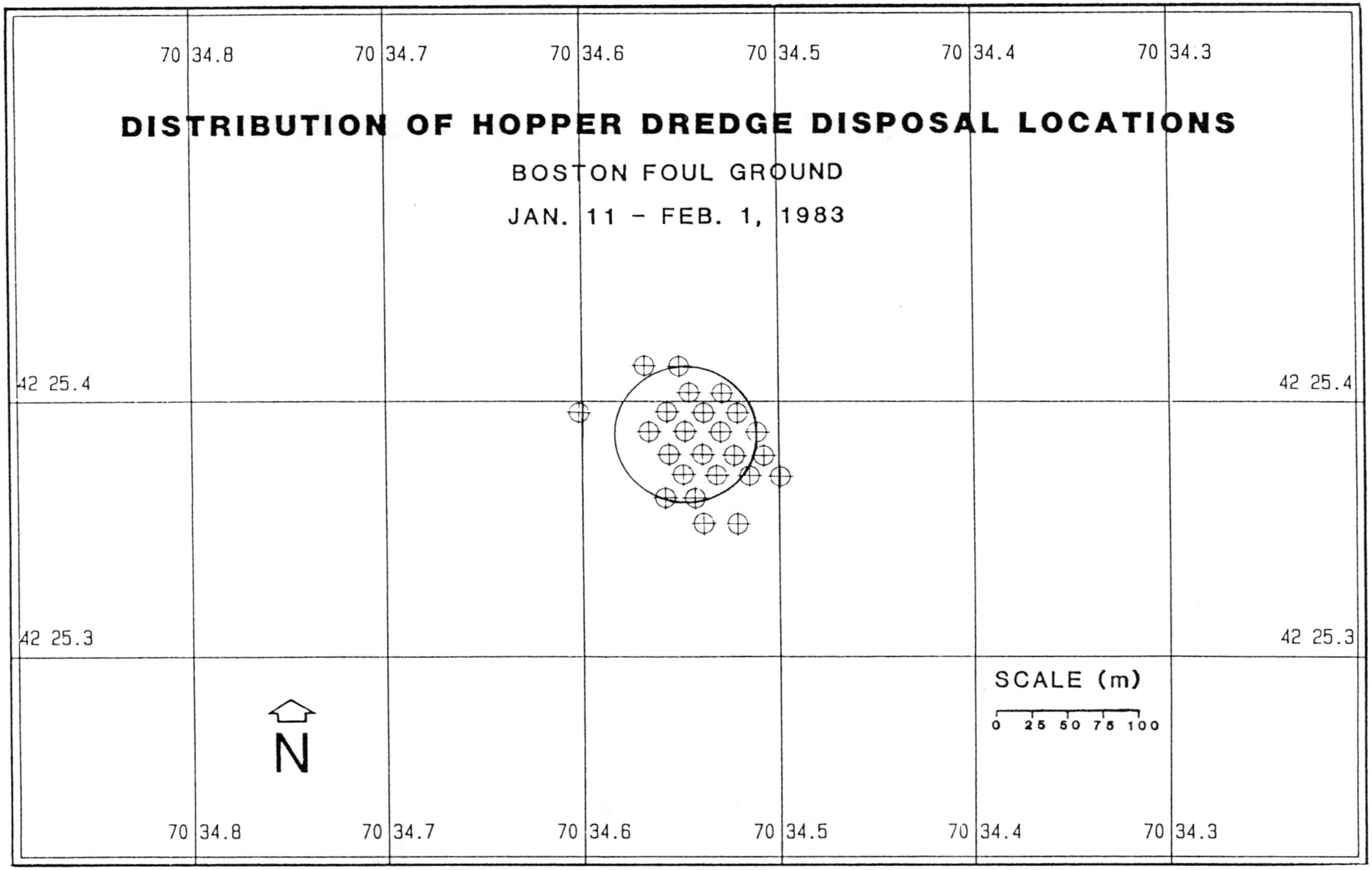

FIGURE 3 Distribution of disposal locations at the Boston Foul Ground. Circle indicates a 50-m radius about the disposal point. SOURCE: SAIC, 1985.

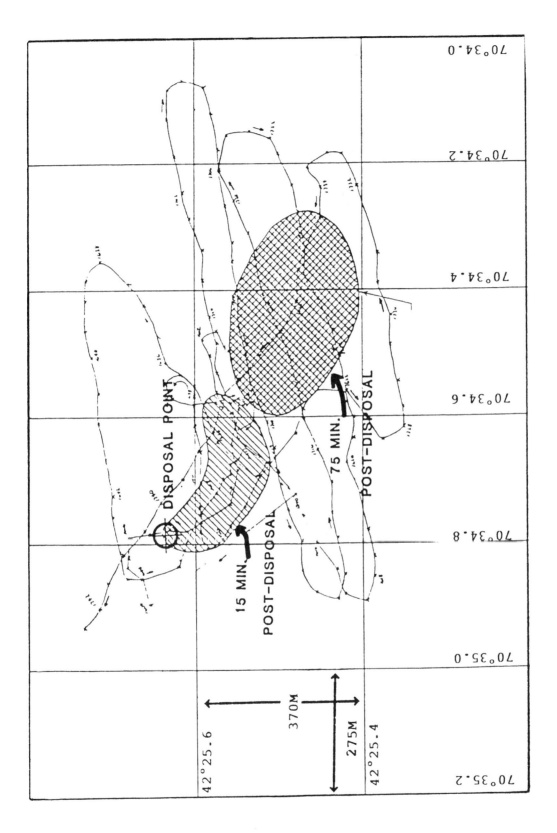

FIGURE 4 Ships track and plume dispersion following hopper dredge disposal at Boston Foul Ground. SOURCE: SAIC, 1985.

Controlled disposal of contaminated dredged material requires a taut-wire moored buoy, which must be accurately deployed. Using such a buoy, disposal can be restricted to less than a 50-m radius and the input of dredged material can be considered as a point source for subsequent capping and monitoring.

On the DAMOS program, buoy deployment and sampling are controlled through an SAIC Integrated Navigation and Data Acquisition System (INDAS), which provides computerized integration of microwave or acoustic positioning systems with environmental sensors and navigation displays to provide accurate ship control and data acquisition. Figure 5 presents an example of the sampling accuracy and precision attained by this system.

- Precision bathymetry: replicate bathymetric surveys provide the basis for sequential monitoring of the volume and distribution of sediment at the disposal site to assess the effectiveness of capping and the long-term stability of the cap. Because of the small changes that occur as a result of erosion or consolidation, this approach requires very precise field measurement procedures and statistical analysis of replicate surveys.
- Sediment profile photography: the REMOTS camera has proven to be a key instrument for assessing the distribution and characteristics of near-surface sediments. In particular, the changes and conditions existing at the fringes of the mound can be examined with a resolution unattainable with acoustic measurements or conventional sampling procedures. Furthermore, this instrumentation examines small-scale effects of physical erosion and bioturbation and provides an efficient method for measuring biological parameters to evaluate the impacts of disposal and capping operations.
- Advanced acoustic measurements: modern acoustic instrumentation such as sidescan sonar, high resolution subbottom profilers and high-frequency plume tracking systems all provide important information on the distribution and physical properties of sediments during and after disposal.
- Specialized instrumentation: development of instrumentation packages such as the Disposal Area In situ System (DAISY) provide information for addressing specific problems associated with dredged material disposal and capping. In particular, the DAISY measures near-bottom current and wave energy associated with sediment resuspension and turbidity to address the long-term stability of capped disposal mounds.

Another example of specialized instrumentation is the Nuclear Density Probe which has been configured with a sediment penetration device and is used with precision bathymetry, REMOTS and subbottom profiling to determine the mass balance of sediment deposited in the capped mound.

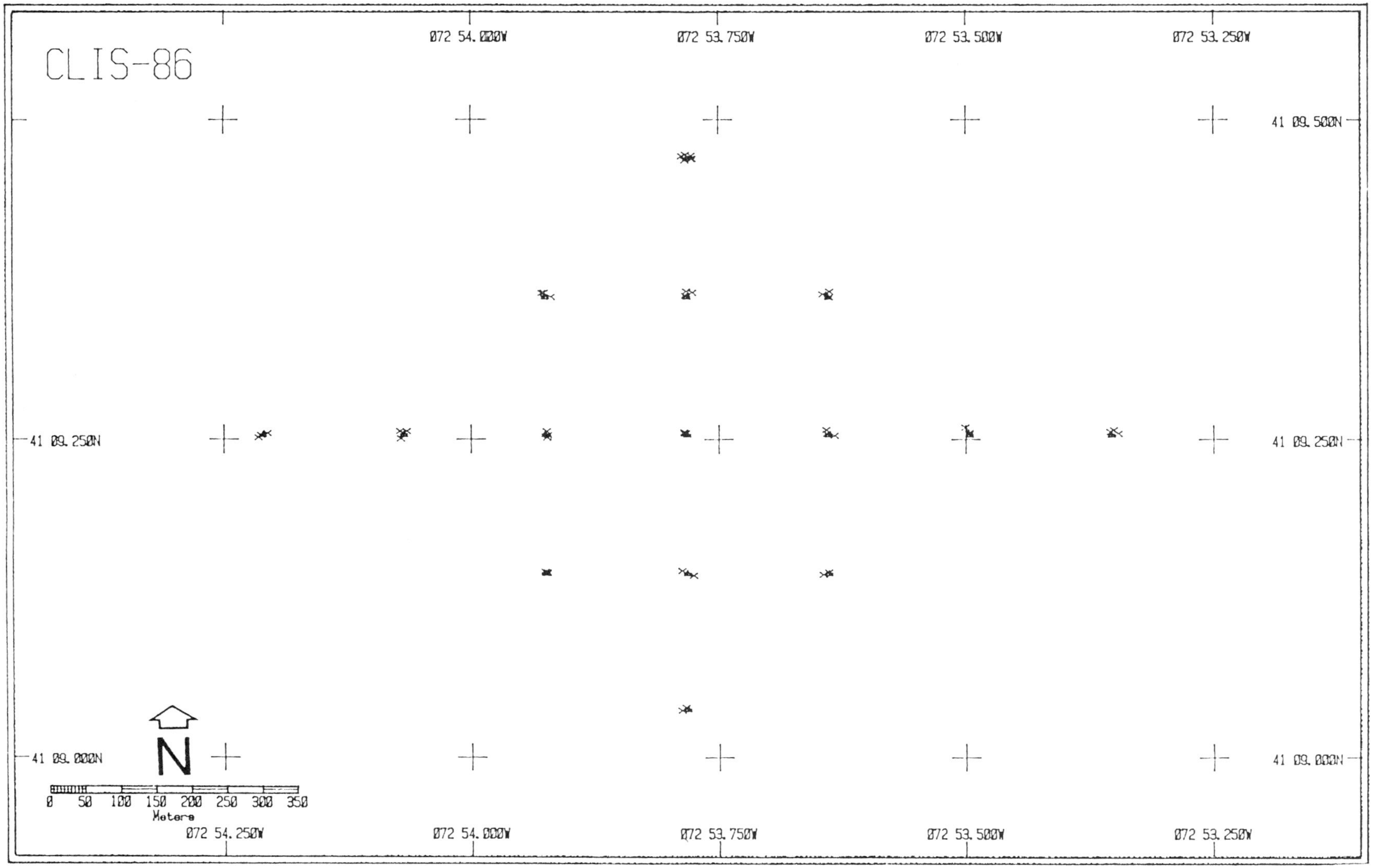

FIGURE 5 Location of replicate sampling at stations at the Central Long Island disposal site.

RECENT CAPPING OBSERVATIONS

The instrumentation and procedures described above have been used extensively on two major field studies recently completed at the New York Experimental Mud Dump site (EMD) and the Foul Area Disposal site (FADS) in Massachusetts Bay. The objectives of the study at the EMD were to assess the long-term (five years) stability of a sand cap deposited over contaminated sediments in the open-shelf environment; while at FADS, the short-term (several months) effects of disposal in water 90-m deep were measured to evaluate the behavior and distribution of sediments and to determine the feasibility of capping in such water depths.

Experimental Mud Dump Site (EMD)

At the EMD the results indicated that following disposal, a cap of approximately 1.5 to 2 m covered most of the contaminated material (Figure 6) and that this cap was essentially unchanged during the subsequent five-year period (Figure 7). These conclusions were supported by the subbottom profile measurements, which indicated a surface deposit of more than 75 percent sand with a mean thickness of 1.5 m in the vicinity of the disposal mound (Figure 8). Subbottom profiles across the disposal site also demonstrated that the cap was continuous.

REMOTS photography supported the subbottom data, but also provided additional information. The "Benthic Process" map generated from the sediment profile photographs (Figure 9) indicated the presence of the same fine-grained, high-reflectance sand in the vicinity of the EMD. However, the photographs also showed bedforms and in some cases alternating layers of sand and mud suggesting that although sediment resuspension and transport can occur on the surface of the mound, the entire region must be in equilibrium, since there has been no significant loss of cap material over time. The recolonization of the disposal mound as measured by REMOTS also indicates that physical disturbance of the sediment surface occurs. State I (opportunistic) species are the redominant infaunal successional stage on the disposal mound and throughout the disposal site, suggesting relatively actibottom conditions. The presence of Stage I species (on the sand cap means that bioturbation will penetrate only a few centimeters into the cap and, therefore, isolation of the contaminated material can be expected. On the flanks of the mound, where cap thickness is not so great and some Stage III organisms are present, some mixing of the contaminated sediment will occur.

Foul Area Disposal Site (FADS)

Extensive disposal site designation studies have recently been conducted at FADS, including investigation of the potential effects of capping operations in water depths of 90 m. Disposal at FADS generally occurs from disposal scows and, occasionally, from hopper dredges;

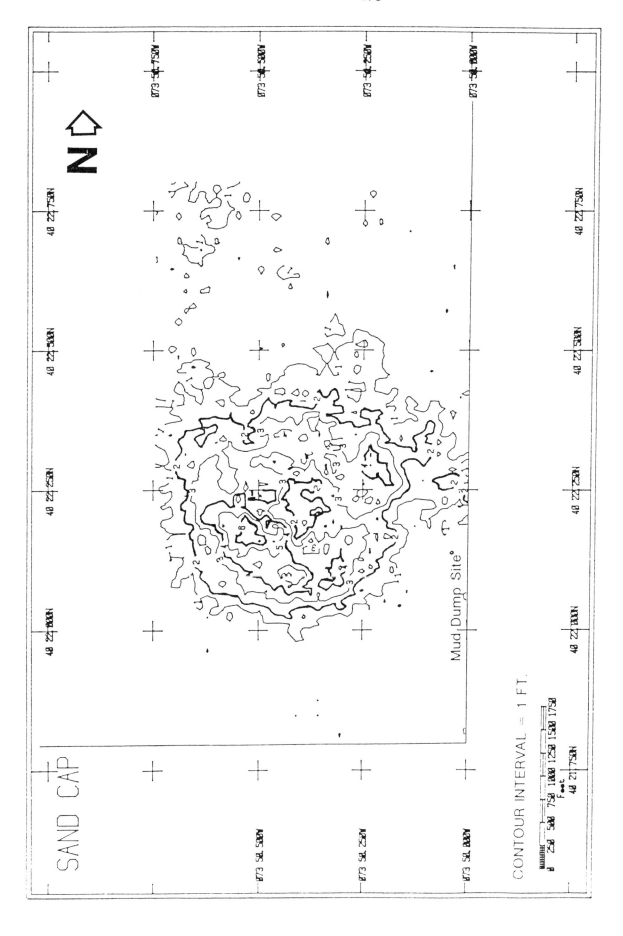

FIGURE 6 Contoured depth (feet) difference chart comparing the June 1980 and August 1981 surveys at the EMD. SOURCE: Parker and Valente, 1987.

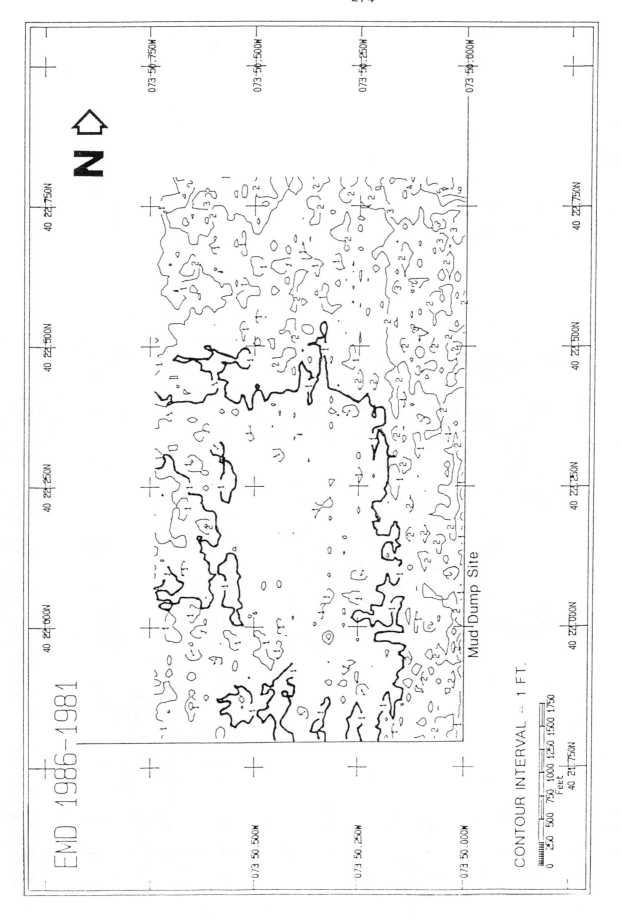

FIGURE 7 Contoured depth (feet) difference chart comparing the August 1981 and November 19 surveys at the EMD. SOURCE: Parker and Valente, 1987.

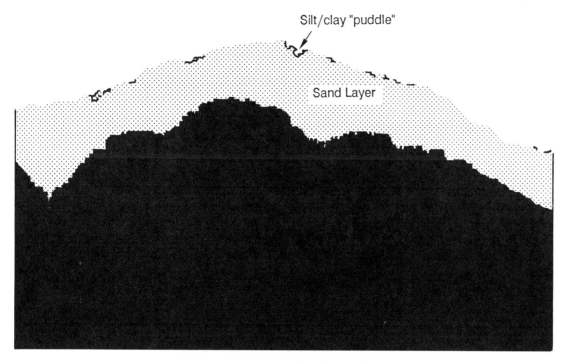

FIGURE 8 Representative trace of acoustic impedance at the EMD.
SOURCE: Parker and Valente, 1987.

however, in all cases, sediment is transported to the bottom through the classical three phases of disposal:

1. the convective descent phase, during which the majority of the dredged material is transport to the bottom under the influence of gravity as a concentrated cloud of material;
2. the dynamic collapse phase following impact with the bottom where the vertical momentum present during the convective descent phase is transferred to horizontal spreading of the material; and
3. the passive dispersion phase following loss of momentum from the disposal operation, when ambient currents and turbulence determine the transport and spread of material.

In shallow water, cohesive sediments disposed under the influence of the above disposal phases create a distinct mound formation with thin flank deposits, while sands, or less cohesive, high-water-content sediments characteristic of the contaminated material produce a broader, more uniform deposit. At FADS, mounding of cohesive sediments was less prevalent even with cohesive sediments; however, the overall spread of material was similar. Regardless of whether the disposal operation was conducted with a hopper dredge or scow, both theoretical and observational data indicate that the majority of the dredged

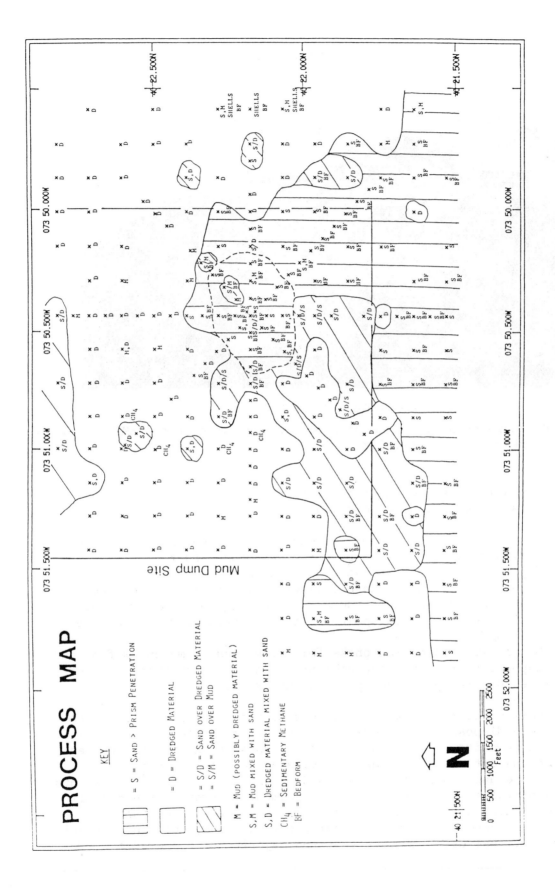

FIGURE 9 Process map illustrating sediment surface characteristics and benthic features around the EMD. The broken line encloses clean, fine-grained sand at the sediment surface.
SOURCE: Parker and Valente, 1987.

material was transported to the bottom at FADS as a discrete plume during the convective descent phase. When this material reached the bottom, the vertical momentum was transferred to horizontal momentum during the dynamic collapse phase. The overall size and thickness of the resulting deposit depended primarily on the amount of material disposed at the site and navigation control exercised during disposal effort.

Recent work, completed during January 1987, using the REMOTS camera has demonstrated that the disposal of dredged material under tight control at FADS resulted in a broad, low deposit spread evenly over an area similar to that covered by disposal in more shallow waters. The major difference in the deposits results from the greater spread of cohesive clumps which inhibits formation of a topographic feature (i.e., disposal mound). Figure 10 indicates the distribution of dredged material as detected by the REMOTS camera following disposal of approximately 200,000 m^3 of cohesive sediment. This operation resulted in a deposit with a thin layer of dredged material extending over a circular area with a radius of 500 m; a deposit comparable to similar volumes deposited in the shallow water of Long Island Sound.

The fact that disposal in deeper water results in a thin, broad deposit can have important implications for capping under such conditions. If careful control of the contaminated material is not exercised, even small amounts of material will cover the same area of the bottom as larger volumes. Consequently, it would take essentially the same amount of material to effectively cap 100,000 m^3 of contaminated material as possibly 250,000 or 500,000 m^3. Assuming the contaminated material covered an area of bottom with a 500-m radius, similar to the deposit created at FADS, then at least 1.1 x 10^6 m^3 of material would be required to produce a deposit one meter thick extending 100 m beyond the edge of dredged material. This is not an unreasonable quantity to cover a substantial project, but would be untenable for a small contamination problem. Therefore, appropriate scheduling of small contaminated projects prior to larger uncontaminated dredging programs must be carefully considered.

SUMMARY

Extensive monitoring of capping projects throughout the New England region under the DAMOS program suggests that capping is a a feasible mitigating measure for disposal of contaminated sediments in the marine environment. However, it must be emphasized that careful control of the operation and comprehensive monitoring of the resulting deposit are required to ensure minimal impact from future projects, particularly if they are conducted in deeper water.

Although capping has not been conducted at FADS, previous disposal operations at that site have demonstrated the effectiveness of disposal control in restricting the spread of material in 90 m of water. This is the single most important factor in a capping operation and, if the disposal location is a containment site, capping should be feasible at those depths with sufficient material. Furthermore, the fact that caps have persisted at the Central Long Island Sound and EMD sites for five

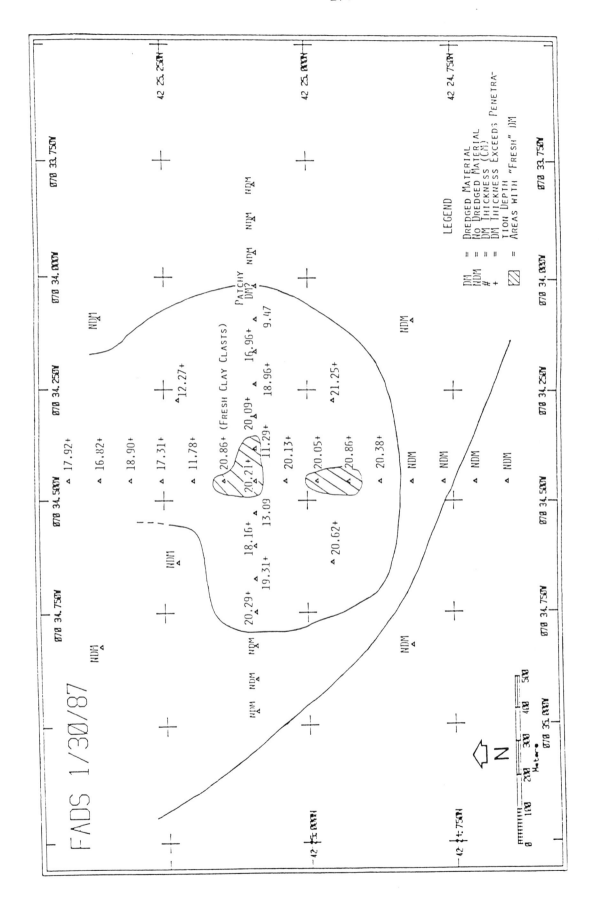

FIGURE 10 Distribution of dredged material at the Foul Area Disposal Site as measured by REMOTS technology. SOURCE: SAIC, 1987.

years or more suggests that containment of contaminated materials can be accomplished even in relatively high-energy environments.

REFERENCES

Morton, R. W. 1980. The management and monitoring of dredge spoil disposal and capping procedures in Central Long Island Sound. In Wastes in the Ocean, Volume 2: Dredged Material Disposal in the Ocean, Kester et al., eds. New York: Wiley and Sons.

Morton, R. W. 1987. Updating the U.S. experience with aquatic capping of contaminated sediments. Presented at 13th U.S./Japan Experts Meeting on Management of Bottom Sediments Containing Toxic Substances, November, 1987. U.S. Army Engineers.

Morton, R. W., C. J. Lindsay, and R. C. Semonian. 1984. Use of Scientific Data for Management of Dredged Material Disposal in New England. Presented at Conference Dredging '84, ASCE.

Morton, R. W. and R. D. Jones. 1985. The importance of accurate navigation in environmental assessment programs. Sea Technology, August.

Parker, J. H. and R. M. Valente. 1987. Long-Term Sand Cap Stability: New York Dredged Material Disposal Site. Vicksburg, Miss.: U.S. Army Engineer Waterways Experiment Station.

Science Applications International Corporation. 1985. Disposal Area Monitoring System (DAMOS) Annual Report, 1984. DAMOS Contribution #46. New England Division, U.S. Army Engineers.

Science Applications International Corporation. 1987. Monitoring Surveys at the Foul Area Disposal Site, February, 1987. DAMOS Contribution #64. New England Division, U.S. Army Engineers.

LIST OF DAMOS CONTRIBUTIONS SUBMITTED TO
NEW ENGLAND DIVISION U.S. ARMY CORPS OF ENGINEERS

7 Stamford/New Haven Disposal Operation Monitoring Survey Report
8 Management and Monitoring of Dredge Spoil and Capping Procedures in Central Long Island Sound
#11 "Capping" Procedures as an Alternative Technique to Isolate Contaminated Dredge Material in the Marine Environment
#12 Precision Disposal Operations Using a Computerized Loran-C S:8ystem
#17 Disposal Area Monitoring System Annual Report, 1980
#22 DAMOS Mussel Watch Program: Monitoring of the "Capping" Procedure Using *Mytilus edulis* at the Central Long Island Sound Disposal Site; 1980-81.
#32 Summary of Disposal Monitoring Methods Used at FVP and Cap Sites #1 & #2 (2 Volumes)
#33 Geotechnical Studies Associated with Capping of Black Rock Sediment
#38 Results of Monitoring Studies @ Cap Sites #1, #2, and the FVP Site in Central Long Island Sound and a Classification Scheme for the Management of Capping Procedures
#46 Disposal Area Monitoring System (DAMOS) Annual Report, 1984
#56 Response to Comments
#57 Observations of the Effect of Hurricane Gloria on the Suspended Material Field in Eastern Long Island Sound

REMEDIAL TECHNOLOGIES USED AT
INTERNATIONAL JOINT COMMISSION AREAS OF CONCERN

Ian Orchard
Environment Canada

ABSTRACT

This paper discusses the need for various technological options for the remediation of contaminated sediments in 39 of 42 nearshore areas in the Great Lakes basin. These areas have been designated as "areas of concern" by the International Joint Commission, established by the United States and Canada for the restoration and enhancement of water quality in the Great Lakes. There are three broad categories of options for dealing with contaminated sediments; leave them alone, remediate in situ, or dredge them. No technologies are presently available within the Great Lakes basin that can be used to remediate; the only real action alternative based on available technology is to dredge up contaminated sediments.

Although considerable experience and technology is available to dredge large volumes of sediment, the disposal of this material remains a problem. Presently, only shore-based, confined disposal facilities and upland landfill disposal of the entire volume of dredge material appear to be possible for large-scale operations. The lack of available space to build confined facilities and to locate landfills has resulted in an urgent need for the development of alternative techniques.

Some concentration (separation), inactivation, and destruction techniques are operational elsewhere in the world but they remain largely at the laboratory (bench) scale or pilot-scale stages in the Great Lakes. Consequently, the focus of this paper will be upon the work of the sediment subcommittee of the International Joint Commission in the identification of appropriate remedial technologies for contaminated sediment.

BACKGROUND

The government of Canada and the United States signed the Boundary Waters Treaty on January 11, 1909, which outlined the rights of each country in the use of Great Lakes waters. In the 1960s, growing public concern about the quality of water in the lower Great Lakes resulted in studies by both governments aimed at assessing the nature and extent of water quality degradation. As a consequence of these studies, Canada and the United States entered into agreements in 1972 and 1978 on Great Lakes water quality aimed at restoration and enhancement of water quality.

An International Joint Commission (IJC) established under the Boundary Waters Treaty functions as the administrative mechanism for ensuring compliance between the parties under the agreement. In 1987 a protocol amending the 1978 agreement was signed between the governments. This protocol placed a greater emphasis on the governments to undertake specific initiatives so as to address the continuing contamination of the Great Lakes. The IJC has assumed a more evaluative role in ensuring that the parties meet the terms of the agreement. The U.S. Environmental Protection Agency (EPA) is the main agency responsible for delivery of U.S. obligations under the agreement and Environment Canada (DOE) is the main Canadian agency.

The development of remedial options for contaminated sediment dates back to the dredging subcommittee of the IJC, which was created under the Great Lakes Water Quality Agreement of 1978. At that time the committee's objectives were to develop guidelines and criteria for dredging activities, maintain a register of dredging projects, exchange information on technology and research, and identify criteria for the classification of polluted sediments in areas that were continually dredged.

In 1986, the sediment subcommittee was created when it became clear that one of the key elements in the development of remedial action plans for areas of concern was the need to refine existing assessment techniques for polluted sediment as well as identify implementable remedial options for polluted sediment. Two work groups were created under the sediment subcommittee. The remedial options work group produced a draft report in November 1987. The report is currently being subjected to scientific and technical review. Its objectives report are to

1. review existing technologies for the remediation of ecosystem related impacts due to sediment contaminants;
2. evaluate the effectiveness and feasibility of existing technologies;
3. develop a system for evaluating the most applicable technology to be used for remediating identified ecosystem impacts due to sediment contamination in the nearshore areas of the Great Lakes basin;
4. identify research needs to further test existing technologies or establish new approaches to mitigate sediment contaminant problems;
5. establish in conjunction with the assessment work group and other committees, work groups or task forces, as necessary, a monitoring program to assess any adverse effects on the ecosystem that may result from moving or otherwise isolating existing contaminated sediments from their present location.

REMEDIAL OPTIONS

The remedial options work group is evaluating remedial options for polluted sediment using the following criteria:

1. Description of Options
   - Stages associated with the implementation of the option (outline of the procedures involved)
   - Is it currently used by jurisdiction(s)?
   - Has it been field validated?
   - Specific case studies associated with the option (refer to title or give short summary of findings)
2. Feasibility
   - Engineering/design feasibility
   - Cost (engineering, not socioeconomic costs)
   - Time frame, from conceptualization through implementation.
   - Limitations--geographic, engineering, scientific, technical, etc.?
   - Can the option be implemented to deal with large-scale contamination (harborwide or can it deal with a small area size limitation)?
   - Does this option deal with different degrees or ranges of contamination?
3. Environmental/Regulatory Criteria
   - What guidelines/criteria exist relative to the implementation of this option?
   - What assessment criteria are available during implementation and for follow-up (how clean is clean)?
   - Does any need exist to develop assessment criteria to address this option's relative success?
4. Long-Term vs. Short-Term Management
   - Is this a quick-fix option that can be implemented in the short-term to deal with a hot spot?
   - Is this a one-shot remedy that could preclude the use of other options in the future?
   - Does this option require long-term management (is its implementation incremental over a period of time)?

The options currently being evaluated are open-water disposal, capping in place, confined disposal, lake filling, agricultural land spreading, strip mine reclamation, decontamination treatment, and solidification. The options can be classified into the following categories based on available data and need for additional testing and field validation:

- remedial options--commonly used techniques,
- remedial options--requiring some testing and assessment,
- remedial options--proposed and tested on a limited scale, and
- remedial options--some limitation and requiring more testing.

A suggested guide for readers to follow has also been developed titled "Sediment Remedial Options" (Figure 1). The guide suggests that one should not consider implementing a remedial option before taking steps to identify the nature and extent of contamination. The causes of contamination can be identified on the basis of available (or collected) baseline data. One must be careful not to propose remediation measures which only address symptoms of contamination. It can be seen from the options listed that we still have a lot of work to do. Most of the options identified require validation and testing.

A short summary of the four categories of remedial options follows. This is a very cursory identification and categorization of options. For a more detailed discussion of remedial technologies the reader is asked to refer to *Technologies for the Remediation of Contaminated Sediments in the Great Lakes; Report of the Sediment Subcommittee and its Remedial Options Work Group to the Water Quality Board of the International Joint Commission*, July 1988.

Commonly Used Techniques

- Dredging and disposal into confined disposal facilities (CDFs) constructed in the nearshore zone: estimated cost, $4,00/yd^3 (capital cost).
- Dredging and disposal on agricultural land: filling of low-lying agricultural land (material) must meet all regulatory requirements). Estimated cost, $0.50/yd^3/mi (pumping), $185,000/mi-pipe (capital cost that can be amortized over the life of several projects), $0.45/yd^3/mi (transport).

Remedial Options Requiring Some Testing and Assessment

- Subaqueous confinement: field verification proved effectiveness of this option in preventing the movement of contaminants into water and biota.
- Capping/covering in-place: suitable for low-energy zones in the lake; contaminants isolated by cap material.
- Beach nourishment: material must match beach substrate and must not be erodable; documented for material containing a limited quantity of nutrients. Estimated cost, $0.50/yd^3/mi, pumping, $185,000/mi (pipe) capital cost.

Remedial Options Proposed and Tested on a Limited Scale

- Depositional zone placement: minimum depth 30 m of 3 x maximum wave height must stay in place. Estimated cost, $0.26/yd^3/mi (transportation).
- Solidification: proven technique for treatment of industrial/municipal wastes. Estimated cost $45-75/yd^3 of wastes (not including cost of removal and disposal).

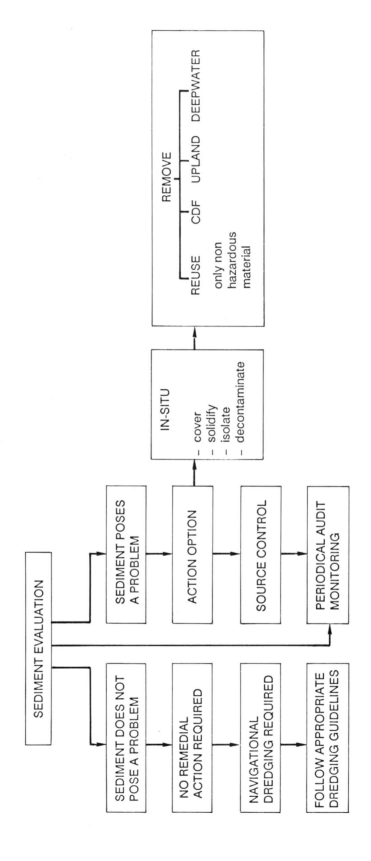

FIGURE 1 Sediment remedial options.

- Dewatering and separation of sediment into coarse (clean) and fine (contaminated) fraction using centrifuges and hydrocyclones.

    1. Further treatment of contaminated fraction by acid leaching/ion exchange (for metals), thermal treatment (for mercury), solvent extraction (for PAHs and oil).
    2. Re-use of decontaminated material or landfill disposal. This option reduces the volume of highly contaminated sediments but involves a high initial cost for equipment.

- Decontamination of PCB-contaminated sediments by low-temperature oxidation, chlorine removal, pyrolysis, removing and concentrating, vitrification, use of microorganisms or different chemical treatments.
- In situ containment of contaminated sediments by synthetic membranes.
- In situ chemical treatment and microbiological treatment.
- Various biological treatments.

Some Limitations and Remedial Options Requiring More Testing

- Hydraulic control navigation relocation: suitable for docking slips/shipping channels; minimizes resuspension of contaminants during shipping activities.
- Reclamation of strip mines and quarries: proximity of site is the major constraint, groundwater impact must be considered. Estimated cost, $12.50/yd^3/for 200 mi transportation.
- Landfill: considered as a poor use of available expensive landfill space. Estimated cost, $3.00/yd^3 (disposal), $0.45/yd^3/mi transport to site).
- Upland fill: Geographic limitations, availability of land, transport to the site. Estimated cost, $0.50/yd^3/mi (pumping), $185,000/mi (pipe-capital cost).

The complexity associated with the selection of suitable remedial options can be demonstrated by Table 1, which summarizes technologies associated with PCB-contaminated sediment. The technologies listed have been excerpted from *PCB Sediment Decontamination: Technical/ Economic Assessment of Selected Alternative Treatment*, Research Triangle Institute for the U.S. EPA, December 1986.

BASELINE MAPPING

Before deciding on an appropriate remedial option, one must know the nature and extent of sediment contamination. Baseline mapping and sampling of an area in question is essential. Mapping identifies the geographic extent of sediment accumulation zones versus erosional areas. This information allows for making loading calculations (once the accumulation rate is known). Also, mapping leads to a subjective

TABLE 1. Technologies Associated with PCB-Contaminated Sediments

Process Type	Process Name	Description	Reaction Temp.	Reaction Time	Efficiency	Development Stage	Feasibility Limitations	Cost
Supercritical Water Oxidation	Modar Supercritical Process	• pressurized to supercritical pressure • oxidized in a controlled rapid reaction	>374 C	NA	>99.9995% residual = <0.1ppb	pilot test	• control of waste material • adaptable • capacity = 570m/day	$250-$733/m^3 (including dredging, transport and treatment
Nucleophilic Substitution	KPEG Terraclean Cl process	• potassium hydroxide (KOH) and polyethylene glycol (PEG) react to form an alkoxide • reagent = PEG, KOH and dimethyl sulfoxide	150 C	30-120 min	>98% residual = <1ppm	pilot tests	• knowledge of analytical chemistry is necessary	4 hour cycle = $160.37 8 hour cycle = $191.16
Radiant Energy Ultraviolet Light	LARC Process	• light activated reduction of chemicals • uses isopropanol, sodium hydroxide and UV light	NA	1.5-2.0 hours	>90% residual = 38-50	lab tests	• concentration of chlorine organics can be controlled • flexibility	$233-$336/m^3
Pyrolisis	Advanced Electric Reactor	• rapid heating with intense thermal radiation • heat transfer by radiative coupling • products = H, Cl, HCl elemental carbon, solid derived waste	2000-2300 C	NA	>99.9999% residual = <1ppb	pilot tests	• pretreatment needed • requires 1,100 Kwh electricity	$829-$942/m^3
Extraction	Acurex Solvent Wash	• solvent is chosen by comparison of absorption isotherm and diffusion rates • each wash has ½ the PCB content of the previous wash	NA	NA	residual = <2ppm	pilot tests	• tolerates up to 40% water	$569/m^3
Extraction	O.H. Materials Process	• slurry soil with methanol	NA	NA	97% residual <25ppm	field test	• predry soil to <5% moisture • solvent resold	$100-$514/m^3
Extraction	Soilex Solvent Extraction	• solvent = kerosene and water • 42 to 45 volume percent water works best	NA	3 hours per batch	95% residual 6-9ppm	pilot stage	• accepts wet sludge • generates RCRA waste	$856-$913/m^3

TABLE 1. continued

Process Type	Process Name	Description	Reaction Temp.	Reaction Time	Efficiency	Development Stage	Feasibility Limitations	Cost
Extraction	N.Y. University Low Energy Extraction Process	• low energy technology • separate into solid and liquid • leach with hydrophilic solvent	NA	NA	NA	confirmed experimentally	• present capacity = 199 m^3/day	$56.67/$m^3$ (treatment only)
Extraction	Propane Extraction Process – Critical Fluid System	• use propane, ambient temperature and 1379 kPa pressure	NA	NA	NA	tested for refinery sludge	• multiple extractions are necessary • capacity = 114L/min	$155-$266/m^3
Extraction	Best Extraction Sludge Treatment	• use triethylamine (TEA) and the natural settling process • mechanically separated	NA	NA	1st extract removes 79.3% subsequent extracts 98.7%	NA	• recovered TEA is recycled • capacity = 520m^3/day (peak = 675m^3/day)	133.30/m^3
Vitrification	Battelle in Situ Vitrification Process	• electrodes are inserted • graphite and glass form a conductive path • high temperatures melt soil	NA	NA	99.9% no residual in vitrified block	pilot stage	• must be dredged • product (solid glass and crystalline) may be more costly to deposit • predrying lowers cost and energy	$293-332/$m^3$
Micro Organisms and Enzymes	Indigenous and Conventional Chemical Mutants	• found naturally in PCB containing soil/sediment	NA	NA	80-100% • fungi less effective than bacteria	has been tested	NA	NA
Micro Organisms and Enzymes	Enzyme Mechanisms	• mixed cultures of bacteria mineralize PCBs with 4 or fewer chlorines/molecule	NA	NA	NA	NA	• unable to degrade chlorinated benzoates	NA
Micro Organisms and Enzymes	Bio-Clean Process	• uses anthrobacteria and/or other naturally adopted microbes • 2 step process 1. extraction 2. destruction	heated to 82 C cooled to 30 C	3 days	• selected PCB cogeners have been reduced to 10ppb in 48-72 hours	lab tests pilot scale available	• naturally occurring organisms • high PCB concentrations inhibit degradation • capacity = 650m^3/day	$156/$m^3$
Micro Organisms and Enzymes	Sybron Bi-Chem 1006 PB/Hudson River Isolates Process	• limited information	NA	NA	NA	lab scale	NA	NA

TABLE 1. continued

Process Type	Process Name	Description	Reaction Temp.	Reaction Time	Efficiency	Development Stage	Feasibility Limitations	Cost
Micro Organisms and Enzymes	Composting	• aerobic and anaerobic • 60% moisture content	55 C	NA	aerobic = 62% in 4 weeks residual = 504-688 anaerobic = 27-47% residual = 825-1120ppm	NA	• lots of work site space needed • lack of control over the weather	NA
Solidification	Solidification	• eliminate free water using a setting agent • physical stabilization	NA	NA	NA	NA	NA	NA
Dredging	Mechanical	• directly apply mechanical force	NA	NA	NA	NA	• capacity = 380m^3/hour • turbidity • can work in close quarters	NA
Dredging	Hydraulic	• remove and transport in liquid slurry • slurry contains 10-20% solids	NA	NA	NA	NA	• >7650m^3/hour • cumbersome	NA
Dredging	Pneumatic	• crane suspended pump	NA	NA	NA	NA	• can handle up to 70% solids • useful around ports	NA
Dredging	Special Purpose Dredge	• mudcat • handheld • land based earth loading equipment	NA	NA	NA	NA	NA	NA
Wastewater Treatment Methods	Mechanical Removal	• absorb onto suspended solids and partition into fats	NA	NA	50-70%	NA	NA	NA
Wastewater Treatment Methods	Activated Sludge	• useful as an adjunct treatment	NA	NA	• non-clorinated degraded in 6 hours • tetrachlorinated degraded only 42% in 15 days	NA	NA	NA

TABLE 1. *continued*

Process Type	Process Name	Description	Reaction Temp.	Reaction Time	Efficiency	Development Stage	Feasibility Limitations	Cost
Wastewater Treatment Methods	Trickling Filter	• crushed rock, slag or stone provide area for biological growth	NA	NA	NA	NA	• aqueous flow medium is necessary • only dilute soluble PCB isomers • no performance data	NA
Wastewater Treatment Methods	Special Biological Treatment Process	• biodisc • bio-surf • ecolotrol • freeze-dried biochemical solution	NA	NA	NA	• haven't been demonstrated for PCBs	NA	NA
Wastewater Treatment Methods	UV/Ozonation/ Ultrasonics and UV/ Hydrogen/ Ultrasonics	• use of ultrasonics and UV Ozone • 3 steps: extraction, solids separation and treatment of PCBs	NA	NA	91-100%	bench scale tests	• based on proven technology • never tried on PCBs	Probably high
Wastewater Treatment Methods	Carbon Absorption	• carbon slurry absorbs impurities	NA	NA	• capable of reducing to <1ppb	NA	• carbon must be replaced	not very cost effective

SOURCE: PCB Sediment Decontamination — Technical/Economic Assessment of Selected Alternative Treatment, Research Triangle Institute for the U.S. EPA December, 1986

understanding of the physical conditions for implementing certain controls within an area of concern.

Baseline maps provide the basis for modeling resuspension and transport in an area and gives additional information on sediment movement into a receiving water body.

Equally important as mapping is the formulation of a comprehensive sediment collection program involving streams and suspended sediments, bottom sediments, sediment resuspension and transport, sediment chemistry, nutrients, organics and sediment toxicity and bioavailability.

OBSERVATIONS

- A knowledge of the ranges of sediment contamination and hot spots determines the nature and feasibility of each option.
- Unless the material is clean, a potential for exposure to the ecosystem already exists. Therefore, the remedial option chosen must minimize or eliminate that exposure pathway.
- Most options can involve removal of the potential risk of exposure to another compartment of the environment.
- There is a need to consider the long-term management of the option chosen (stewardship).
- When faced with a range of contamination one must investigate the utility of a variety of remedial options having both short-term and long-term feasibility.
- There is rarely a quick-fix option for areas possessing a range of contamination.

CONCLUSIONS AND RESEARCH RECOMMENDATIONS

1. Technologies for remediating contaminated sediments, which are already being employed outside the Great Lakes basin require immediate, first-hand evaluation, and where feasible, they should be used in the Great Lakes.
2. These technologies presently at or near pilot-scale operation need to be comparatively tested in one or more large-scale demonstration projects.
3. The more rapid development of technologies currently under investigation in the laboratory needs to be supported and encouraged.
4. Further research associated with in situ treatment techniques, in particular, and techniques for inactivation and destruction are necessary.

ECONOMIC CONSIDERATIONS OF MANAGING CONTAMINATED MARINE SEDIMENTS

Thomas A. Grigalunas and James J. Opaluch
University of Rhode Island

ABSTRACT

Contaminated marine sediments pose highly uncertain but potentially serious threats to public health and the environment. However, cleanup of these sites is very expensive and costs increase rapidly with level of effort. Thus, important tradeoffs are faced in the social decision concerning appropriate cleanup level. This paper discusses the application of economic analysis as input to the social decision process for managing contaminated marine sediments. Two general approaches are outlined: cost-benefit and cost-effectiveness analysis. Both approaches can provide valuable input into remedial action decisions at a site and in allocating efforts among multiple sites (fund balancing). However, significant difficulties and uncertainties characterize all approaches for managing contaminated marine sediments and economics is no exception. The difficulties and potential for application of the two economic approaches are discussed, along with the potential role of strict liability for damages in providing incentives for source control to avoid creation of new sites.

INTRODUCTION

Contaminated sediments occur in marine coastal areas throughout the United States and in the Great Lakes (A. D. Little, 1987; U.S. Environmental Protection Agency [EPA], 1985; Office of Technology Assessment [OTA], 1987; U.S. Dept. of Commerce [DOC], 1988). This contamination stems from a variety of point and nonpoint sources including day-to-day releases from industry, sewerage treatment plants, urban runoff, rivers, federal facilities, shoreline erosion, atmospheric sources, and periodic spills from vessels, pipelines, and shoreside facilities. The substances contained in sediments include heavy metals, synthetic organic compounds, petroleum hydrocarbons, and other materials.

Concern with contaminated sediments stems from the threat they pose to public health and the environment. At sufficient concentrations, toxic substances give rise to health threats to individuals exposed either directly through contact with contaminated materials or much more likely, indirectly via the food web. Risks to health from consumption of contaminated shellfish or finfish have caused public officials to close fishing grounds or restrict the catch of certain species, thereby imposing economic losses on commercial and recreational users of the affected species (e.g., Freeman, 1987; OTA, 1987; A.D. Little, 1987). Indeed, even the perception of a possible but uncertain threat to health from consuming fish exposed to toxic substances can impose losses (Schwartz and Strand, 1981).

Other losses also can result from contaminated marine sediments. Direct lethal effects on adult fish and shellfish, juveniles, eggs and larvae can cause short- and long-term commercial and recreational fishery losses (Grigalunas et al., 1987, 1988). Indirect or ecological effects can lead to losses of commercial and recreational fisheries, waterfowl, and other marine resources through loss of habitat (Kahn and Kemp, 1985) and via the food web (Grigalunas et al., 1987, 1988). Concern about exposure to toxic materials also may impose losses on recreational beach users, and a reduction in property values can occur as a consequence of a loss in amenity services at or near a contaminated site (e.g., Freeman, 1987).

Although the presence of contaminated sediments can impose a variety of losses, remedial actions typically are very costly and in many cases the cost increases rapidly as additional levels of remediation or treatment are sought. For example, remedial actions described or proposed for New Bedford Harbor (EBASCO Services, Inc., 1987); the Hudson River (Mark Brown, New York State Department of Environmental Conservation, personal communication); and Commencement Bay (Lukjanowicz et al., 1988) could cost millions of dollars. An analysis of alternative levels of cleanup of Hudson River polychlorinated biphenyls (PCBs) show rapidly increasing unit costs with increasing levels of cleanup (National Research Council [NRC], 1979). Hence, additional degrees of public health protection and environmental benefits can be achieved--but typically only at far greater cost. Given limited funds to use in cleanup of the rapidly increasing number of contaminated sites throughout the country, higher expenditures on cleanup at one site implies smaller remaining budget for cleanup at other sites. Thus, important tradeoffs are faced in determining the level of cleanup that should be carried out at any particular site.

Given the potential significance of public health effects and environmental costs at contaminated marine sites, on the one hand, and the potential high costs of taking corrective action at all sites, on the other hand, issues relating to the management of contaminated marine sediments are of major national importance. This importance is reflected in the passage of the Comprehensive Environmental Response, Compensation and Liability Act of 1980, CERCLA (PL 96-510) and is given additional emphasis by the enactment of the Superfund Amendments and Reauthorization Act of 1986, SARA (PL 99-499). SARA adds a new criterion to the hazardous ranking system, which is used to assess sites to determine whether they should be included on the National Priorities List (NPL). Under the act, consideration must be given to "the damage to natural resources which may affect the human food chain" (Sec. 105 (a)(2)). As a consequence, SARA increases the likelihood that marine contaminated sediment sites will be included on the NPL and thereby be eligible for use of fund-financed remedial actions.

Deciding how, to what extent, and whether to remediate at a site on the NPL unavoidably confronts decision makers with very difficult decisions and tradeoffs. SARA mandates that the remedies selected must

1. protect human health,
2. be cost-effective,
3. meet federal and state standards, and
4. utilize permanent remedies and alternative technologies "to the maximum extent practical" (Sec. 121 (b)).

However, the level of risk to health will depend on the amount committed to an action and the technology used, and remedial actions will differ in their cost and degree of permanence. SARA expresses a clear preference for permanent remedies, but permanent actions are not required and have been avoided because of their high cost, at least in the short run. Hence, implicit--if not explicit--tradeoffs among goals cannot be avoided in making a decision at any site. The difficulties involved are underscored by the evolving state of the art for remedial action technologies; the formidable problems inherent in quantifying risks to human health and the environment (Lave, 1987); the need to make remedial action decisions among multiple sites in the context of fund balancing; and the requirement under SARA that the public and potentially responsible parties be actively involved in the decision process.

Clearly, site-specific factors are critical considerations in remediation decisions at a given location. The case studies presented at the workshop illustrate the particular concerns that drive the desire for remedial actions, the strategies considered, and the institutional factors that influenced the decisions made or the actions considered in specific cases. Nonetheless, there are general principles which transcend particular applications.

This paper examines some of the economic principles and issues that arise in deciding whether, how, and to what extent to remediate at a site. These approaches could be used to complement current approaches to evaluating sediment management alternatives (EPA, 1985b, 1985c, 1986). A generally applicable, economics-based methodology using concepts from benefit-cost analysis and cost-effectiveness analysis is developed, and suggestions are made concerning how this methodology could be applied so as to capture the special characteristics of particular sites. The potential usefulness and limitations of the economics methodology for assisting in decisions concerning the management of contaminated marine sediments is described.

To make economic considerations more applicable to contaminated marine sediment issues, it is particularly important that the various components be quantifiable. Hence, particular emphasis will be placed on measurement, and the potential for quantifying these arguments for particular applications will be discussed. It is recognized that quantification of benefits from remedial actions is exceedingly difficult, and may not be possible in all cases. Hence, the use of economic analysis is limited in some cases. However, it also is recognized that considerable uncertainty surrounds the use of any approach--whether based on concepts from the natural or the social sciences--used to make decisions at and among sites, and thus substantial use must be made of imprecise information and informed judgment.

In keeping with the scope of the workshop, particular attention is given to the role of cost-effectiveness analysis in making management decisions at particular sites. However, remediation decisions at individual sites are made within a broader framework. This broader framework involves, for example, fund balancing among sites and implicit if not explicit judgments concerning relative public health and environmental benefits and costs. In another vein, the liability provisions established under CERCLA and the Clean Water Act, as amended, have potentially important implications for encouraging source control to help avoid the creation of new contaminated sediment sites (Grigalunas and Opaluch, 1988). Recognizing the importance of this broader framework, this paper goes beyond consideration of cost-effectiveness and outlines

1. the potential contribution and problems which arise from the use of benefit-cost analysis to help guide remediation decisions, and
2. the use and limitations of liability as an approach for source control.

GENERAL ECONOMIC CONSIDERATIONS

As noted, contaminated marine sediments are widespread and can impose a number of public health and environmental costs, and in particular cases, these costs can be substantial. For example, New Bedford Harbor, an area heavily contaminated with PCBs, is the marine Superfund site that has been most carefully studied by economists. Research funded by the National Oceanographic and Atmospheric Administration (NOAA) estimated the present value of damages to marine resources (using a 3 percent real rate of discount) to range from a total of $39.6 million to $52.4 million in 1985 dollars (Freeman, 1987). These damages resulted from injury to

1. the lobster fishery,
2. public beaches and recreational fishing, and
3. reduced amenity services experienced by people living near the harbor alleged to arise from high concentrations of PCBs.

Direct losses to striped bass recreational fishermen alleged to have resulted from Hudson River PCB contamination have been estimated by New York State to be more than $4 million annually (OTA, 1987, p. 43).

New Bedford Harbor and the Hudson River are among the most dramatic examples of contaminated marine sediment sites. However, numerous other cases exist of restrictions imposed on harvesting marine resources from areas contaminated with metals and organic chemicals, particularly for shellfish and bottom species near urban and industrial centers (see, e.g., OTA, 1987; A. D. Little, 1987; Haberman et al., 1983).

On the other hand, remedial action alternatives can require a major commitment of resources, and the costs are very sensitive to the removal and disposal option selected. To illustrate, in the case of Everett, Washington on Puget Sound, it was found that the construction cost of $55 million for upland disposal was almost four times larger than the $14.5 million construction cost of the selected alternative, contained aquatic disposal in the deep waters of Puget Sound (Lukjanowicz et al., 1988, p. 38). Further, available estimates suggest rapidly increasing costs with additional degrees of remediation. For example, an NRC report of PCB contamination (NRC, 1979) concludes that cost per pound of PCBs removed from the Hudson River vary from $65 per pound PCBs removed for initial cleanup of hot spots to $3,153 per pound for complete removal of low-concentration river sediments, as shown in Table 1.

Ideally, one would like to measure the economic damages at a site prior to remediation, and the reduction in damages (the resultant benefits) expected at different levels of remediation. With this information, it would be possible to compare the increments in benefits from greater levels of remediation with the associated increments in cost. It then would be possible to assess whether remedial action at a site was worthwhile on economic grounds and to use economic principles to help guide the extent of remediation--a textbook solution to the "how clean is clean" dilemma. Difficulties inherent in measuring the full spectrum of damages make such an ideal approach beyond the reach of the state of the art in many cases. Nonetheless, in a number of instances some of the potential benefits from remediation can be quantified, and this information can be used as part of the decision process concerning proposed remedial actions.

As part of any analysis, attention must be given to several important factors. Particularly important is the potential for uncertain future costs. For example, landfilling or disposal of contaminants in marine waters may cause adverse environmental impacts at some future, uncertain date if the substances become re-released. SARA recognizes

TABLE 1 Incremental Costs of PCB Control in Hudson River (in 1978 Dollars)

Policy	Incremental quantity controlled (kg)	Incremental control costs ($ million)	Incremental cost per kg (dollars)
A. Maintenance dredging	23,100	$ 2.5	$ 108
B. Removal of remnants	7,700	$ 0.5	$ 65
C. Removal of stabilized remnant deposits	15,100	$ 3.3	$ 219
D. Hot spot dredging	77,000	$ 22.4	$ 291
E. Removal of all river sediments	55,600	$175.3	$3,153

this possibility and requires that in addition to short-term costs, long-term costs must be considered when selecting a remedial action. Such long-term costs include potential adverse health effects, long-term maintenance costs, and potential future remedial action costs should the selected alternative fail (Sec. 121 (b)). A rigorous examination of long-term costs thus would include an analysis encompassing all feasible options and an assessment of the probability of their failure at points in time. Further, the analysis must include the probability that damages would result, given failure of each alternative; the chance that further remedial action would be taken; and the cost of such action and maintenance costs at each point in time. Clearly, the data requirements for such an analysis impose a truly major research burden on scientists and others charged with assessing remedial action alternatives.

Other factors also must be considered. Potential beneficial effects in addition to health and environmental benefits could result in particular cases. For example, it may be possible to use removed materials to construct islands or provide other natural resource enhancement (e.g., Landin, 1988) or to recover materials for reuse, as is planned for uranium contained in sediments at Port Hope, Canada (Orchard, 1988). Another factor to be considered is the availability and capacity of upland disposal sites. Given the limited availability of landfills and the difficulty in siting new facilities, the opportunity costs of use of the disposal site must be considered in evaluating the social impacts of alternative remediation strategies when upland disposal is being considered.

Finally, it is important to recognize that in some cases, sediment contaminants may degrade/dilute over time or become covered with clean material as a result of natural deposition. In these cases, the benefits to be achieved through remedial action can be negligible--or, in fact, severe health or environmental costs could result should resuspension occur. The James River kepone case is an important example. Natural sedimentation and dilution have reached the point that commercial fishing in the James River will be allowed for the first time in more than a decade (Huggett, 1988).

Thus, the simplified contaminated sediment management problem can be depicted in two stages;

1. public health and environmental effects--sediments are perceived to cause public health and other social losses.
2. removal and remediation--these actions result in health and environmental improvement but are costly, in terms of monetary costs of the action and possibly environmental costs associated with ecosystem disruption from physical removal and disposal in case of nonpermanent actions. Landfilling, nearshore, and offshore disposal may imply further uncertain costs or possible benefits.

Although it is easy to enumerate possible costs and benefits that might arise in particular cases, it is difficult to provide quantitative economic information to assess net social impact. In this regard the potential use of uncertain economic information is on the same footing as the use of uncertain information from the natural sciences for making management decisions at sites. Hence, the choice is not use of one approach that provides precise information versus another that yields inexact information. Rather, all available approaches necessarily involve important elements of imprecision, subjectivity, and judgment. Important social decisions must necessarily be made within this uncertain environment since no amount of research can completely resolve the uncertainty faced by society.

Clearly, a great deal of scientific and technical information is needed to describe the problem and the tradeoffs that result from the management alternatives for addressing the problem. However, the final choice among the alternatives is necessarily a social decision based on a weighing of these tradeoffs. The next section describes two general frameworks that have been used by economists to address such issues.

ALTERNATIVE ECONOMIC FRAMEWORKS FOR ADDRESSING THE CONTAMINATED MARINE SEDIMENTS PROBLEM

Benefit-Cost Analysis

Benefit-cost (B-C) analysis attempts to quantify all important beneficial and detrimental impacts of a proposed action in dollar terms. This approach potentially can be very valuable because it is very flexible, and, moreover, it is the only approach that can indicate whether or not remediation is a good investment of society's resources. Hence, to the extent B-C analysis can be used as part of remediation decisions, it puts investments in this area on the same economic footing as public investments for environmental improvement in other areas and for public projects in general.

However, B-C analysis is limited in its potential applicability due to the difficulty of providing a monetary measure of damages when evaluating commodities that are not sold in established markets, as is typically true when evaluating many environmental damages. Despite the extreme difficulty in quantifying environmental damages in dollar terms, a great deal of progress has been made in measuring these nonmarket effects.

Economic Methods for Evaluating Nonmarket Environmental Goods

Many environmental damages involve nonmarket goods and services--that is, goods and services that are not traded in the marketplace, such as sports fishing or public beach use. Values of nonmarket goods are sometimes viewed as "subjective," and it is often claimed that these values cannot be measured in monetary terms, as is the case for market goods. However, values for market goods in many respects are no less subjective than those for nonmarket goods. For example, consumer preferences for taste, texture, color or other attributes of salmon are no less subjective than preferences for viewing wildlife; consumer willingness to trade off price differences for salmon attributes could reveal the value of these attributes (Anderson, 1988). The difficulty in measuring values for nonmarket goods does not derive from the fact that they are more subjective, but rather from the fact that these preferences are not directly revealed in market transactions.

Two general approaches have been developed for evaluating nonmarket commodities: the revealed preference approach and contingent valuation approach. A brief description of each approach follows.

Revealed Preference Approaches

The basis of the revealed preference concept is that through an individual's actions, his or her preferences are revealed. Use of revealed preference is most straightforward for economic valuation of goods and services sold in the marketplace since it is relatively easy to infer preferences from these market decisions. However, the concept of revealed preference also can be used for nonmarket goods by using related-market techniques. This approach is based on the concept that the value a good or service which is not traded on a market can be inferred from closely-related goods which are sold on the market (Freeman, 1979). Two related-market approaches are outlined and illustrated in the following paragraphs.

Travel cost approach. Suppose the problem is to evaluate a non-market recreational experience, such as a fishing trip to a particular site. If a fishing experience could be bought in the marketplace, such as through an entrance fee, then the observations on the decisions individuals made, given this market price of participation, could be used to value the fishing experience. Although there may be no entrance fee for most fishing sites, the cost of participating can be measured, since in order to fish at a particular site one must travel to the site and incur certain costs in the process. Thus, the travel cost can be viewed as the price of participating, and usual market valuation approaches can be applied by examining participants revealed behavior. A recreational fisherman's value for changes in catch rates, for example, can be measured in terms of willingness to travel longer distances to more remote sites, which have higher catch rates (Brown and Mendelsohn, 1984).

An example of the application of this related-market approach to the problem of contaminated marine sediments is provided by the economic damage assessment study of New Bedford Harbor PCBs. It was hypothesized that public awareness of the PCB contamination in the harbor affected recreational beach users and recreational fishermen. Damages would be reflected in a decrease in the demand for these recreational activities relative to the no pollution situation. For recreational beach use, the study focused on three beaches adjacent to waters or sediments with significant PCB concentrations. Telephone interviews were conducted with a random sample of residents of nearby communities. The results revealed that "among those aware of the PCB pollution, up to twice as many households would have visited the beach in 1986 if the PCBs had been cleaned up" (Freeman, 1987, p.9). The estimated total damages to beach users alleged to have resulted from the PCB pollution of the sediments was $8.3 million (Freeman, 1987, p.10).

For recreational fishing, it was estimated that the PCB pollution resulted in the diversion of 41,935 trips to other sites with an average cost of diversion per trip of $1.60. Using this approach, annual damages alleged to result from the PCB pollution were estimated to be $67,100, and the present value of these alleged damages was $3.1 million in 1985 dollars.

Hedonic-price approach. This related-markets approach is based on the concept that a particular market good is composed of various nonmarket characteristics, and that given market prices for similar goods with differing levels of the characteristics, one could calculate the implicit "price" people are willing to pay for the characteristics. For example, the characteristics of a house can be described in terms of square footage, style, age, number of bathrooms, yard size, and neighborhood characteristics. Given data on a large number of house sales, statistical techniques could be used to relate price differentials to each of the characteristics in order to identify the implicit price of these characteristics. If one of the characteristics is, for example, sediment quality of the adjacent marine waters, then the value of sediment quality to an individual can be estimated in terms of the additional amount the individual would be willing to pay for a house in an area of high sediment quality, as compared to a house that has identical characteristics, or after correcting for other differences in characteristics, except for the one "neighborhood" characteristic, sediment quality. Note that willingness to pay in this context is based on actual or revealed behavior as reflected in the sales price of homes.

This approach is particularly useful for valuing the effects of contaminated marine sediments and, in fact, was applied in the New Bedford Harbor PCB damage assessment study. Any reduction in the amenity services of the harbor should be reflected in relative decreases in the prices of nearby residential properties. It was hypothesized that the effect of the pollution on housing prices would be stronger for those houses near the more contaminated harbor waters. The study area was divided into three zones of diminishing levels of pollution, and data were assembled on residential sales prices for single family dwellings located within two miles of the New Bedford harbor shoreline. The results indicated that the estimated total damages (reduction of property values) alleged to result from PCB contamination of marine sediments was between $26.2 million and $39.0 million in 1985 dollars (Freeman, 1987, p.17).

Contingent valuation approach. This survey-based approach asks individuals how they would behave under some set of given circumstances in an attempt to elicit the individual's preferences. A contingent valuation study of a fishing experience, for example, may ask whether an individual would participate in fishing at a particular site if it costs $X to do so; or the individual could be asked whether they would pay $X to experience (or avoid experiencing) a specified increase (decrease) in the catch rate. The responses to the questionnaire are then used as though they were actual market behavior to assess the value of the resource issue in question. Note that this approach, in contrast to the approaches outlined in preceding paragraphs, is not based on revealed behavior. A detailed assessment of the state of the art in contingent valuation, including the importance of sources of potential bias and "reference operating conditions" for controlling bias, is contained in Cummings et al. (1985).

Valuing Public Health Risks

Public health effects are a primary concern in assessing remedial actions, and SARA establishes a number of important health-related authorities to provide a better understanding of the health effects from exposure to toxic substances (Sec. 110). From an economic viewpoint particularly difficult issues arise when contaminated sediments result in a threat to human health. Revealed preference is probably not useful for valuing an individual's life, since this would be tantamount to determining what one is willing to pay to continue living, or willing to accept to die, neither of which are reasonable concepts. However, most environmental impacts increase the *risk* of death for some population, rather than directly incurring death of an particular individual. Revealed preference may be useful for evaluating health risks since individuals make decisions that change risks every day, for example, through the decision to wear seatbelts, drive a car, smoke cigarettes, etc. It is possible to elicit values related to increments in risk by observing how individuals behave when making decisions that determine the level of risk.

One particularly useful related-markets approach to valuing mortality risks is to examine behavior in choosing risky occupations (Fisher et al., 1988). For example, bridge painters are often paid differing amounts depending on whether they paint the more risky, top parts of the bridge, or the relatively safer, lower parts. Thus, the individual trades off wages for risk of death, and behavior revealed in the labor market can provide information on the individual's preferences for safety, or lack of risk.

The key concepts used in economic analysis on mortality risk are the value of a statistical life and cost per life saved, which will be discussed below. Valuing a statistical life is an example of a cost-benefit approach that can be illustrated as follows. Suppose that contaminated sediments lead to human ingestion of a toxic substance through the food chain and that risk analyses show that this level of concentration implies the probability of death through cancer is increased by .001 percent for each of one million people. The number of statistical lives lost by this pollutant is

$$.00001 \times 1,000,000 = 10$$

If, on average, individuals are willing to give up $100 in wages in order to reduce the risk of death on the job by .001 percent, then the value of a statistical life is $10 million, as revealed by actions of the individuals. This implies that a remedial action that would remove this pollutant from sediments, hence removing the health risk, would result in $100 million in benefits. This concept can be applied to the problem of contaminated sediments, given a risk analysis of the increased mortality which result from various exposure pathways.

It should be noted that the approach discussed in this section assesses statistical lives saved, or mortality risks, not reductions in sublethal effects, or morbidity. The concepts to be used to evaluate the benefits from reduced morbidity--reduced medical costs, smaller amount of time out of work due to illness, and so forth--are relatively straightforward. Note, however, these approaches do not value the illness per se, such as the discomfort or suffering of the individual, but only the associated monetary costs, and thus would strictly understate the value of reduction in morbidity rates. Additionally, it can be very difficult to establish the incremental improvements in health from total or partial remediation of contaminants at a given site due to the inherent difficulties in isolating the effect of exposure to one or more contaminants from all other effects that influence health.

An alternative approach that has been employed to value mortality is the so-called human capital approach. This approach uses the earning potential of the individual over his or her future life as the value of human life. The human capital approach has been widely used in the courts in cases of wrongful or accidental death. While this approach measures potential future earnings, it does not place a value on the loss of life, per se, nor does it measure the individual's willingness to accept risk of death. In addition, the approach has an innate bias against those who have little or no direct wage income, such as a housewife whose services are not valued through the market. Thus, the human capital approach would be expected to significantly understate the value of life. Indeed in practice, the value of a statistical life determined from revealed preference studies tends to be significantly higher than the value determined from the human capital approach. For example, the EPA uses figures of $400,000 to $7 million as a range of reasonable values for a statistical life based on revealed preference. A present value of this size implies a perpetual income of $32,000 to $560,000 at an 8 percent discount rate.

Cost-Effectiveness Analysis

General Considerations

Generally, health and environmental benefits of remediation at a site are not explicitly measured in dollar terms, but implicitly may be judged to be worth the costs of remedial measures. In these cases, cost-effectiveness (C-E) techniques can be used to guide the selection of remediation alternatives by helping to assess the relative cost and the effectiveness of the alternative removal and disposal strategies. SARA requires that C-E is to be considered in the evaluation of remedial actions, and that long-term as well as short-term costs be taken into account.

C-E analysis provides a systematic way for determining (1) the least-cost approach(es) for achieving a given objective, or (2) the maximum level of the objective which can be achieved for a given cost. Correctly applied, C-E can be a powerful tool because it potentially allows decision makers to screen remedial action alternatives on the basis of a common measure. All else being equal, if remedial action alternative A is less expensive than B, then A clearly is preferred; or, if A and B are equally expensive but A results in a greater public health and environmental improvement than B, then A would be selected. However, all else generally is not equal, comparisons rarely are so straightforward, and as a result C-E analysis is subject to several potentially important shortcomings.

Potential Problems with Cost-Effectiveness Analysis

C-E analysis will lead to misleading results if (1) important costs are ignored or (2) costs for alternative actions at a site or between sites are not estimated using a consistent approach. As noted, SARA expresses a clear preference for permanent actions, involving thermal, biological or chemical treatment to reduce the volume, toxicity or mobility of the substance(s). Permanent remedies typically cost more initially than actions that do not involve treatment (e.g., capping). For example, a recent report found that for the ten cases studied (none marine), the average cost of the five cases involving permanent remedies through treatment of the removed materials was $16 million as compared to $7.5 million for nontreatment (impermanent) remedies (OTA, 1988). Assessing whether permanent remedies are more cost-effective when long-term as well as short-term costs are considered is extremely difficult, as noted above. Nonetheless, long-term costs must be considered if C-E analysis is to be a meaningful guide for remedial action policy.

Clearly, for C-E analysis results to lead to appropriate decisions, it also is vital that costs be assessed consistently. A recent report by OTA found that in some cases (none marine) different approaches were used by contractors to estimate costs. To the extent the use of inconsistent approaches at marine sites causes large differences in apparent costs, the usefulness of cost-effectiveness analysis can be severely compromised.

Application of Cost-Effectiveness to Public Health Risks

The C-E approach does not place a value on health or other benefits and hence cannot be used to provide an economic argument in support of, or against, remediation at a site. However, C-E analysis can play an important role in helping to guide the extent of remediation at a site and among sites (fund balancing). To illustrate this, the concept of cost-per-life-saved is used. Rather than placing a value on a statistical life, the cost-per-life-saved approach ranks options according to the number of statistical lives saved per dollar spent. Similarly, the approach could be applied for nonlethal affects, for example by evaluating cost per reduced cancer case.

To illustrate, suppose there is a fund of $50 million, and six non-mutually exclusive cleanup alternatives are being compared, as shown in Table 2. The alternatives may represent, for example, cleanup of various subareas within some particular contaminated site, such as the four subareas of New Bedford harbor discussed by Ikalainen and Allen (1988). Additionally, some alternatives may represent subareas from differing contaminated sites, such as four subareas in New Bedford Harbor and two subareas in the Hudson River.

For the cases depicted in Table 2, assuming no impacts other than reduced health risks, alternatives one, two, and three would be chosen, resulting in 14 statistical lives saved. Given these alternatives, this is the greatest reduction in risk that would result from this fixed expenditure of the $50 million fund. This approach implies a social willingness-to-pay per statistical life saved between $6.7 million, the highest cost per life saved for the alternatives chosen, and $8 million, the lowest cost per life saved for the alternatives *not* chosen. Given the same alternatives, the cost-benefit criterion outlined in the preceding section would justify alternatives one, two, three, and four, using the $10 million hypothetical value per statistical life derived above. This would result in 15 statistical lives saved and would result in expenditures of $58 million, which would require $8 million in addition to the $50 million contained in the fund.

SOURCE CONTROL: THE ROLE OF LIABILITY UNDER CERCLA AND THE CWA

Under CERCLA and the Clean Water Act, as amended, polluters are liable not only for cleanup and reasonable assessment costs, but also for "damages for injury to, destruction of, or loss of natural resources" (Sec.107.(a)(4)(C) (hereafter, injury to natural resources) resulting from a spill. Briefly, CERCLA provides for two types of

TABLE 2 Depiction of Cost-Effectiveness Strategy Using the Cost-per-Life-Saved Approach

	Alternative					
	One	Two	Three	Four	Five	Six
Cost ($ Million)	$10	$20	$20	$8	$12	$15
Number of statistical lives saved	6	5	3	1	1	1
Cost per statistical life saved	$1.7	$4	$6.7	$8	$12	S15

damage assessment regulations. The type A regulations provide a simplified approach, involving minimal field observation to be used for minor incidents of short duration, while the type B regulations describe methods for site-specific natural resource damage assessments with potentially extensive field observations, to be used for major incidents. Since contaminated sediment problems generally arise as a result of chronic releases over an extended period, the type B approach almost always will be the appropriate approach for measuring damages. For example, the economic damage assessment study of New Bedford harbor PCBs was a type B study, the first carried out under CERCLA (e.g., Freeman, 1987).

The two-tiered damage assessment approach mandated by Congress recognizes that undertaking a damage assessment can be very expensive. For example, the economic study of the damages alleged to result from the presence of PCBs in New Bedford Harbor cost $0.5 million (Meade, NOAA, personal communication). For some cases, these assessment costs can exceed the value of the damages that can be ascertained. For example, 5,600 barrel *ARCO Anchorage* oil spill cost $250,000 for damage assessment, while the resultant damage estimate was $33,000. Clearly, it only makes sense to spend the large amounts of money necessary to carry out field significant investigation when assessing damages for very large incidents, such as the New Bedford Harbor case.

The intent of CERCLA is to compensate governments for damages to publicly controlled natural resources in their role as trustees of these resources. Thus, the primary goal of the act is to encourage fairness by compelling the responsible party to pay compensation for the damages resulting from their actions; the amount recovered is to be "available for use to restore, rehabilitate, or acquire the equivalent of such natural resources by the appropriate agencies..." (Sec.107(f)). However, as an unintended side-effect, the liability provisions of CERCLA create a legal framework for what is akin to a "tax" on pollution incidents covered under the act. As such, the damage assessment regulations introduce what could be an important new approach for using economic incentives to avoid pollution for a wide range of incidents (Opaluch and Grigalunas, 1984; Grigalunas and Opaluch, 1988). For example, in discussing the liability provision, the 1982 version of the Clean Water Act requires that "the Administrator shall . . . conduct a study and report to Congress on methods, mechanisms, and procedures to create incentives to achieve a higher standard of care in all aspects of the management and movement of hazardous substances."

The potential importance of incentives embodied in the liability provisions in the CERCLA regulations for source control is made clear by examining its applicability and unique characteristics. The regulations apply to virtually all publicly controlled natural resources, and encompass a wide span of pollution discharges. Also, CERCLA holds polluters strictly liable for their actions, so that following an incident, there is no need to establish negligence in a prolonged and costly court trial. Moreover, CERCLA establishes joint and several liability. Thus, any one polluter can be held liable for *all* cleanup or remediation costs, even if they contribute only a small share of the total amount released into the environment. Note, however, that these incentives are applicable only to accidental spills, and not, for example, to routine discharges permitted under the National Pollution Discharge Elimination System (NPDES) of the Clean Water Act.

Another unique and very important characteristic of the CERCLA natural resource damage assessment regulations merits emphasis. The regulations provide an advantage for trustees in that they carry the force of rebuttable presumption (Sec.111 (h)(2)). That is, if the process set out in the regulations is correctly applied by the authorized official following a spill, the resulting measure of damages is presumed to be correct, unless the potentially liable party can show otherwise by a preponderance of the evidence. In most cases it will be very difficult and costly to prove that the results of a damage assessment carried out under the act are incorrect, especially for the type A approach, by virtue of the fact that it is intended to be simplified and based on minimal field observation. Hence, the rebuttable presumption provision of the CERCLA can have important implications for the effectiveness of the damage assessment regulations, in general, and especially for the type A approach.

It should also be noted that under CERCLA, liability extends beyond custody of the material to include materials spilled by those under contract, directly or indirectly, with the firm. Thus, liability under CERCLA provides incentives not only for careful handling of materials, but also for careful choice of parties with whom to contract for waste removal and final disposition. Hence, CERCLA recognizes the importance of choice of contracted parties in order eliminate the obvious financial incentive to hire inexpensive "fly-by-night" contractors who practice midnight dumping of hazardous wastes.

There is some evidence that liability provisions have been effective in providing incentives for damage reduction. For example, in a study of industries which produce hazardous wastes, Killory (1987) concludes "[source reduction] has become an increasingly attractive environmental policy for the organic chemical industry because of the high costs of waste disposal and, more importantly, the greater liability that producers now incur for generated wastes." Further, the only empirical study of economic behavior under liability finds some evidence that is consistent with provision of incentives (Opaluch and Grigalunas, 1984).

Thus, the liability provisions of CERCLA may provide incentives both for source control and for careful handling and disposal of hazardous materials. This is an extension of the so-called "polluter pays principle" which holds the polluter financially responsible for costs associated with the harmful effects of pollution emissions. Note, however, that despite the success of incentive-based approaches in Europe, U.S. environmental legislation does not maintain this same incentive system for other sources of environmental pollutants, such as pollution emission under NPDES permits. For many years economists have argued for environmental policy based at least in part on financial incentives. One alternative would be a mixed system of direct regulation and financial responsibility that would require the firm to attain some stated treatment percentage, but would also require the firm to pay a fee for the remaining pollutants emitted or would pay a subsidy if pollution were reduced below that required by the regulation (Baumol and Oates, 1975; Roberts and Spence, 1976), perhaps accounting for potential locational differences in impacts (Tietenburg, 1978).

Although the act is widely applicable, the type A approach cannot be used in a number of important cases; and CERCLA itself is of limited applicability in some cases. For example, the type A approach cannot be used to assess damages from chronic releases, although the type B approach can be employed. CERCLA does not apply to releases under NPDES permits, although it does apply when the permitted release is exceeded, nor can the act be used to assess damages from releases of fertilizer or pesticides resulting from normal use. Where CERCLA does not apply, other laws and approaches will have to be used to control releases of contaminants into marine waters. However, these acts do not generally provide incentives, such as those implied by the liability provisions of CERCLA, the Clean Water Act, or the Outer Continental Shelf Lands Act.

The CERCLA natural resource damage assessment regulations are relatively new and, in many respects, novel. How effective they will prove to be depends importantly upon on several factors, including how active states are in implementing this approach. To be effective, trustees must be appointed, staff must become familiar with the regulations, and efforts to apply the regulations to releases under the act must be pursued. If the act is not implemented, than it will be of little use. It is not clear that trustees have fully explored the potential usefulness of this approach. Further, liability can only be applied in cases where one or more responsible parties can be identified. For many cases of illegal dumping this may not be possible.

A final note is in order. A unique part of CERCLA is the requirement that the damage assessment regulations be reviewed every two years and updated, as appropriate. Hence, there is the important opportunity to suggest new techniques or data to be included in updated natural resource damage assessment regulations. Also, the biennial review mandated for CERCLA may provide an opportunity to explore the feasibility of developing a simplified approach which could be applied to those contaminated marine sediment cases which may not warrant the high cost of a type B study but which cannot be encompassed within the present type A framework.

SUMMARY AND CONCLUDING COMMENTS

Contaminated marine sediments are of concern because they can impose a variety of adverse public health effects and environmental losses. At the same time, remediation can be very costly. Hence, whether, how, and the extent to which sites should be remediated are important national issues. However, quantification of public health and environmental effects unavoidably involves considerable scientific uncertainty and social tradeoffs. In light of the many uncertainties involved and the lack of clear criteria, these decisions concerning remediation at a site necessarily are based on imprecise information and important elements of uncertainty and subjectivity. Hence, these issues are principally social decisions, and not merely scientific, technological or economic issues.

Two economic frameworks are available for contributing to social management decisions at contaminated marine sediment sites. The two economic frameworks presented--benefit-cost analysis and cost-effectiveness analysis--can contribute to the decision process by making explicit the costs and benefits (in the case of cost-benefit analysis) of the alternatives. Despite the many difficulties inherent in quantifying some of the important factors, these approaches can be used to complement the use of scientific information in making decisions at or among contaminated sediment sites.

These economic frameworks provide a means of organizing the information in ways that can be helpful to decision makers. The techniques are particularly useful for identifying alternatives that achieve goals at excessive costs. For example, the cost-per-life-saved approach would identify policy options that save few lives at relatively high costs, in favor of alternatives that result in a greater reduction in risk per unit expenditure. To the extent that the benefits from alternative cleanup policies can be quantified, benefit-cost analysis can provide a perspective on relative benefits and costs to help determine whether goals appear to be reasonable. This can be particularly useful when "conservative" high or low estimates can be consistently utilized and imply an unambiguous solution. For example, if conservative, low estimates of benefits of a remedial action exceed costs, then certainly that level of remediation would be warranted. On the other hand, if costs greatly exceed benefits, even when overstated benefit estimates are used, then it is likely that somewhat less ambitious levels of action may be warranted, with reallocation of funds for expenditure at some alternative site. For example, complete remediation of all tainted sediments from a particular site may be prohibitively expensive and the costs of doing so would be beyond any reasonable level of benefits which may result. Finally, if the results of a cost-benefit analysis are not conclusive, then the project can neither be justified nor rejected on a C-B basis, and other considerations would dominate the decision.

It must be recognized, of course, that quantifying benefits and costs can be exceedingly difficult. This is particularly true when examining long-term costs associated with nonpermanent solutions. In evaluating costs associated with impermanent solutions, such as in situ capping, long-term costs associated with failure must be considered to make comparisons with costs associated with permanent solutions, such as incineration. EPA calls for a screening of the alternative actions to eliminate those that cost more but do not provide a "commensurate" public health or environmental benefit. However, determining whether these benefits are "commensurate" places a significant burden on scientists and economists. To do so the probability of failure for impermanent solutions must be determined, in addition to the consequences of failure considering the potential for, and costs of, any associated remedial action.

However, it is possible in many cases to measure the benefits from improvement, as was illustrated by the results presented for the New

Bedford Harbor damage assessment study. Moreover, it is important also to recognize that noneconomics-based approaches must also consider the same uncertainties and tradeoffs, but will do so in a way that leaves the tradeoffs implicit. In contrast, an economics-based approach specifies these tradeoffs explicitly, so that when choosing an action, the decision maker can see the alternatives being given up--the opportunity cost.

C-E analysis begs the question of whether remediation ought to take place at a site. However, given that a decision to remediate has been made, C-E analysis can be an important part of remediation decisions, and under SARA C-E is given a central role in designing remedial actions. This section reviewed the C-E approach and illustrated its application to public health effects. Properly applied, C-E analysis can provide a powerful tool for choosing among alternative approaches for remediation at a given site and for allocating resources among sites. However, there is the danger that the concept of cost effectiveness may be confused with the least costly remedial action. Various actions can only be compared on a cost-effective basis if the benefits are equal, but the costs differ; if the costs are equal and the benefits differ; or if the least costly alternative also results in the highest level of benefits. Cost-effectiveness analysis cannot be used to compare two alternatives where one is more costly but leads to greater environmental benefits.

Additionally, the role of liability as an approach for encouraging source control was examined. CERCLA holds polluters strictly liable for their actions. The prospect of paying potentially very considerable sums for damages, assessment, and remediation actions creates a powerful incentive to reduce the amount and the toxicity of materials potentially spilled, as well as to handle more carefully and dispose of the materials that remain. Another unique, and very important characteristic of the act is that the natural resource damage assessment regulations carry the force of rebuttable presumption. Thus, the damage assessment process is greatly facilitated by shifting the burden of proof.

Given the characteristics of the act and its broad potential applicability, the damage assessment regulations established by CERCLA clearly are a major development in environmental policy. CERCLA's damage assessment regulations also may represent a major, and perhaps unprecedented, expansion of the use of economic incentives to control pollution. However, to be effective trustees must be appointed and the act must be enforced. It is not clear that states have fully exploited the potential of the act for assessing damages and remediating contaminated marine sediment sites.

REFERENCES

Anderson, J. L. 1988. Analysis of the U.S. Market for Fresh and Frozen Salmon. Staff Paper. Kingston, RI: Department of Resource Economics, University of Rhode Island. 65 pp. (plus app.).

A. D. Little, Inc. 1987. An Overview of Sediment Quality in the United

States. Final report to Monitoring and Data Support Division, Office of Water Regulations and Standards, U.S. Environmental Protection Agency, Washington, D.C. (June).

Barnett, H. C. 1985. The allocation of Superfund, 1981-1983. Land Economics 61(3):255-262.

Baumol, W. J. and W. Oates. 1975. The Theory of Environmental Policy: Externalities, Public Goods and the Quality of Life. Englewood Cliffs, N.J.: Prentice Hall.

Brown, G. and R. Mendelson. 1984. The Hedonic travel cost method. Rev. Econ. and Stat. 66(3):427-433.

Cummings, R. G., D. S. Brookshire, and W. D. Schulze, eds. 1986. Valuing Environmental Goods. Totowa, N.J.: Rowman and Allanheld. 270 p.

EBASCO Services. 1987. Detailed Analysis of Remedial Technologies for the New Bedford Harbor Feasibility Study. EPA Contr. 68-01-7250. Washington, D.C.: U.S. EPA.

Fisher, A., L. G. Chestnut, and D. M. Violette. 1988. The value of reducing risks of death: A note on new evidence. J. Policy Analysis and Management 8(1).

Freeman, A. M. 1979. The Benefits of Environmental Improvement: Theory and Practice Resources for the Future. Baltimore: Johns Hopkins Press.

Freeman, A. M. 1987. Assessing damage to marine resources: PCBs in New Bedford Harbor. Paper presented at the meetings of the Assoc. of Environmental and Resource Economists, Chicago, Dec. 27. 25 p.

Grigalunas, T. A., J. J. Opaluch, D. French, and M. Reed. 1987. Measuring Damages to Coastal and Marine Natural Resources: Concepts and Data Relevant for CERCLA Type A Damage Assessments. Springfield, Va.: National Technical Information Service (2 vols.).

Grigalunas, T. A., J. J. Opaluch, D. French, and M. Reed. 1988. Measuring damages to marine natural resources from pollution incidents under CERCLA: Application of an integrated ocean systems/economic model. Mar. Res. Econ. 5(1).

Grigalunas, T. A. and J. J. Opaluch. 1988. Assessing liability for damages under CERCLA: A new approach for providing incentives for pollution avoidance? Nat. Res. J. 28(3).

Haberman, D., G. B. Mackiernan, and J. Macknis. 1983. Toxic compounds. Chapter 4 in Chesapeake Bay: A Framework for Action. Philadelphia: U.S. Environmental Protection Agency.

Huggett, R. J. 1988. Kepone and the James River. Paper presented at the National Research Council Workshop on Contaminated Marine Sediments, Tampa, Fla.

Ikalainen, A. J. and D. E. Allen. 1988. New Bedford Harbor Superfund Project. Paper presented at National Research Council Workshop on Contaminated Marine Sediments, Tampa, Fla.

Kahn, J. R. and W. M. Kemp. 1985. Economic losses associated with the degradation of an ecosystem: The case of submerged aquatic vegetation in Chesapeake Bay. J. Environ. Econ. and Manage. 12(3).

Killory, H. C. 1987. Getting to the source of hazardous waste. Resources No. 89 (Fall).

Lave, L. B. 1987. Improving quantitative health risk assessment techniques. In Environmental Monitoring, Assessment and Management,

Draggan et al., eds. New York: Praeger Publishers.

Lukjanowicz, E., J. R. Faris, P. F. Fuglevand and G. L. Hartman. 1988. Strategies and technologies for dredging and disposal of contaminated marine sediment. Presented at the National Research Council Workshop on Contaminated Marine Sediments, Tampa, Fla.

Landin, M. C., ed. 1988. Beneficial Uses of Dredged Material: Proceedings of the North Atlantic Regional Conference, 12-14 May 1987. Baltimore Maryland. Vicksburg, Miss.: U. S. Army Engineer Waterways Experiment Station.

National Research Council (NRC). 1979. Polychlorinated Biphenyls. Washington D.C.: National Academy Press.

Office of Technology Assessment (OTA). 1987. Managing dredged materials. In Wastes in the Marine Environment. Washington, D. C.: OTA.

Opaluch, J. J. and T. A. Grigalunas. 1984. Controlling stochastic pollution events with liability rules: Some evidence from OCS leasing. The Rand J. Econ. 15(Spring).

Orchard, I. 1988. Oral statements presented at National Research Council Workshop on Contaminated Marine Sediments, Tampa, Fla., May.

Roberts, M. J. and M. Spence, 1976. Effluent charges and licenses under uncertainty. J. Pub. Econ. Vol. 5.

Schwartz, D. G. and I. E. Strand, 1981. Avoidance costs associated with imperfect information: The case of kepone. Land Econ. 57(2).

Tietenburg, T. H. 1978. Spatially differentiated air pollutant emission charges: An economic and legal analysis. Land Econ. 54(3).

U.S. Congress. 1980. Comprehensive Environmental Response, Compensation and Liability Act of 1980. Public Law 96-510, 96 Cong. 94 (Dec. 11, 1980) Stat. 2767-2811.

U.S. Congress. 1982. The Clean Water Act (as amended through December, 1981). Public Law 97-117, 97th Cong. (Feb., 1982). Washington, D.C., U.S. Government Printing Office.

U.S. Congress. 1986. The Superfund Amendments and Reauthorization Act of 1986. Public Law 99-499, 99th Cong. Washington, D.C.

U.S. Department of Commerce (DOC), National Oceanographic and Atmospheric Administration, Ocean Assessments Division, 1988. A Summary of Data on Chemical Contaminants Collected During 1984, 1985, 1986, and 1987. Rockville, Md.: NOAA.

U.S. Environmental Protection Agency (EPA). 1985. Remedial Action Costing Procedures Manual. Cincinnati, Ohio: Hazardous Waste Engineering Research Laboratory, Office of Research and Development.

U.S. Environmental Protection Agency. 1985b. Guidance on Feasibility Studies Under CERCLA. Office of Emergency and Remedial Response and Offices of Waste Programs Enforcement and Office of Solid Waste and Emergency Response, Washington, D.C.

U.S. Environmental Protection Agency. 1985c. Guidance on Remedial Investigation Under CERCLA. Office of Research and Development, Office of Emergency and Remedial Response and Office of Solid Waste and Emergency Response, Washington, D.C.

U.S. Environmental Protection Agency. 1986. Superfund Public Health Evaluation Manual. Office of Emergency and Remedial Response, Washington, D.C.

CASE STUDIES

NEW BEDFORD HARBOR SUPERFUND PROJECT

Allen J. Ikalainen and Douglas C. Allen
E.C. Jordan Company, C.E. Environmental

ABSTRACT

This case study about the ongoing remedial investigation (RI) and feasibility study (FS) for the New Bedford Harbor Superfund Site discusses events and prior studies leading to the current RI/FS. It includes discussion of multiple sampling and analytical programs to describe contamination and to develop and calibrate physical-chemical and food-web models to evaluate contaminant movement. Engineering feasibility, pilot-scale dredging and disposal studies, bench- and pilot-scale testing of innovative treatment technologies and public health and environmental risk assessment are utilized to evaluate the feasibility of a range of alternatives to meet site-specific clean-up objectives.

OVERVIEW OF THE SITE

New Bedford Harbor, a tidal estuary, is situated between the city of New Bedford on the west and the towns of Fairhaven and Acushnet on the east at the head of Buzzards Bay, Massachusetts. For administrative purposes, the site can be divided into three geographic areas, as shown in Figure 1. The northernmost portion of the site extends from the Coggeshall Street Bridge north to Wood Street in Acushnet. The remainder of the site extends south from the Coggeshall Street Bridge through the New Bedford Hurricane Barrier and into Buzzards Bay as far as the southern limit of PCB Closure Zone 3. Geographic boundaries include the shoreline, wetlands, and peripheral upland areas.

The New Bedford Wastewater Treatment Plant, the combined sewer system outfalls, the Aerovox plant, and the Cornell-Dubilier plant, all documented discharge points of PCBs, are within the areas of concern for the site. The New Bedford and Sullivan's Ledge landfills are repositories of PCBs and are being addressed separately from the harbor.

The estuary and harbor/bay area within the limits of the New Bedford Harbor Superfund Site is over 5,000 acres. Water depths range from 1 ft at the northern limit of the site to over 30 ft at the last shellfish closure line in Buzzards Bay. Freshwater discharge from the Acushnet River to the harbor is 30 ft^3 per second, average annual flow.

Significant features of the estuary and harbor include

- the 50-acre Fairhaven Marsh on the eastern shore of the Acushnet River;
- three bridge crossings that form constrictions and define boundaries for the feasibility study (FS) areas;
- a very active commercial fishery in both Fairhaven and New Bedford (commercial fish landings in 1987 were the largest of any U.S. port); and
- the New Bedford hurricane barrier dike, 4,600-ft long with a top elevation of 22 ft and navigation opening 150-ft wide, that forms the lower limit of the harbor.

Description of the Problem

Selecting and implementing a cost-effective remedial action for New Bedford Harbor requires that the nature and extent of contamination by PCBs and metals be determined and that environmental effects, including impacts on public health, be evaluated. Conducting a remedial investigation/feasibility study (RI/FS) to select the remedial action currently involves five federal agencies or departments and six private consultants or institutions. The following subsections contain discussions of the environmental problem.

The Environmental Problem

Since the initial survey of the New Bedford area in 1974, a much better understanding of the extent of PCB contamination has been gained. The entire area north of the hurricane barrier, an area of 985 acres, is underlain by sediments containing elevated levels of PCBs and heavy metals. PCB concentrations range from a few parts per million (ppm) to over 100,000 ppm. Portions of western Buzzards Bay sediments along the New Bedford shoreline south of the hurricane barrier are also contaminated, with concentrations occasionally exceeding 50 ppm. The water column in New Bedford Harbor has been measured to contain PCBs in the parts per billion (ppb) range well in excess of the U.S. Environmental Protection Agency's (EPA) 30 parts per trillion (ppt) guideline for protection of saltwater aquatic life from chronic toxic effects. Much of the PCB sampling performed before 1980 was analyzed for Aroclor 1254. Woods Hole Oceanographic Institution scientists have presented evidence suggesting that, as a result, PCB contamination is often understated by factors of three to five. Sampling and analyses performed since 1980 have included PCB isomers. Sediment copper concentrations were reported in 1977 to range from more than 6,000 ppm near the head of the harbor, to less than 100 ppm at the edge of Buzzards Bay. Other metals are also present at lower concentrations. The direct discharge of PCB-contaminated wastewater from Cornell-Dubilier and Aerovox plants has been significantly reduced, as a result of EPA's amendments to their wastewater discharge permits. However, uncontrolled releases from the tidal mudflats beneath Aerovox's discharge have continued unabated. Studies have shown that 200 to 700 lbs of PCBs were previously

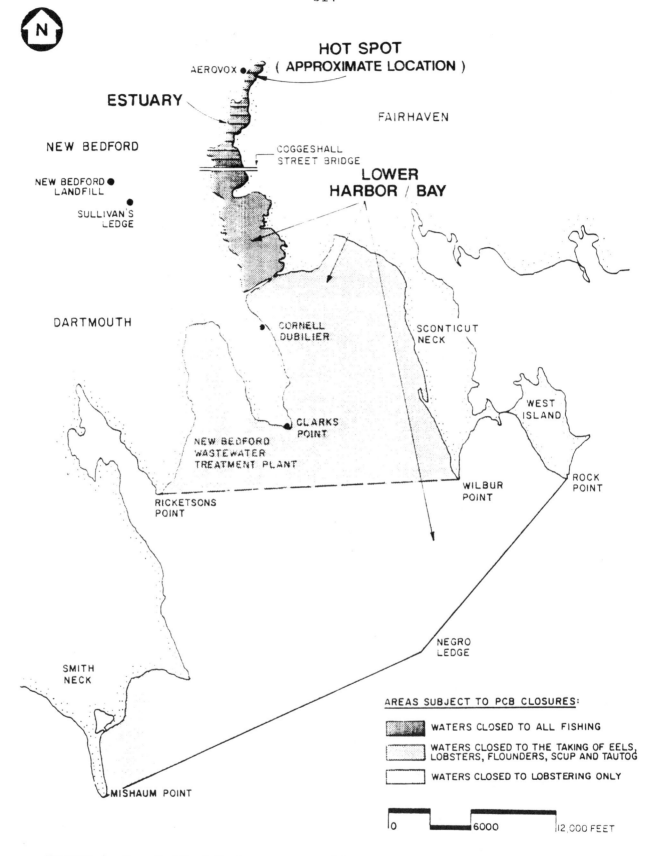

FIGURE 1 New Bedford Harbor areas subject to PCB closures.

discharged per year to Buzzards Bay via the Clark's Point outfall. The
magnitude of PCB discharge from the sewer system and treatment plant is
being addressed by EPA in its review of New Bedford's application for a
waiver from secondary treatment under Section 301(h) of the Clean Water
Act, as amended.

In addition to these known PCB disposal sites, EPA has investigated
a number of other potential sources and disposal sites. Of 30 areas
investigated initially, five or fewer sites appear to warrant further
investigation. These sites are being addressed by EPA's pre-remedial
program.

The environmental impacts at the New Bedford Harbor site due to PCB
and heavy metal contamination include both human health and effects on
fishing in the area. The most probable link of PCBs to human intake is
the consumption of contaminated fish and shellfish from the Acushnet
River estuary. Widespread contamination of the Acushnet River estuary
environs has resulted in the accumulation of PCBs in many marine species. Although thousands of acres have been closed to the harvesting
of shellfish, finfish, and lobsters, residents are known to still harvest both finfish and shellfish, thus exposing themselves to ingestion
of PCBs. In addition, many individuals regularly consumed contaminated
fish before the extent of environmental contamination by PCBs was
known. The chronic toxicity effects on these people have not been
evaluated.

The closure of the harbor and sections of Buzzards Bay to fishing
has resulted in an estimated capital loss of $250,000 per year to the
lobster industry alone. Shellfish and finfish industries, as well as
recreational fishing, have also been negatively affected.

Figure 1 shows the three closure areas established by the
Massachusetts Department of Public Health on September 25, 1979. Area
1 (New Bedford Harbor) is closed to the taking of all finfish,
shellfish, and lobsters. Area 2 is closed to the taking of lobster and
bottom-feedng fish (eels, scup, flounder, and tautog). Area 3 is
closed to the taking of lobster. Responsibility for enforcement of
these closures is entrusted to the Massachusetts Office of
Environmental Affairs Division of Law Enforcement.

Contaminated sediments have also affected proposed harbor
development projects, most of which require dredging. Dredging in New
Bedford Harbor is restricted by the difficulties encountered in
fulfilling state and federal regulatory requirements for the disposal
of contaminated dredge spoils.

ASSESSMENT OF CONTAMINATION

As previously noted, contamination of New Bedford Harbor sediments
has been assessed since 1974. The most recent studies conducted in
1984-1987 have formed the basis for performing physical-chemical and
food-web modeling, public health and environmental risk assessments,
and identifying specific locations for sediment cleanup.

Following are summaries and results of the sampling programs.
Section 2.2 describes assessment of public health and environmental

risk. Information on the distribution of contamination and the risk assessment help to define clean-up objectives (remedial response objectives) described in Section 2.3.

Sampling Programs for the Acushnet River Estuary and New Bedford Harbor

Acushnet River Estuary

The U.S. Army Corps of Engineers (COE) New England Division (NED) and Waterways Experiment Station (WES) in Vicksburg, Mississippi, conducted two sampling programs in the Acushnet River estuary. The first was designed to characterize sediment contaminant concentrations throughout the estuary and included

- sediment cores on a 250-foot grid north of the Coggeshall Street Bridge (180 locations, Figure 2);
- 30 cores selected for testing at a 3-ft depth, plus other depths as required, that were analyzed for PCBs, metals, oil and grease, and physical tests;
- 10 cores selected for the EPA hazardous substances list analyses.

In addition, a more concentrated sampling program was conducted in the area of the highest concentrations of PCBs in sediments, the "hot spot." This program consisted of

- sediment cores on a 150-ft grid,
- 49 cores selected for testing at two depths,
- five cores selected for testing at 36- and 48-in depths, and
- 13 cores selected for physical testing.

Within the harbor/bay area south of the Coggeshall Street Bridge, NUS conducted a sampling program to characterize sediment contaminant levels (Figure 3). It consisted of

- sediment cores on a 500-ft grid between I-195 bridge and the area just south of the hurricane barrier (180 locations);
- the top 6 in of each core, selected for PCB analysis;
- multiple depth samples selected for PCBs and metals analyses at approximately 30 locations;
- multiple depth samples selected for full hazardous substances list at 5 to 10 locations; and
- physical testing of selected samples.

For purposes of calibration and verification of the physical-chemical and food-web model by Battelle/HydroQual, Battelle New England developed and completed a sampling program consisting of

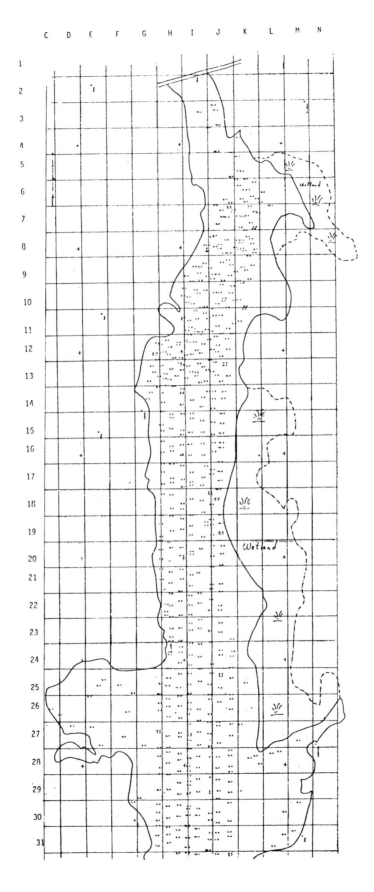

FIGURE 2 Arbitrary sampling grid for Acushnet River estuary and New Bedford Harbor, north of Coggeshall Street Bridge. SOURCE: U.S. Army COE, 1985.

FIGURE 3 Location map and sampling grid for New Bedford Harbor.
SOURCE: NUS Corp.

- water, sediment, and biota sampling at 25 stations, extending from the Acushnet River estuary to Buzzards Bay, on three occasions, September and December 1984, and June 1985;
- a sampling program after a storm event;
- PCB analysis to measure isomers (chlorination levels) using GC/MS (re-analyses were done to achieve a detection limit in the range of ng/liter using GC/ECD and GC/MS);
- metals analysis for copper, cadmium, and lead; and
- support by Woods Hole Oceanographic Institution's (WHOI) tide, current, and drifter studies.

Type of Sample	Number of samples
Sediment	233
Tissue	366
Filtrate	300
Particulate	300
Pore water	3
TOC	233
Grain size	136
POC	300
TSS	300

Concurrent with these sampling programs aimed at providing data for the overall New Bedford FS, EPA's Narragansett Laboratory conducted a sediment toxicity and characterization study to investigate the toxicity of New Bedford Harbor sediments on two amphipods, and the effects of contaminants--including PCBs--accumulated in sediments on sheepshead minnow reproduction.

Sediment for the bioassay tests was collected at stations in the Acushnet River estuary through the lower harbor to the hurricane barrier. Following the bioassay studies and analyses of sediments for PCBs, analyses were also performed for polynuclear aromatic hydrocarbons (PAHs). Sampling locations and results are shown in Figures 4-6.

The most recent sampling in New Bedford harbor was done to provide an environmental baseline on water quality conditions prior to the pilot dredging and disposal study by COE. This sampling, conducted by EPA's Narragansett Laboratory, included physical measurements of currents, tides, temperature, salinity, and suspended solids. Water column samples were composited over two tidal cycles and analyzed for PCBs, cadmium (Cd), copper (Cu), and lead (Pb). Toxicity tests were comprised of mussel (*Mytilus edulis*) deployment to evaluate growth and survival, sea urchin (*Arbacia punctulata*) sperm cell fertilization, red alga (*Champia parvula*) reproduction, and fish (*Cyprinodon variegatus*) growth and survival.

Results of Assessment of Contamination

The accumulated data from the various New Bedford Harbor sampling programs are extensive. Summaries of the data are provided here to

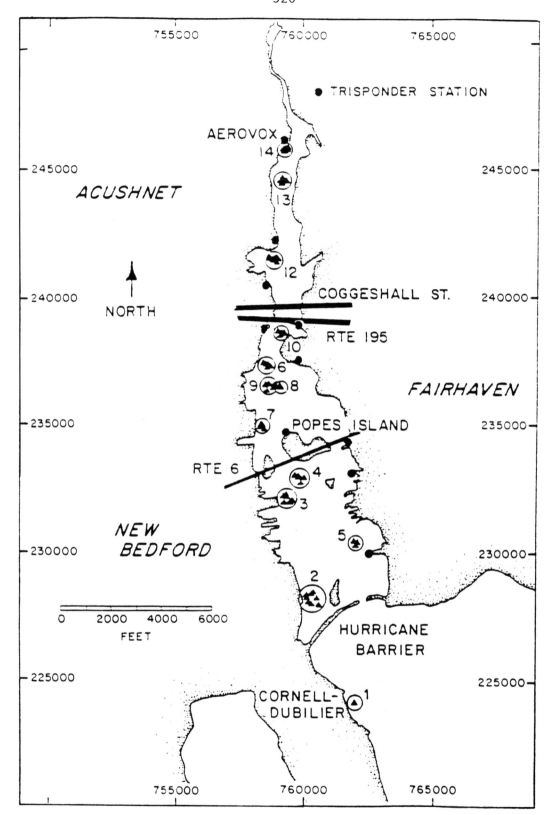

FIGURE 4 EPA Narragansett sediment sampling stations for New Bedford Harbor. SOURCE: U.S. EPA, 1987.

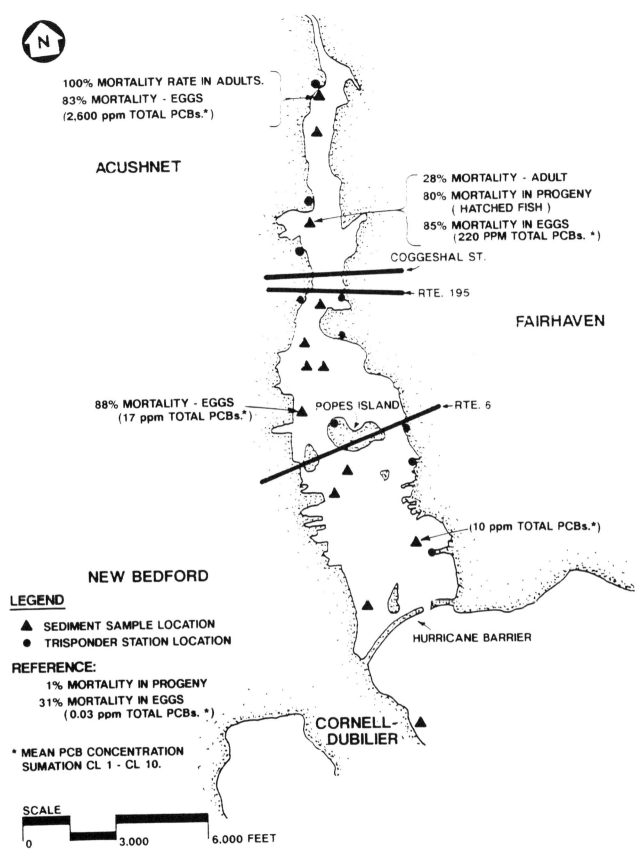

FIGURE 5 Toxicity of New Bedford Harbor sediments to the fish *Cyprinadon variegatus*. SOURCE: U.S. EPA, 1987.

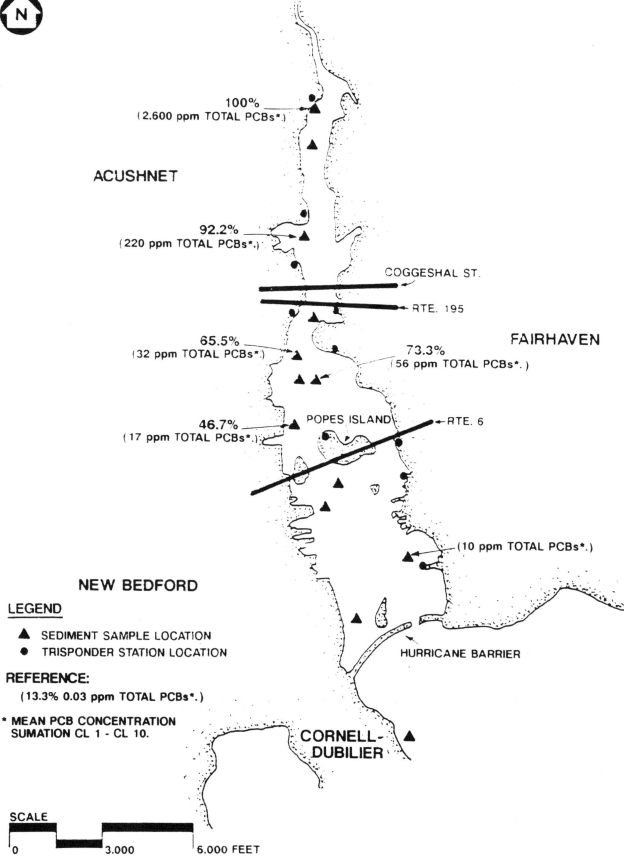

FIGURE 6 Mortality of the amphipod *Ampelisca abdita* (Hansen, 1986) in New Bedford Harbor sediments. SOURCE: U.S. EPA, 1987.

indicate some general patterns resulting from the evaluation of the data.

Sediment PCB Concentrations

Figures 7 and 8 show PCB sediment concentrations by regions of the harbor extending from the Acushnet River to the outer harbor and by north-south location. Concentrations of total PCBs in sediments range from over 100,000 ppm in the Acushnet River to less than 10 ppm in outer New Bedford Harbor. Table 1 shows a range of organic compound concentrations in the Acushnet River and lower New Bedford Harbor.

Water Column and Biota PCB Concentrations

Table 2 shows mean and maximum PCB values by areas designated for assessment of risk (Figure 9). Ranges of PCBs in biota are shown in Table 3.

Risk Assessment

As part of the FS, E.C. Jordan is conducting the risk assessment for the New Bedford Harbor site. The risk assessment serves to establish the actual and/or potential threat to public health and welfare and the environment. It is based on the sampling data contained in the New Bedford Harbor data base, various site investigation reports, the Greater New Bedford Health Effects Study, and ongoing studies being conducted for this site.

The risk assessment evaluates both the public health and environmental risks associated with exposure to contaminants from the New Bedford Harbor area under baseline (existing) and future potential conditions. The results of the risk assessment will be used to determine the need for and extent of remediation. It serves as the public health and environmental evaluation of the no-action remedial alternative for the site, and will be used in the effectiveness evaluation of each remedial alternative.

The risks to human health and environmental ecosystems were determined by two primary factors: exposure to and toxicity of the contaminants detected at the site. Exposure is defined as contact with a chemical or physical agent and toxicity is the inherent ability of the chemical to cause harm. Combined, these two factors determine risk. The risk assessment for the New Bedford Harbor Superfund site consists of both a public health and an environmental assessment of the risks associated with exposure to the contaminants of concern: PCBs, cadmium, copper, and lead. A summary of these risks follows.

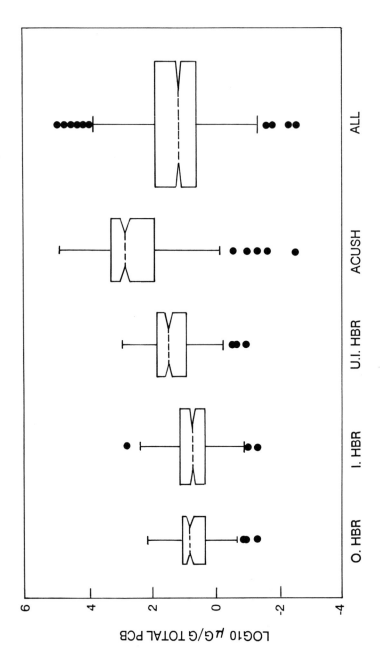

FIGURE 7 PCB concentration by region, New Bedford Harbor.

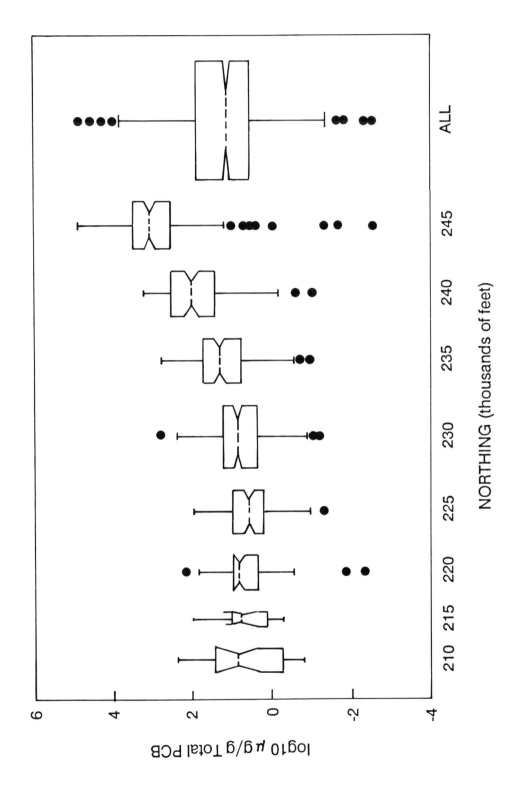

FIGURE 8 PCB concentration versus north-south location, New Bedford Harbor.

TABLE 1 New Bedford Harbor Organic Compounds in Sediment

Location	Compound	Concentration
Acushnet River estuary	phenol	140 ppb
	2 methylphenol	56 ppb
	bis(2-ethylhexl)phthalate	120-4,500 ppb
	di-n-octyl phthalate	60 ppb
	di-n-butyl phthlate	250 ppb
	acenaphthene	1,900 ppb
	acenaphthylene	52 ppb
	anthracene	110-13,000 ppb
	benzo(a)anthracene	240-7,400 ppb
	benzo(b)fluoranthene	440-5,600 ppb
	benzo(k)fluoranthene	53 ppb
	benzo(g,h,i)perylene	330-3,700 ppb
	benzo(a)pyrene	320-4,100 ppb
	chrysene	290-4,700 ppb
	dibenz(a,h)anthracene	100-710 ppb
	flouranthene	420-2,900 ppb
	flourene	2,300 ppb
	indeno(1,2,3-cd)pyrene	390-940 ppb
	napthalene	73-980 ppb
	2-methylnapthalene	350 ppb
	phenanthrene	57-8,500 ppb
	pyrene	490-7,100 ppb
	benzoic acid	260 ppb
	dibenzofuran	1,500 ppb
Harbor/bay	acetone	21-360 ppt
	4 methyl-phenol	100-560 ppb
	bis(2-ethylhexyl)phthalate	75-4,400 ppb
	di-n-butylphthalate	1,200-1,100 ppb
	diethylphthalate	380-1,400 ppb
	acenaphthene	320 ppb
	acenaphthylene	180 ppb
	anthracene	230-270 ppb
	benzo(a)anthracene	87-2,600 ppb
	benzo(b)fluoranthene	580-1,100 ppb
	benzo(k)fluoranthene	580-1,100 ppb
	benzo(g,h,i)perylene	1,300 ppb
	benzo(a)pyrene	690 ppb
	chrysene	860-1,700 ppb
	fluoranthene	81-2,300 ppb
	fluorene	250 ppb
	indeno(1,2,3-cd)pyrene	550-990 ppb
	napthalene	110-270 ppb
	2-methylnapthalene	140 ppb
	phenanthrene	180-2,500 ppb
	pyrene	110-3,200 ppb
	benzoic acid	68-520 ppb

TABLE 2 Mean and Maximum Concentrations of PCBs (µg/liter) in New Bedford Harbor Water

	Area I[a] (n)[b]	Area II (n)	Area III (n)	Area IV (n)
PCBs total				
Mean[c]	7.93 (6)	0.23 (97)	0.057 (72)	0.00461
Max	36.94	1.41	0.64	0.06

NOTES:
[a] Areas correspond to those labeled in Figure 9
[b] 2n = sample size
[c] Mean and maximum from HydroQual, 1988

Public Health Risk Summary

The public health risk evaluation was composed of three sections: an exposure assessment, a toxicity evaluation, and a risk assessment. The exposure assessment focused on site-specific demographic and land use information to identify the subpopulations (children, adults, teens, etc.) considered to be at risk from contaminant exposure and to determine the principal pathways of contaminant exposure at the New Bedford Harbor site.

Analyses of the demographic and land use information, as well as activity and behavior patterns, indicated that exposure to contaminants in the New Bedford Harbor area could occur through dermal contact with sediments and water, ingestion of water and biota, and/or inhalation of airborne contaminants. A quantitative screening analysis of these exposure pathways was performed to identify the principal pathways of exposure. These were

- ingestion of aquatic biota,
- direct contact with sediments,
- ingestion of sediments, and
- inhalation of airborne contaminants.

These exposure pathways accounted for 99 percent of the potential exposures within the New Bedford Harbor area, and were the focus of the remaining risk assessment efforts.

Exposure dose levels were estimated for each exposure pathway and based on a variety of exposure conditions, including the mean and maximum contaminant levels detected in the various media. To provide realistic exposure concentrations for sediment and water, the New Bedford Harbor site was divided into three areas (Figure 9), as follows:

TABLE 3 Range[a] of Total PCBs in New Bedford Harbor Biota

Organism	Area[b]	PCBs (ppm)	n[c]
Lobster	1		
	2	0.195-1.234	15
	3	0.042-0.351	14
	4	0.016-0.176	21
Winter flounder	1	5.658-8.235	23
	2	0.514-6.348	17
	3	0.925-4.504	27
	4	0.122-2.615	22
Mussel	1	1.467-2.962	9
	2	1.461-6.204	9
	3	0.254-0.278	5
	4	0.008-0.039	6
Quahog	1	0.282-2.121	22
	2	0.010-1.181	18
	3	0.025-0.478	21
	4	0.002-0.137	11
Green Crab	1	0.070-0.725	5
	2	.067-0.031	4
	3	0.624-1.329	2
	4	0.020-0.076	4
Polychaetes	1	12.97[d]	
	2	16.54[d]	
	3	96.24-689.46	
	4	0.182-0.789	

NOTES:
[a] Each mean represents the mean of several organisms within one size class taken during Cruises 1, 2, or 3.
[b] Areas correspond to Figure 10.
[c] Total number of organisms sampled in each area.
[d] Only one value available.

- Area I, between the Wood Street Bridge and the Coggeshall Street Bridge;
- Area II, between the hurricane barrier and the Coggeshall Street Bridge; and
- Area III, south of the hurricane barrier.

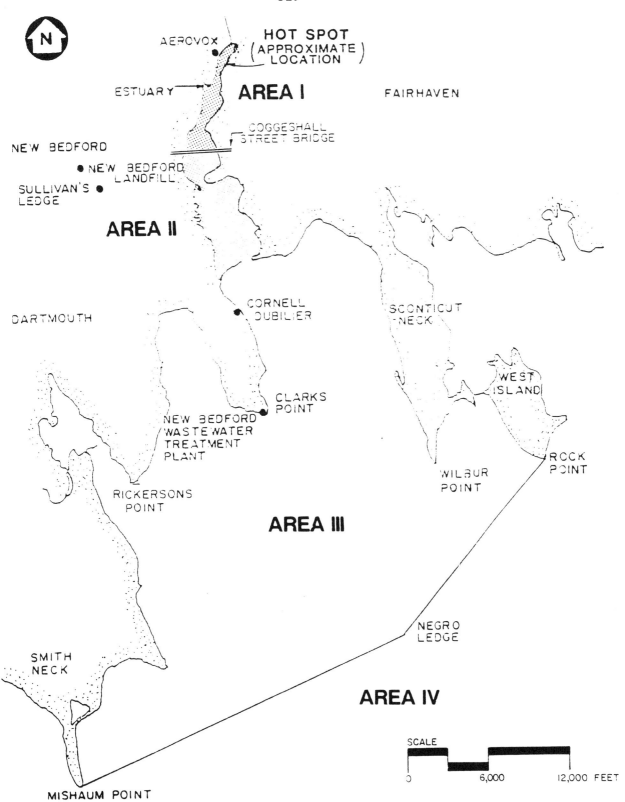

FIGURE 9 Areas for ecological risk assessment for New Bedford Harbor.

Exposure through the ingestion of biota was assessed separately for the following four areas (Figure 10):

- Area I, between the Wood Street Bridge and the Hurricane Barrier;
- Area II, between the Hurricane Barrier and Wilbur and Ricketsons Point;
- Area III, between Wilbur and Ricketsons Point; and Rock Point, Negro Ledge, and Mishaum Point; and
- Area IV, outside Area III.

A standard 8-oz (227 g) fish meal was considered to be a reasonable estimate of fish consumption for older children and adults from this area. Ingestion frequencies of daily (365 meals/year), weekly (52 meals/year), and monthly (12 meals/year) were assumed. Body dose calculations for each route of exposure were evaluated using the most appropriate health-based standard, criteria, or guideline value derived for PCBs, cadmium, copper, and lead.

Noncarcinogenic risk estimates were generated for acute and chronic exposure to PCBs, cadmium, copper and lead for direct contact and ingestion of sediments (PCBs only) and the ingestion of biota. Carcinogenic risk estimates were generated for subchronic, chronic, and lifetime exposure to PCBs for direct contact and ingestion of sediments, ingestion of biota, and inhalation of air.

Ecological Risk Summary

The ecological risk assessment for the New Bedford Harbor site examined the potential risks to marine biota from exposure to PCBs, cadmium, copper, and lead. This evaluation was composed of three sections: an exposure assessment, an ecotoxicity assessment, and a risk assessment.

The exposure assessment described the potential exposure of aquatic biota to PCBs, cadmium, copper, and lead. Three routes of exposure (direct contact with the water, direct contact with sediments, and ingestion of contaminated biota) were evaluated. To represent the potential aquatic receptors in this area, 33 species were selected and used to describe exposure. These species also represented five different classes of organisms, including each major trophic level. In addition, a food web was constructed to illustrate the potential for contaminant movement within an ecosystem.

Direct contact with contaminated sediments was evaluated for the first three of the four areas of the New Bedford Harbor site shown in Figure 9. These areas are the same as those used to assess public health risks.

The concentration of PCBs detected in the sediments from these areas were used to assess exposure. A comparison of these PCB concentrations to site-specific and appropriate toxicity data and/or criteria values was used to assess the potential risks to aquatic biota.

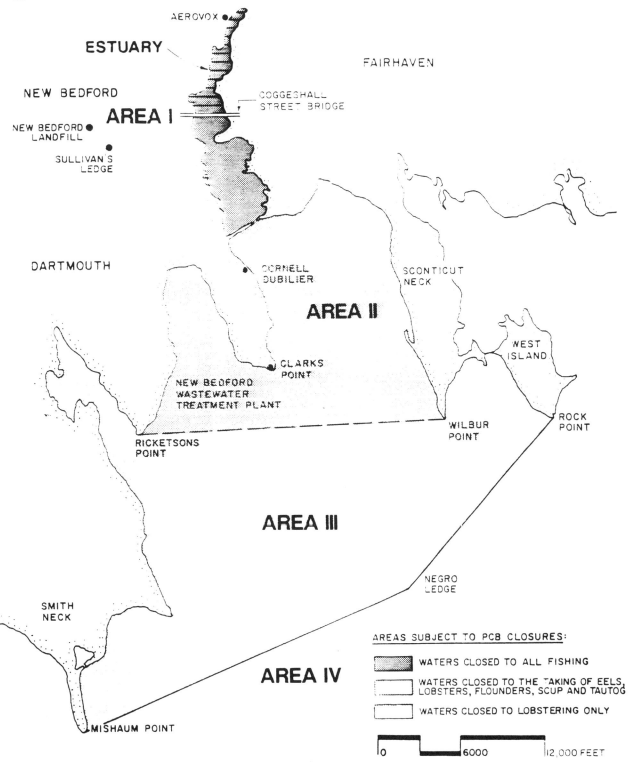

FIGURE 10 Areas used to assess exposure to biota in New Bedford Harbor.

Direct contact exposure to water-borne contaminants was assessed for all areas previously described, including Area IV, which extends further out into Buzzards Bay. Water concentrations in these four areas were compared to both EPA's Ambient Water Quality Criteria (AWQCs) and benchmark concentrations (laboratory-derived toxicity values) to evaluate the potential risks to aquatic organisms.

The observed contaminant concentrations in aquatic biota were evaluated against species-specific toxicity data. The ecological risk assessment focused on site-specific toxicity data when available. The AWQC and other applicable toxicity data were also used to evaluate PCB exposure from the water column and were derived by comparing exposure concentrations to benchmark concentrations (laboratory-derived toxicity values).

Remedial Response Objectives

Under EPA's Superfund process, the next step following assessment of the distribution of contamination at a site is to establish remedial response (clean-up) objectives. Response objectives for the New Bedford Harbor FS will be established to guide the development and evaluation of remedial alternatives for this site. Based on the results of the risk assessment, the response objectives for this site will likely consider methods to

- reduce PCB biota concentrations to levels considered protective of public health, and/or protective of aquatic life;
- reduce total PCB sediment concentrations in the Acushnet River estuary to levels considered protective of public health and/or protective of aquatic life;
- reduce total PCB sediment concentrations in the Acushnet River estuary to levels considered protective of public health and/or protective of aquatic life, and reduce total PCB sediment concentrations in New Bedford Harbor to levels considered protective of aquatic life; and
- reduce water-column PCB and copper concentrations to below the AWQC in the Acushnet River estuary and New Bedford Harbor.

REMEDIAL TECHNOLOGIES AND ALTERNATIVES

Overview of the New Bedford Harbor FS Process

The FS for New Bedford Harbor is being conducted under the EPA Superfund program. The goal of this FS is to present EPA with remedial alternative(s) that address the cleanup of PCBs and metals in New Bedford Harbor. The New Bedford Harbor FS is a multistep process formulated to meet the legislative requirements of the Comprehensive Environmental Response, Compensation, and Liability Act of 1980 (CERLCA); the Superfund Amendments and Reauthorization Act of 1986 (SARA); and the National Oil and Hazardous Substances Pollution Contingency Plan

(NCP 1985, Final Rule, Section 300.68). Figure 11 shows the FS process for New Bedford Harbor. The methodology for conducting the FS was developed in accordance with guidance documents including CERCLA FS guidelines (EPA OERR/OSWER) and EPA directives (EPA OWSER).

General response or clean-up actions were identified to provide a framework for subsequent identification and evaluation of component remedial technologies selected to address site problems and meet clean-up goals and objectives. The range of clean-up actions identified for New Bedford Harbor is shown in Figure 12.

Remedial Technologies

Identification

From a remedial perspective, the sediments in New Bedford Harbor constitute the focal point for the FS. Therefore, for each of the response actions, technologies were identified to address the PCBs and metals in the sediments. Nonremoval technologies contain, isolate, or treat the PCBs and metals in the sediments via biological, chemical, or physical means, without removing the sediments. Removal technologies remove the PCBs and metals from the harbor bottom by removing the sediment where the contaminants are located. Treatment technologies destroy PCBs or render the PCBs and metals less toxic and/or less mobile by chemical or physical alteration. Disposal technologies contain and isolate untreated or treated sediments that have been removed from the harbor.

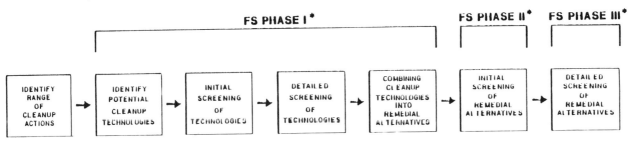

* BASED UPON EPA GUIDANCE SUPERFUND AMENDMENTS AND REAUTHORIZATION ACT (OCTOBER 1986)

FIGURE 11 FS process for New Bedford Harbor.

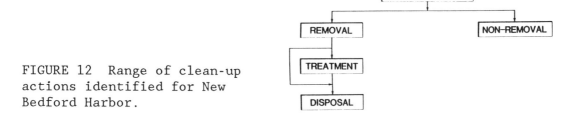

FIGURE 12 Range of clean-up actions identified for New Bedford Harbor.

Figure 13 shows the clean-up technologies that were identified for New Bedford Harbor. Sources of information and data that were used to identify these technologies included: vendor literature surveys, other Superfund FSs, and remedial technology reports from the government, academic, and private sectors.

Initial Screening

An initial screening of the identified technologies was conducted to reduce the number of technologies for further detailed analysis. Waste-specific and site-specific factors were used to assess the effectiveness, implementability, and cost of each technology identified for New Bedford Harbor. The screening step eliminated technologies that

- were not effective in remediating marine sediments contaminated with PCBs and metals,
- could not be implemented due to an insufficient level of development or constraints on existing capacity or accessibility to the site;
- would cost an order of magnitude more than a technology that provided similar results.

Figure 14 shows the technologies that were retained from the initial screening step and were determined to be potentially applicable for New Bedford Harbor.

Detailed Evaluation

An evaluation of each of the potentially applicable technologies was conducted to compile detailed information and data relating to effectiveness, implementation, and cost. The effectiveness of each technology was assessed on the basis of reliability and whether it would significantly and permanently reduce the toxicity, mobility, or volume of hazardous constituents. The implementation of a technology considered factors relating to the technical, institutional, and administrative feasibility of installing, monitoring, and maintaining that technology. The costs associated with a technology were estimated on the basis of direct and indirect capital costs, and operation and maintenance expenses. Figure 15 summarizes the evaluation process and the criteria.

Treatment Technology Bench Test Program

SARA emphasizes a preference for remedial actions employing treatment technologies that permanently and significantly reduce the mobility, toxicity, or volume of hazardous constituents. EPA directives require consideration of innovative (and often unproven in the field)

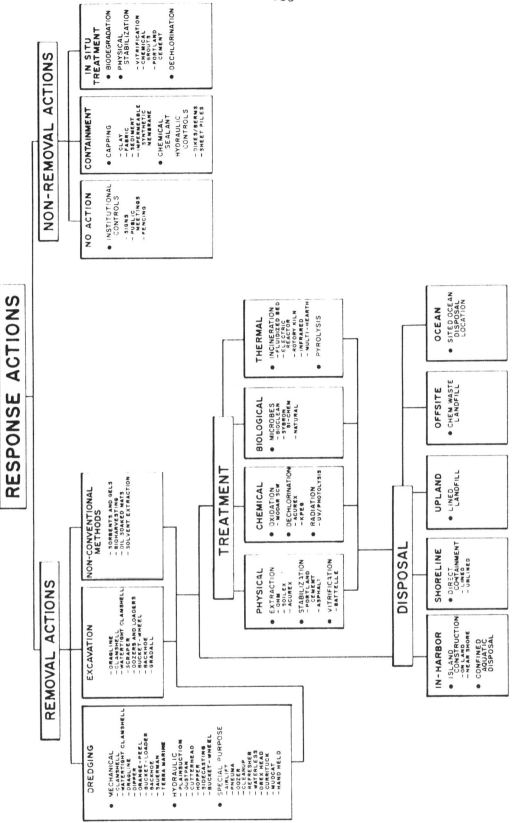

FIGURE 13 Categories of response action, New Bedford Harbor FS.

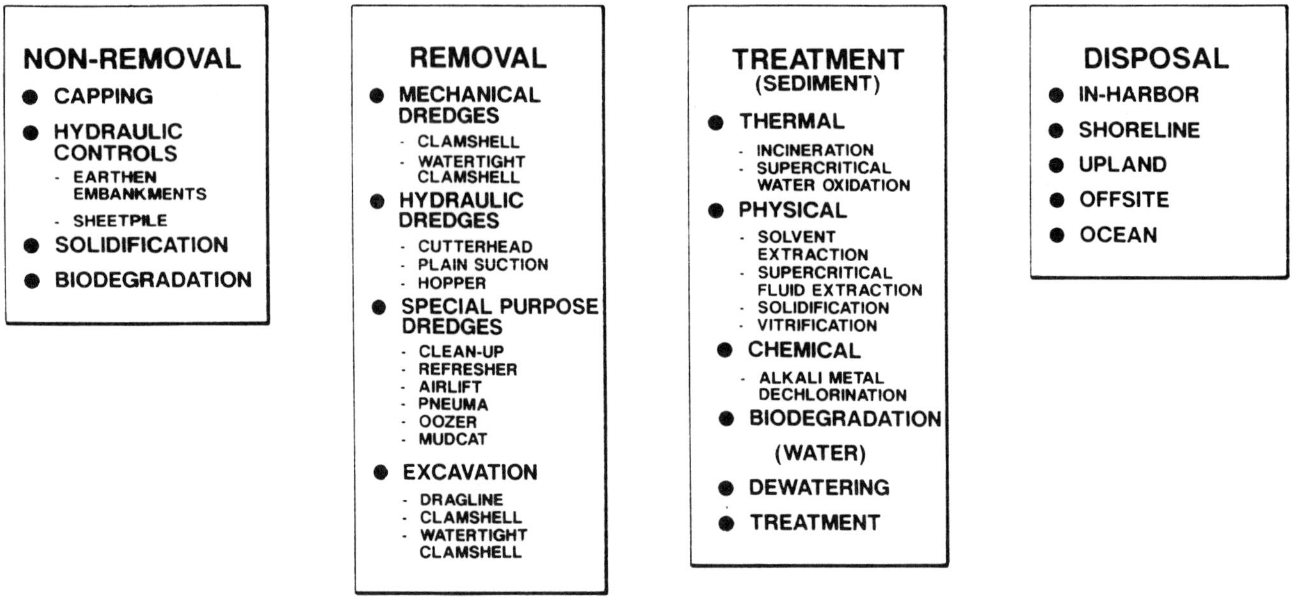

NON-REMOVAL
CAPPING
HYDRAULIC CONTROLS
- EARTHEN EMBANKMENTS
- SHEETPILE
SOLIDIFICATION
BIODEGRADATION
REMOVAL
MECHANICAL DREDGES
- CLAMSHELL
- WATERTIGHT CLAMSHELL
HYDRAULIC DREDGES
- CUTTERHEAD
- PLAIN SUCTION
- HOPPER
SPECIAL PURPOSE DREDGES
- CLEAN-UP
- REFRESHER
- AIRLIFT
- PNEUMA
- OOZER
- MUDCAT
EXCAVATION
- DRAGLINE
- CLAMSHELL
- WATERTIGHT CLAMSHELL
TREATMENT (SEDIMENT)
THERMAL
- INCINERATION
- SUPERCRITICAL WATER OXIDATION
PHYSICAL
- SOLVENT EXTRACTION
- SUPERCRITICAL FLUID EXTRACTION
- SOLIDIFICATION
- VITRIFICATION
CHEMICAL
- ALKALI METAL DECHLORINATION
BIODEGRADATION
(WATER)
DEWATERING
TREATMENT
DISPOSAL
IN-HARBOR
SHORELINE
UPLAND
OFFSITE
OCEAN

FIGURE 14 Technologies for detailed evaluation, New Bedford Harbor.

REMEDIAL TECHNOLOGY
EFFECTIVENESS
RELIABILITY
PUBLIC HEALTH
ENVIRONMENT
IMPLEMENTATION
TECHNICAL FEASIBILITY
LEVEL OF DEVELOPMENT
SUPPORT REQUIREMENTS
AVAILABILITY
INSTALLATION
TIME
SAFETY
MONITORING AND MAINTENANCE
PERMITTING
LEGAL CONSTRAINTS
IMPACTS ON HISTORICAL AND CULTURAL RESOURCES
COST
DIRECT/INDIRECT CAPITAL COSTS
OPERATION AND MAINTENANCE
APPLICABLE TECHNOLOGY

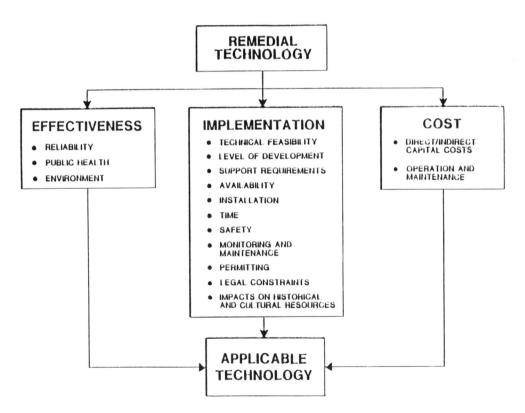

FIGURE 15 Criteria for the detailed analysis of remedial technologies for CERCLA feasibility studies as impacted by SARA, 1986.

technologies that could be implemented to provide better, more effective treatment at lower cost than demonstrated technologies.

A bench test program was developed and implemented to obtain performance data on the treatment of PCB-contaminated New Bedford Harbor sediments. Vendors of treatment technologies were identified through vendor literature and presentations, and detailed questionnaires were sent to vendors. The information and data were compiled and used to evaluate the level of development and potential effectiveness of these various treatment technologies. A select list of vendors was solicited for bench test proposals. Contracts were awarded to vendors of five different treatment technologies:

1. alkali metal dechlorination (KPEG),
2. solidification,
3. biological treatment,
4. solvent extraction (B.E.S.T.-triethylamine), and
5. vitrification.

Each vendor is bench testing small volumes (1 to 3 gal) of sediment containing two concentration levels of PCBs: less than 500 ppm and greater than 1,000 ppm.

COE Dredging and Disposal Evaluation Program

COE is conducting a detailed analysis of dredging and disposal of contaminated sediments in New Bedford Harbor. This effort includes

1. numerical modeling of sediment and contaminant transport during dredging;
2. laboratory studies of estuary sediment characterization, leachate and surface runoff from confined disposal facilities (CDFs), subaqueous capping, solidification/stabilization technologies, and settling and chemical clarification; and
3. conceptual designs of CDFs and confined aquatic disposal (CAD) areas.

In addition to the laboratory studies, COE conducted a pilot-scale study of dredging and disposal of a small volume (approximately 10,000 yd^3) of contaminated sediment during the fall and winter of 1988. The pilot study evaluated three dredges and two disposal techniques. Contaminated sediments were dredged using a mudcat, a cutterhead, and a matchbox dredge. The sediments were deposited in a shoreline CDF and in a CAD, both located in the immediate vicinity of the dredging operation.

The results of this study are an important part of the New Bedford Harbor FS and will be used to determine

- the effectiveness of the three dredges on contaminant removal and migration, and impacts on water quality;
- the feasibility of disposing contaminated sediments in

underwater CADs; and
- the degree and cost of treating the water removed with the sediments.
- The ability of the CDF to contain contaminated sediment and leachate.

Remedial Alternatives

Development of Alternatives

Combinations of technologies from each of the response actions were assembled into remedial alternatives for New Bedford Harbor. These alternatives represent a range of treatment and containment combinations and an emphasis on permanent solutions mandated by SARA. The alternatives include a no-action alternative that serves as a baseline for comparison with other remedial actions, a containment option involving little or no treatment, treatment alternatives that would eliminate the need for long-term management at the site (included in this group are off-site treatment and disposal options), and treatment alternatives that would reduce the toxicity, mobility, and volume of hazardous constituents as their principal element.

Remedial alternatives were developed for three specific areas within New Bedford Harbor. The three areas were delineated on the basis of contaminant levels in the sediment and differences in topography and bathymetry. Although the overall response objectives are the same, the differences between these three areas warranted the development of remedial alternatives tailored for each area. The approach also allows EPA to consider remediation of the three areas separately as well as remediation of the overall site.

The hot spot (Area I) is located along the western bank of the Acushnet River. This 4-acre area contains PCB concentrations in silts and marine clays ranging from 4,000 ppm to over 100,000 ppm. Mean low water depths range from 1.6 to 2.2 ft. The estuary (Area II) is defined as a 200-acre area between the Wood Street Bridge to the north and the Coggeshall Street Bridge to the south. PCB concentrations in the organic silts and silty sands of the estuary range from < 1 ppm to 5,000 ppm. Water depth at mean low water ranges from 18 ft at the Coggeshall Street Bridge to 6 ft at the Wood Street Bridge. The lower harbor (Area III) is a 785-acre area that extends from the Coggeshall Street Bridge to the hurricane barrier (lower harbor). PCB concentrations in the sediments range from < 1 ppm to 100 ppm. Water depths range from 6 to 12 ft in nearshore areas, and 30 to 35 ft in the ship channel transecting the harbor. An additional area of approximately 5,000 acres (considered part of Area III) constitutes the upper Buzzards Bay portion of the New Bedford Harbor site. This area, extending from the hurricane barrier to an imaginary line between Mishaum Point, Negro Ledge, and Rock Point, contains isolated areas of PCBs in the range of 100 ppm and below. Water depths in the bay vary from tidal flats nearshore to 35 ft in the channel.

Screening of Alternatives

Screening of the remedial alternatives was conducted to reduce the number of alternatives for further detailed evaluation. Each of the alternatives was evaluated against the same three criteria used for screening remedial technologies: effectiveness, implementation, and cost. However, for the screening of alternatives, greater emphasis was placed on cost/benefits and technical/administrative implementability. Table 4 shows the remedial alternatives for each of three FS areas that were retained after screening.

Detailed Evaluation of Alternatives

The detailed evaluation of alternatives provides the basis for identifying a preferred alternative (selection of remedy) and preparing the proposed remedial action plan. The information and data compiled during the alternative development and screening phases is supplemented with: evaluations of the reduction of potential risk to human health and the environment; attainment of federal and state applicable, relevant, and appropriate requirements (ARARs); and the results of treatability studies, if conducted.

As a result of recent EPA directives and revisions to the CERCLA guidance for conducting FSs, a detailed evaluation of remedial alternatives must address nine criteria, shown in Figure 16. Particular emphasis is placed on evaluating each alternative against the five criteria

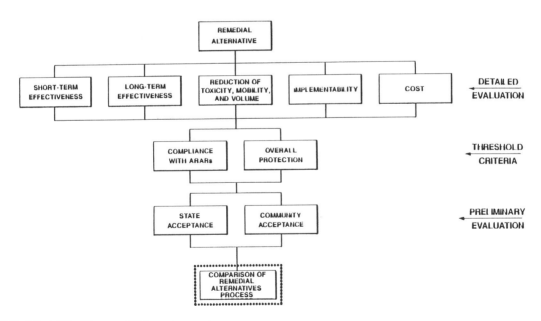

FIGURE 16 FS Phase III, detailed analysis of alternatives under EPA CERCLA guidance (1988 draft).

TABLE 4 Remedial Alternatives for Three New Bedford Study Areas

REMEDIAL ALTERNATIVES
HOT SPOT

		RESPONSE ACTIONS		
ALTERNATIVE	REMOVAL	TREATMENT	DISPOSAL	NON-REMOVAL
HS-1				No Action
HS-2	Dredge*	Sediments -Dewater -Solidification Water treatment	RCRA landfill (out-of-state)	
HS-3	Dredge*	Sediments -Dewater -Incineration -Solidify ash Water treatment	Shoreline/Ocean	
HS-4	Dredge*	Sediments -Solvent extract. -Solidification Solvent-PCBs -Incineration (off-site) Water treatment	Shoreline/Ocean	

*Analysis of dredging with and without containment of area to be dredged.

REMEDIAL ALTERNATIVES
ESTUARY

		RESPONSE ACTIONS		
ALTERNATIVE	REMOVAL	TREATMENT	DISPOSAL	NON-REMOVAL
EST-1				No Action
EST-2	Dredge	(Sediments) Water treatment	CDF/CAD	
EST-3	Dredge	Sediments -Dewater -Solidification/ vitrification Water treatment	Shoreline/Ocean	
EST-4	Dredge	Sediments -Dewater -Solvent extract/ Incineration -Solidify residue (Solvent-PCBs) -Incineration (off-site) Water treatment	Shoreline/Ocean	

REMEDIAL ALTERNATIVES
LOWER HARBOR/BAY

		RESPONSE ACTIONS		
ALTERNATIVE	REMOVAL	TREATMENT	DISPOSAL	NON-REMOVAL
LHB-1				No Action
LHB-2	Dredge	Sediments -Dewater -Solidification Water treatment	Shoreline/Ocean	Cap (select areas)
LHB-3	Dredge	Sediments -Dewater Water treatment	CDF	Cap (select areas)
LHB-4	Dredge	Sediments -Dewater -Incineration -Solidify ash Water treatment	Shoreline/Ocean	
LHB-5	Dredge	Sediments -Dewater -Solvent extract. -Solidify residue Solvent-PCBs -Incineration (offsite) Water treatment	Ocean/shoreline	
LHB-6	Dredge	Sediments -Dewater Water treatment	Lined/Unlined CDF	
LHB-7	Dredge		CAD	

shown in the top line of Figure 16, since these criteria consider technical, cost, institutional, and risk concerns. The threshold criteria, compliance with ARARs, and overall protection, are used to assess whether a remedial alternative achieves compliance with ARARs and whether it provides overall protection of human health and the environment. The final two criteria (state and community acceptance) assess the state's and community's preferences or concerns about the alternatives.

Evaluation of the effectiveness of remedial alternatives in achieving the response objectives is the major environmental analysis step in the FS. Effectiveness evaluations consider time until protection is achieved, environmental impacts, magnitude of residual risk, and human health and environmental protection. Table 5 summarizes these factors and the analysis to be done under each.

Following assessment of individual alternatives against the nine criteria, a comparative analysis will be conducted for alternatives developed for each area. This analysis, which will identify advantages and disadvantages of each alternative relative to one another, will assist EPA in selecting the preferred remedial alternative for each area. Alternatives from each area will also be combined to form remedial action scenarios for the overall New Bedford Harbor site. This approach of area-specific and site-wide remedial alternatives will provide EPA with the options of partial or complete remediation.

IMPLEMENTATION AND MONITORING CONSIDERATIONS FOR REMEDIAL ACTION

The New Bedford Harbor FS is scheduled for completion in the spring of 1989. Following completion of the FS, EPA will decide on the remedial alternative to be implemented. This will not occur until summer 1989. Thus, implementation and monitoring information for the selected remedial alternative is not available for discussion at this time.

The New Bedford Harbor FS does consider implementation and monitoring aspects in the detailed evaluation of each remedial alternative. Factors considered are grouped in EPA's most recent guidance on performing feasibility studies under headings of technical feasibility, administrative feasibility, and availability of services and materials. Technical feasibility includes considerations of

- technical difficulties and unknowns with applying a new and innovative technology or a known technology under new conditions;
- technology capabilities in meeting specified operation rates and performance standards, along with consideration of materials handling and time needed to solve operational problems;
- ease of performing further remedial action, particularly, if the site is remediated in stages; and
- monitoring considerations, including the ability to measure how effective the remedy is during and after implementation, including risks of exposure if monitoring does not detect a release of contaminants or the technology not performing to standards should be assessed.

TABLE 5 Remedial Alternatives Effectiveness Evaluation

Analysis Factor	Analysis
SHORT-TERM EFFECTIVENESS	
Time until protection is achieved	Time until effect of hot spot removal is seen in water column, sediment, and biota, and change in risk is achieved
Environmental Impacts	Hot spot containment construction, impact of release on water column, sediment, and biota Evaluation of mitigative measures Impact after application of mitigative measures
Environmental Impacts	Hot-spot dredging, no containment, impact of release of 0.6-1.5 kg PCBs per tidal cycle for 3-4 weeks through Coggeshall Street Bridge on water column, sediment, and biota and change in risk achieved Duration and impact of removal of hot spot to less than 10 ppm total PCB on estuary, harbor-bay on water column, sediment, and biota and changes in risk achieved Duration and impact of treated water discharge on estuary, harbor-bay, water column, sediment, and biota Evaluation of mitigative measures; hydraulic controls at Coggeshall Street Bridge, hurricane barrier closure, flood tide dredging, impact after application of mitigation measures
LONG-TERM EFFECTIVENESS	
Magnitude of residual risk	Risk from residual PCBs after hot spot dredging, no containment Biota ingestion, direct contact water, sediment, exposure Risk from estuary, harbor-bay after hot spot dredging Biota ingestion, direct contact water, sediment, exposure
Adequacy of controls	Hydraulic controls at Coggeshall Street Bridge Hurricane barrier closure Flood tide dredging

TABLE 5 Continued

Analysis factor	Analysis
OVERALL PROTECTION	
Site alternatives	Protection over time, reducing PCB and metal concentrations in water, sediment, biota as reductions in harmful levels and risks
	Combined alternative simulations with model results in overall site predicted concentrations

Under administrative feasibility, permitting and regulatory agency coordination time and costs need to be considered.

Factors under availability of services and materials that are important are

- treatment and storage capacity at the time of implementation,
- availability of equipment and experienced operators, and
- extent that new and innovative technologies are proven in full-scale operation.

INSTITUTIONAL AND MANAGEMENT CONSIDERATIONS

A number of institutional considerations are major factors in the New Bedford Harbor FS. They include siting and land space available for locating sediment handling, treatment, and disposal facilities; future use of the harbor and shoreline; and community and state acceptance. Management considerations to be addressed by EPA when selecting a clean-up alternative for the site are fund-balancing of the costs of the New Bedford Harbor remedy with other high priority sites and the SARA's emphasis on on-site permanent remedies.

The State of Massachusetts is considering their responsibility under SARA to manage the long-term remedial alternative as they review and comment on the clean-up options.

Siting of Handling, Treatment, and Disposal Facilities

Alternative sites are generally of two types: off-site beyond the confines of the Acushnet River Estuary and New Bedford Harbor, or on-site within the confines of the Acushnet River and New Bedford Harbor. EPA's National Contingency Plan (40 CFR 300.68(f)(1)(i)) requires that remedial alternatives include treatment or disposal at an off-site facility.

The siting evaluations consider various options, including: offsite disposal in existing PCB-approved landfills; disposal at upland sites in the vicinity of New Bedford; onsite disposal, including shoreline sites and sites within New Bedford Harbor (islands and CADs); aquatic disposal; and ocean disposal.

NUS has conducted the major disposal site studies completed to date. During the process of conducting the 1984 fast-track FS for cleanup of PCB-contaminated sediments in the Acushnet River Estuary, NUS completed an interim report titled "Initial Evaluation of Potential Disposal Sites for Contaminated Dredged Materials" (June 1984). The report included an initial identification, evaluation, and ranking of potential sites, both upland and shoreline. Both EPA and the Massachusetts Interagency Task Force for New Bedford Harbor participated in establishing criteria for screening the identified sites. These criteria are listed in the previously mentioned report. In addition to the Interagency Task Force, state and local governmental information on siting was obtained from previous solid-waste disposal and regional planning studies.

For upland sites, 37 potential disposal sites remained following the first phase screening by NUS. In the second phase, the five highest ranking sites were determined. The first phase screening identified sites with "critical flaws" to eliminate such sites from further screening. These flaws included sites being located in close proximity to developed/populated areas, state parks, or wildlife management areas; public drinking water supply watersheds; highly productive stratified glacial deposits, including aquifers used for public water supplies; and wetlands. These five sites are not being evaluated further at this time due to SARA's preference for onsite disposal and the state of Massachusetts' policy of not establishing new hazardous waste disposal facilities in "nondegraded" areas.

For onsite disposal, NUS, in the 1984 Interim Siting Report, identified 12 sites, which were then screened to a subset of five using the factors listed in Table 6.

Following public comment on the fast-track FS in 1984, EPA decided that further evaluation of potential in-harbor disposal sites was warranted. NUS completed the evaluation in April 1986; the results are described in the report "Investigation and Ranking of Potential In-Harbor Disposal Sites." The April 1986 report identified 15 potential in-harbor disposal sites as the most promising. The identification was based on a quantitative ranking procedure similar to that used in the 1984 siting study by NUS. For purposes of the overall New Bedford FS, the 15 in-harbor sites identified by NUS will receive further evaluation when the in-harbor disposal alternative is studied.

As the siting results are being used in the FS, it is becoming apparent that shoreline and land areas for disposal are limited and that it will be necessary to preserve land areas for addressing other environmental needs in the area. For example, if all of New Bedford Harbor, from the estuary to the Hurricane Barrier, needs to be dredged to achieve a residual PCB concentration of < 10 ppm total PCBs in the sediment, and it is disposed of without volume reduction, there will not be sufficient capacity for disposal in lined shoreline disposal facilities. A major aspect of the capacity of such facilities is

TABLE 6 In-Harbor Disposal Siting Characteristics

Factor		Site #1 Western Cove	Site #1A Cove Extension N
I.	Current site conditions		
	Current use	*Undevel. land; open water	*Undevel. land; open water
	+Footage of waterfront lost	2.83	3.28
	Property ownership	Municipal; private	Municipal; private
II.	Engineering Feasibility		
	Depth to bedrock	37.5'	65'
	Depth to water	0.5'	1.25'
	Sediment characteristics	Deep muck	Deep muck
	% Land	35%	25%
	Existing discharges	One present	One present
	+Length of embankment	5.87'	*5.22'
III.	Site Access		
	Distance to dredge	3,000'	3,000'
	Route conditions	*River transport (pipeline)	*River transport (pipeline)
IV.	Storage capacity		
	Capacity to +10 MSL	459,850 yd^3	671,700 yd^3
	Change per 1' increase	44,800 yd^3	61,100 yd^3
V.	Environmental factors		
	+Acres of open water lost	0.04	0.04
	+Acres of wetlands lost	0.006	>0.004
VI.	Public Health Considerations		
	Short-term exposure (ind/comm)	High	High
	Long-term exposure (resident)	High	High
	Buffer zones	Poor	Poor

Factor		Site #1B Cove Extension N	Site #2 Sycamore Road Lowland
I.	Current site conditions		
	Current use	*Undevel. land; open water	Salvage yard; forest; wetland
	+Footage of waterfront lost	1.79'	4.87'
	Property ownership	Municipal; private	Municipal; private
II.	Engineering Feasibility		
	Depth to bedrock	45'	50'
	Depth to water	1.7'	0.65'
	Sediment characteristics	Varies; muck, silts/sands	Onshore; silts/sands
	% Land	24% present	83%
	Existing discharges	Multiple present	*None known
	+Length of embankment	11.83'	8.17'
III.	Site Access		
	Distance to dredge	*500'	1,500'
	Route conditions	*River transport (pipeline)	*River transport (pipeline)
IV.	Storage capacity		
	Capacity to +10 MSL	*726,910 yd^3	688,000 yd^3
	Change per 1' increase	65,800 yd^3	*68,340 yd^3
V.	Environmental factors		
	+Acres of open water lost	0.04	0.01
	+Acres of wetlands lost	>0.004	0.06
VI.	Public Health Considerations		
	Short-term exposure (ind/comm)	High	Medium
	Long-term exposure (resident)	High	High
	Buffer zones	Poor	Poor

+ = Per 1,000 C.Y. of capacity; * - Optimal feature; (Ind/Comm) - Industrial/commercial

SOURCE: Data from NUS, 1986. Investigation and Ranking of Potential In-Harbor Disposal Sites. Appendix A

lining them to prevent leaching of disposed contaminants in dredged material.

Other environmental needs in the New Bedford area are space for an expanded and upgraded wastewater treatment plant and future solid-waste disposal, since the existing landfill is nearing capacity.

State and Community Acceptance

EPA Region I has very actively sought state and community involvement in the FS process for New Bedford. This has been accomplished by holding monthly progress meetings and presenting information on the study process and results to the Greater New Bedford Environmental Community Work Group.

At monthly progress meetings, representatives of state agencies involved in regulatory review of the FS, a city of New Bedford representative, and the Community Work Group receive progress updates and results and have the opportunity to comment to and question EPA, COE, and the contractors performing the FS. Also, as various parts of the FS are completed, EPA and its contractors present results to the members of the Community Work Group to enable them to discuss and comment on all phases of the work as it proceeds.

Through this process to date, it has become apparent that there are two major concerns shared by the community. The first is that remedial action should allow areas closed to fishing to be opened. This is particularly important because of misimpressions that fish landed are sold through New Bedford are related to the PCB contamination-based fishing closures. In actuality, New Bedford fish are caught in the Georges Bank, some 200 mi east of New Bedford Harbor.

The second major concern is availibility of shoreline and land areas for locating treatment and disposal facilities. The primary interest is to maintain shoreline areas as suitable for development of port facilities. Other land needs competing for space in New Bedford are an expanded wastewater treatment facility, a planned waste-to-energy, solid-waste disposal facility and continued expansion of commercial port facilities.

ACKNOWLEDGMENTS

The information in this case study has been funded by the U.S. EPA under REM III contract 68-01-7250 to Ebasco Services Incorporated. It has been subject to EPA review and has been approved for publication as preliminary information from an ongoing study. Mention of trade names or commercial products does not constitute endorsement or recommendation for use.

REFERENCES

Battelle New England Marine Research Laboratory. 1984. Work Plan for

Modeling of the Transport, Distribution and Fate of PCBs and Heavy Metals in the Acushnet River/New Bedford Harbor Buzzards Bay System. Duxbury, Mass.: Battelle.

E. C. Jordan/Ebasco Services Incorporated. 1988. Baseline risk assessment for the New Bedford Harbor site. Preliminary Draft. Portland, Me.: E. C. Jordan Company.

E. C. Jordan/Ebasco Services Incorporated. 1986. Project Management Plan for New Bedford Harbor, Massachusetts. Portland, Me.: E. C. Jordan Company.

Massachusetts Department of Pulbic Health. 1987. The Greater New Bedford PCB Health Effects Study 1984-1987, Executive Summary, Massachusetts Department of Public Health, The Massachusetts Health Research Institute, the U.S. Centers for Disease Control.

Teeter, A. M. 1988. Sediment and Contaminant Hydraulic Transport Investigations. Report 2 of 12, New Bedford Superfund Project: Acushnet River Estuary Engineering Feasibility Study Series, Technical Report EL-88. Vicksburg, Miss. U.S. Army Engineer Waterways Experiment Station. In preparation.

U.S. Army Corps of Engineers (U.S. Army COE). 1988. New Bedford Harbor Superfund Project, Acushnet River Estuary Engineering Feasibility Study of Dredging and Dredged Material Disposal Alternatives. Draft Final Report. Vicksburg, Miss.: U.S. Army Engineer Waterways Experiment Station.

U.S. Army Corps of Engineers. 1987. Pilot Study of Dredging and Dredged Material Disposal Alternatives. Waltham, Mass.: COE New England Division.

U.S. Environmental Protection Agency. 1988. New Bedford Harbor pilot study, pre-operational monitoring progress report. Draft Report, U.S. EPA, Narragansett, R.I.

U.S. Environmental Protection Agency. 1988. Guidance for conducting remedial investigations and feasibility studies under CERCLA. Draft. U.S. EPA, Washington, D.C.

U.S. Environmental Protection Agency. 1986. Preliminary Data Report, New Bedford Harbor Project, Draft Report. U.S. Environmental Protection Agency, Narragansett, R.I., and Science Applications International Corp. Narragansett, R.I.: U.S. EPA.

ADDENDUM

Completed Project Reports

The following is a list of New Bedford Harbor Superfund project reports that have been issued to date:

<u>Task 7</u>. Draft Technical Review Report; Evaluation of the New Bedford Wastewater Treatment Plant and Sewage System for PCB Discharges to the Acushnet River Estuary, New Bedford Harbor and Buzzards Bay, Bristol County, Massachusetts. September 1986. E. C. Jordan/Ebasco Services Incorporated.

Tasks 18, 23, 24. Draft Initial Screening Report: Non-removal and Removal Technologies, April 1987. E. C. Jordan/Ebasco Services Incorporated.

Tasks 18, 23, 24. Final Draft Initial Screening Report: Non-removal and Removal Technologies, November 1987. E. C. Jordan/Ebasco Services Incorporated.

Tasks 18, 19, 21, 23, 24. Draft Detailed Analysis of Remedial Technologies for the New Bedford Harbor Feasibility Study, August 1987. E. C. Jordan/Ebasco Services Incorporated.

Task 13. Upper Estuary Sediment Characterization, Field Investigation and Analytical Testing by Woodward Clyde Consultants, January 1987.

Task 14. Contaminant Migration Analysis:

A. Baseline Conditions for Contaminant and Sediment Migration, January 1987. COE, Waterways Experiment Station.
B. Estimated Contaminant Release from Pilot Study Operations, July 1987. COE, Waterways Experiment Station.
C. Controls for Dredging, January 1987.
D. Numerical Modeling of Sediment Migration from Dredging and Disposal, May 1987.
E. Suspended Material Transport at New Bedford Harbor (ASCE paper by Al Teeter [not on your reference list--please include]), May 1987. COE, Waterways Experiment Station.

Task 16: Composite Sample Sediment Testing (COE, Waterways Experiment Station):

A. Chemical Analysis of Composite Sediment and Site Water, January 1987.
B. Surface Runoff Water Quality from New Bedford Harbor Sediment, June 1987.
C. Interim Results from Leachate Testing, May 1987.
D. Capping Effectiveness Testing, June 1987.
E. Dredged Material Settling Tests for New Bedford Sediment, January 1987.
F. Chemical Clarification Testing, May 1987.

Task 21: Field Operations Plan, E. C. Jordan/Ebasco Services Incorporated.

A. Site Management Plan.
B. Field Sampling and Analysis Plan.
C. Health and Safety Plan.

Task 19: Disposal Site Selection, E. C. Jordan/Ebasco Services Incorporated.

A. Final Draft Report, Alternative Disposal Site Selections, February 1987.
B. Statement of Work for Drilling Services for Preliminary Geotechnical Investigation of Engineering Properties, August 1987.
C. Statement of Work for Survey Services for Preliminary Geotechnical Investigation of Engineering Properties, August 1987.

Task 20: Preliminary Flood Plain Assessment Investigation, September 1987. COE New England Division.

Task 21: Detailed Evaluation of Detoxification/Destruction Technologies, Initial Screening Report, January 1987. E. C. Jordan/Ebasco Services Incorporated.

Task 21: Detailed Evaluation of Detoxification/Destruction Technologies, Initial Screening Report (Final Draft), September 1987. E. C. Jordan/Ebasco Services Incorporated.

Task 21: Technical Memorandum, Pilot Testing of Detoxification/Destruction Technologies, February 1987. E. C. Jordan/Ebasco Services Incorporated.

Task 21: Requests for Proposals - May 1987. E.C. Jordan/Ebasco Services Incorporated.

A. Bench Testing of Selected Technologies for PCB Detoxification/Destruction.
B. Bench Scale Testing of Biodegradation Technologies for PCBs in New Bedford Harbor Sediment.

Task 22: Draft Report: Exposed Species Analysis, July 1987. E. C. Jordan/Ebasco Services Incorporated.

Draft Report: Selection of Contaminants of Concern, July 1987. E.C. Jordan/Ebasco Services Incorporated.

Task 23: Technical Memorandum, Hot Spot Feasibility Study, March 1987. E. C. Jordan/Ebasco Services Incorporated.

Task 26: Pilot Study of Dredging and Dredged Material Alternatives - New England Division, USACE. September 1987.

Technical memo on proposed target levels for PCB concentration in air, October 1987.

Task 50: Project Management Plan for New Bedford Harbor, Massachusetts, August 1986. E.C. Jordan/Ebasco Services Incorporated.

Task 52: Technical Review Report of Comments to the NUS Report, Draft Feasibility Study of Remedial Action Alternatives, Acushnet River Estuary above Coggeshall Street Bridge, New Bedford Harbor, Bristol

County, Massachusetts, June 1986. E. C. Jordan/Ebasco Services Incorporated.

Task 63: Regulation Assessment (ARARs) for New Bedford Harbor, October 1986. E. C. Jordan/Ebasco Services Incorporated.

Task 20: Hydrology of Floods, NBH by NED, September 1987. (Prelim. Floodplain).

PHYSICAL TRANSPORT INVESTIGATIONS AT NEW BEDFORD, MASSACHUSETTS

Allen M. Teeter
U.S. Army Engineer Waterways Experiment Station

ABSTRACT

Migrations of sediment, sediment-associated contaminant, and dissolved materials released by proposed dredging and disposal operations were predicted as part of an engineering feasibility study of dredging cleanup. Highly contaminated sediments blanket most of upper New Bedford Harbor (the Acushnet River estuary), and were found to be escaping from the upper harbor toward Buzzards Bay. Field measurements, laboratory tests, and computer models, each necessary to support the others, comprised the study. Field measurements characterized hydraulic conditions and transport mechanism for salt, sediment, and contaminants for a series of surveys. Suspended material was found to migrate from Buzzards Bay upstream in the estuary at concentrations generally below 10 ppm, and settle in the upper harbor at about 2,200 kg per tidal cycle. The flux of PCB-Aroclors was found to be seaward and averaged 1.55 kg per tidal cycle. Laboratory tests for settling, deposition, and erosion of sediment material were carried out. The most mobile sediment fraction was found to make up 28 percent of the sediment. Models for hydrodynamics and sediment transport were applied to the upper New Bedford Harbor, and used to predict sediment and contaminant migration for dredging and disposal scenarios. Results indicated that the flux of sediment materials from the upper harbor would be 15 to 20 percent of the rate of sediment resuspension.

INTRODUCTION

An engineering feasibility study (EFS) of a possible Superfund dredging cleanup for upper New Bedford Harbor was conducted by the U.S. Army Corps of Engineers (COE). The COE's Waterways Experiment Station (WES) Hydraulics Laboratory evaluated hydraulic conditions and sediment migration as part of the WES Environmental Laboratory dredging and disposal EFS conducted for the U.S. Environmental Protection Agency (EPA), Region I, under the direction of COE's Missouri River Division, and in cooperation with its New England Division. Highly contaminated sediments blanket most of the upper New Bedford Harbor (Acushnet River Estuary), threatening to spread to other harbor areas and adjoining

Buzzards Bay, and adversely impacting fisheries resources. The EFS is one component of EPA studies that will lead to a Superfund cleanup of the harbor and upper harbor.

The Acushnet River estuary EFS had components addressing physical and chemical testing of sediments to determine appropriate dredging limits, acquisition of bathymetric and geotechnical information, and study of contaminant behavior under simulated field conditions. An EFS report series is in preparation. Averett (1988) gives an overview of the study. The WES Hydraulics Laboratory investigated hydraulic conditions and transport mechanisms for salt, sediments, and contaminants as part of the EFS (Teeter, 1988). The approach of the study was to integrate prototype measurements, laboratory data, and model results to quantify present conditions and predict dredging and disposal effects. The objectives were to evaluate

1. contaminant and sediment migrations away from resuspension points, and out of the upper harbor,
2. the hydraulics of the present and dredged upper harbor, and
3. concentrations of sediment and contaminant in the upper harbor during dredging and disposal releases.

Confined disposal facilities (CDF) are diked settling basins that can include treatment methods to enhance contaminant settling. Confined aquatic disposal (CAD) is a controlled operation using excavated cells or chambers in the estuary bed, which are capped after filling. Suspended sediments (total suspended material [TSM]) will be discharged with CDF effluent, and released during CAD filling. The rate of sediment release, sediment characteristics, and ambient conditions will control the amount of sediment that will escape from the proximity of the dredge and disposal facilities, and from the upper harbor during dredging. This paper describes methods used to predict sediment and contaminant escape.

FIELD DATA COLLECTION

Figure 1 shows the layout and approximate dimensions of New Bedford Harbor, which is located on the north shore of Buzzards Bay and is the estuary of the Acushnet River. The mean tide range at New Bedford Harbor is 1.1 m, and the spring range is 1.4 m. Surface-to-bottom salinity differences are generally less than 0.5 ppt. The Acushnet River has a mean annual freshwater discharge of about 0.85 cubic meters per second. The upper harbor is shallow, with an average depth of only about 1 m at mean low water.

Survey dates, tides, freshwater flows, and winds were as follows:

Survey date	Freshwater inflow m^3/sec	Tidal range at tide gauge 3, m	Wind direction, speed, km/hr	Water temperature
Mar 6	1.17	1.04	S, 24-32	4°C
Apr 24	1.50	1.65	NE, 8-12 then 32-48	11°C
Jun 5	0.25	1.04	SW, 16-24	17°C

Nine stations were sampled repeatedly over tidal cycles (Figure 1). Current speed and direction, salinity, and TSM were sampled 0.6 m up from the bed and below the surface, and at mid-depth. Station 9 was sampled only at mid-depth.

Bridge Flux of TSM

The Coggeshall Street Bridge is a key point at which transport measurements and predictions were made. Bridge fluxes of TSM were estimated by integrating half-hour measurements of velocity (u) and TSM

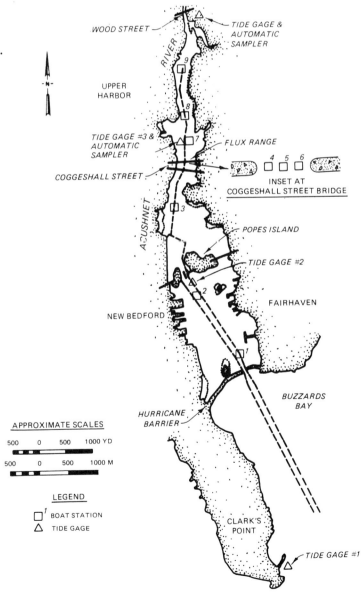

FIGURE 1 Sampling and gauging locations for New Bedford Harbor.

concentration over the tide-corrected, cross-sectional area, and integrating in time. Results are shown in Table 1.

The net flux of TSM was always found to be upstream, although fluxes in either direction were at least twice the net flux values. Average flux of TSM into the upper harbor, corrected for tidal asymmetry, was about 2,200 kg per tidal cycle. The freshwater inflow added some additional sediment, on the order of a several hundred kg per tidal cycle. Shoaling resulting from the deposition of 2,500 kg per tidal cycle amounts to 3 mm per year when spread over the entire surface area of the upper harbor at a bulk wet density of 1.5 g/cm^2 (775 dry-g/liter). The average sedimentation rate for the harbor has been estimated to be 7 mm per year (Summerhayes et al., 1977).

Estuarine TSM Flux Components

Figure 2 shows a plot of the longitudinal distribution of depth- and tidal-averaged TSM concentration. TSM concentrations were lowest at the most seaward stations, and increase upstream. A turbidity maximum occurred in the upper harbor. Differences in TSM concentration between spring- and neap-tide surveys were small, indicating that redispersion of near-bed suspended material rather than erosion of bed sediments contributed to tidal variations in TSM.

The most important flux components for TSM were tidal pumping. TSM transport by steady vertical shear closely associated with transport by gravitational circulation, was small for New Bedford Harbor. Mechanisms for upstream tidal pumping were evaluated. Maximum TSM resuspension produced by the highest tidal currents, usually occurring near low water, were transported in the flood direction, producing upstream tidal pumping. Water column redispersion time scales were much shorter than settling time scales, and produced phase differences between tidal velocities and TSM concentrations (tidal pumping).

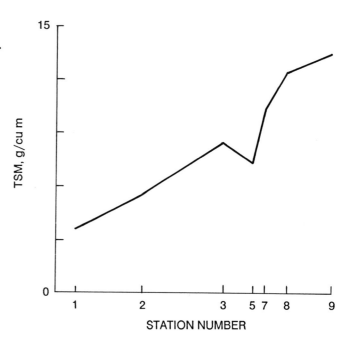

FIGURE 2 Longitudinal distribution of TSM concentration.

TABLE 1 Fluxes of PCBs and TSM

Survey date	Tidal phase	Tidal volume E9l	PCB, Aroclor ppb	TSM ppm	PCB/TSM Aroclor, ppm sed. dry-wt	PCB flux kg	TSM flux kg	Tide-corrected[a] PCB flux kg	Tide-corrected[a] TSM kg
3/6/86	ebb	-1.13	1.3	3.9	333	-1.47	-4,400		
	flood	0.89	1.3	7.2	180	1.15	6,400		
	net	-0.25				-0.32	2,100	-0.07	3,100
4/24	ebb	-1.47	2.0	5.9	339	-2.94	-8,700		
	flood	1.57	0.5	8.1	62	0.79	12,800		
	net	0.10				-2.16	4,000	-2.36	2,900
6/5	ebb	-0.67	5.8	6.6	879	-3.90	-4,400		
	flood	0.88	3.0	7.4	405	2.63	6,500		
	net	0.21				-1.27	2,100	-2.22	605

NOTE
[a] See text for explanation.

Vertical mixing was found to be more intense during flood tidal phases and was less damped by density effects on flood tidal phases.

PCB Fluxes

Bridge fluxes of PCBs are shown in Table 1. Flow-proportioned ebb and flood PCB Aroclor concentrations were multiplied by the tidal volumes to obtain ebb and flood PCB fluxes. The difference between ebb and flood fluxes is the tidal net flux. Observed net fluxes were always seaward (negative) with a mean net flux of -1.25 kg per tidal cycle.

Tidal biases were removed from the raw tidal net fluxes by summing net-flow fluxes (freshwater volume times mean concentration) and tidal pumping fluxes (the difference between ebb- and flood-mean concentrations times the mean tidal volume). Corrected flux values are also shown in Table 1, and were also seaward with a mean flux of -1.55 kg per tidal cycle.

Floatable material samples at the bridge were low in PCBs. Accurate transport rates could not be estimated for floatable material, but were at least several orders of manitude less than that for suspension.

PCBs attached to sediment particles at the surface of the bed could be exchanged into the overlying sediment suspension by a physical particle exchange mechanism, and thus be mobilized for transport. Such an exchange could take place without a significant mass flux of sediment. A particle exchange theory, based on aggregation and disaggregation of cohesive particles at the sediment/bed interface was developed during this study. That analysis used laboratory data on another estuarine sediment. The results indicated that particle exchange could be an important transport mechanism.

LABORATORY TESTING

Depositional and erosional characteristics of fine-grained sediments vary greatly, and are critical to the prediction of sediment and contaminant migration. Direct laboratory testing on sediments less than 74 μm from the study area was therefore undertaken.

Deposition (D), or flux of sediment material to the bed, is the sum over a number of fractions of settling flux multiplied by deposition probability:

$$D = \sum_{i=1}^{n} P_i W_{s_i} C_i \qquad (2)$$

where W_s is the settling velocity, P is the probability that an aggregate reaching the bed will remain there, C is the concentration just above the bed, the subscript i indicates a sediment fraction, and n is the number of fractions. P varies linearly from 0 at a critical shear stress for deposition, τ_{cd}, to 1 at zero bed shear stress, $\tau_b = 0$.

The functional form $1 - \tau_b/\tau_{cd}$ where $\tau_b < \tau_{cd}$ is used for P.A suspension of uniform material in a steady, uniform flow will either deposit completely or remain entirely suspended depending on whether the bed shear stress is below or above τ_{cd}, according to Eq. 2. The objective of the deposition testing was to determine τ_{cd}, and the magnitude of the product $P\,W_s$ for each sediment fraction identified.

The mode of resuspension (used synonymously with erosion) considered important to potential contaminant migration at New Bedford Harbor is particle erosion. At τ_b above a critical value, particles individually dislodged from the sediment bed as interaggregate bonds are broken. Particle resuspension (E) is related to the shear stress in excess of a critical value, and to an erosion rate constant (M), thus:

$$E = M\left(\frac{\tau_b}{\tau_c} - 1\right), \quad \tau_b > \tau_c \qquad (3)$$

where τ_c is the critical erosion shear stress (Ariathurai et al., 1977). Observed erosion does not follow Eq. 3 indefinitely. Suspension concentrations above experimental eroding beds often reach equilibrium values that depend on the bed shear stress. Equilibrium suspensions form as erosion rates decrease with time to zero, while the flow remains constant. Eqilibrium suspensions have been related to vertical inhomogeneity in the bed (either particle characteristics or bed density) or to armoring by selective erosion at the bed surface.

See Figure 3 for the configuration of the sediment water tunnel. This testing device was developed for this study to safely test contaminated sediments. It was a closed-conduit sediment water tunnel, open to the air only at a small expansion chamber. The water tunnel had a uniform cross-section area, which changed from rectangular in the horizontal, deposition/resuspension sections to circular in the vertical settling and pumping sections. The water tunnel was calibrated so that propeller speed could be related to average velocity and bed shear stress.

Three sediment fractions were identified, and designated 1, 2, 3. The τ_{cd}'s developed from the analysis of the data were 0.42, 0.33, and 0.043 N/m^2 for fractions 1, 2, and 3, respectively. Considering all results, the approximate composition of the sieved composite sample was 30, 30, and 40 percent for fractions 1, 2, and 3, respectively. W_s's were about 2, 1, and 0.006 mm/sec for fractions 1, 2, and 3, respectively. Median Ws values are shown in Figure 4. For the most erodible fraction, τ_c was found to be 0.06 N/m^2. The most easily eroded fraction is the same fraction identified as the slowest to deposit (fraction 3). Only about an additional 15 percent of the total bed material, or half of fraction 2, eroded between 0.06 and 0.6 N/m^2, and the remainder of the material had τ_c's greater than 0.6 N/m^2.

MATHEMATICAL MODELING

Near-field Plume and CAD Models

Suspended-sediment plume calculations were performed to evaluate

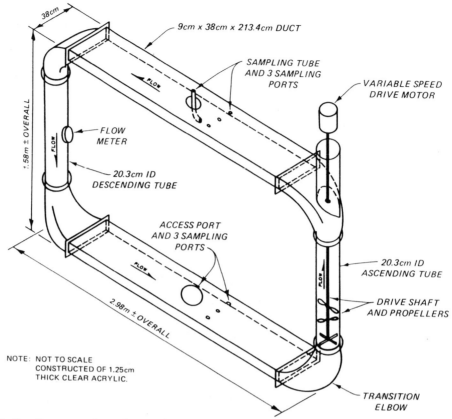

FIGURE 3 Isometric view of sediment water tunnel.

the escape of sediments and contaminants from proposed dredging and disposal site outfalls in upper New Bedford Harbor. Near-field analyses of the escape of sediments from a CAD during the filling phase were also performed.

The near-field plume and CAD models assumes an infinitely small, vertically well-mixed, suspended sediment source. Model suspended sediment plumes are advected away from the source in the X direction, spread or diffused in the Y direction, and allowed to settle. A diffusion velocity formulation was used to introduce a length scale-dependence into model plume spreading, similar to those observed in field experiments.

The required data for the plume model were Q_s, H, U, and W_s. A site depth of 1 m was assumed, the average depth of the upper harbor. The remaining two variables, U and W_s, were assigned distributions.

Plume calculations were made for a test matrix of 16 conditions formed by four values of current speed, and four values of settling velocity. Plume predictions for dredging in upper New Bedford Harbor indicate that on average about 35 and 29 percent of the material released at the dredge head will escape from 50 and 100 m of the site, respectively. The remainder settled within this radius. Sediment that escaped did so at the highest current speed, and had the lowest settling rate. However, escape totals were highest for the moderate

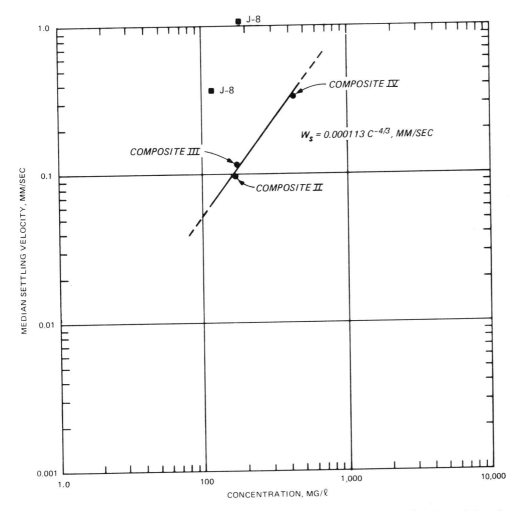

FIGURE 4 Median settling velocity/concentration relationship for settling test phases and field samples at grid cell J-8.

current speeds, and for the lowest settling rate. Moderate current speeds had the greatest frequency of occurrences. Figure 5 shows an example plot of plume concentration contours.

Average CAD release ratios were calculated for a three-dimensional test matrix in U, W_s, and H. Four depositional classes and current speed ranges were identical to those used in the plume calculations. Results indicated that only the finest or slowest settling fraction escaped from the CAD, and that the escape of this fraction was almost complete.

Estuarine Numerical Modeling

Computer codes RMA-2V and RMA-4 of the TABS-2 numerical modeling system (Thomas and McAnally, 1985) were used to schematically model vertically averaged hydrodynamics and sediment transport, respectively. To properly describe boundary conditions, the model domain was extended downstream to the hurricane barrier. A numerical mesh of 219 elements

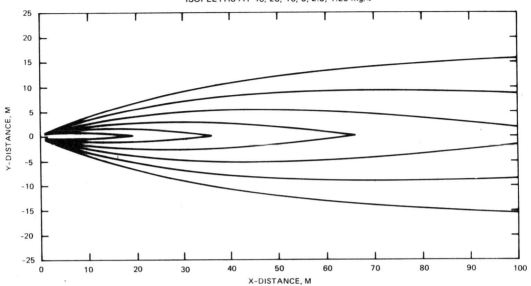

FIGURE 5   Example plume concentration isopleths for Q_s = 5 g/sec, U = 0.03 m/sec and W_s = 0.01 mm/sec.

was developed to cover the study area for use by both RMA-2V and RMA-4. The upper harbor portion of the mesh is shown in Figure 6.

A mean-tide sequence was applied to the seaward boundary of the model. At the upper end, velocities corresponding to a constant freshwater inflow of 0.85 m³/sec were specified.

The numerical hydrodynamics model was verified to field data by adjusting friction coefficients and turbulent exchange coefficients. Hydrodynamic computations were performed by "spinning up" the model from a steady, flat, water surface condition. Results from the hydrodynamics model were used to construct an 8-tidal-cycle sequence for input to the sediment transport model.

Sediment transport modeling was performed to estimate escape probabilities from the upper harbor for various sediment materials which might be resuspended as a result of dredging. Transport of resuspended material was modeled as a steady mass loading at specified points. Only sediments released from the mass-loading point were included in computations. Three mass-loading locations were employed to represent various locations where sediment releases might occur along the axis of the estuary. Five sediment fractions were modeled to characterize the range of sediments that might be actually released. The effective sediment deposition coefficient used was

$$\alpha = \frac{W_s P}{H}$$

The five depositional fractions tested covered the range 0.10 < α < 25.6, where α has the units of 1/day.

An example contour plots of the concentration field with α = 0

FIGURE 6 New Bedford upper Harbor mesh showing CAD exclusion zone (hashed) for 5.5-ft spring tide.

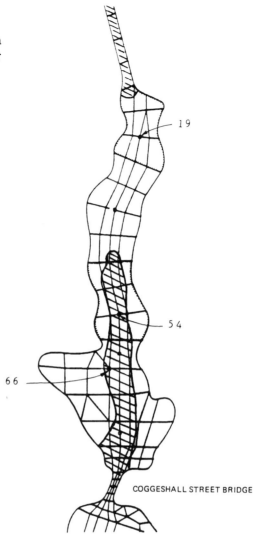

for flood and ebb tide phases are shown in Figure 7. Maximum concentrations were about 3 mg/liter for a release rate of 15 g/sec. However, the estuarine model over-estimates spreading near the source, and near-field predictions should be applied here. Concentrations were proportional to release rates.

Sediment transport results were used to calculate the escape probabilities of resuspended sediments rleased in the upper harbor. Average transport rates under the Coggeshall Street Bridge were computed for flood and ebb tidal phases after the model had reached dynamic equilibrium. The difference between ebb and flood transport rates, normalized by the mass loading rate, represents the escape probability. Escape probabilities were calculated for each depositional sediment fraction at three source locations in the upper harbor, and with three variations of geometry representing dredging changes.

A plot of escape probability versus a-infinity for three source locations is shown in Figure 8 for the existing estuary geometry. Source locations are shown in Figure 6.

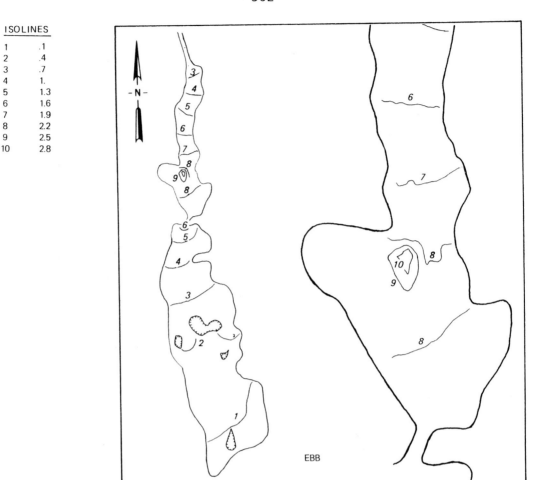

FIGURE 7 Sediment concentration field, ebb tide.

SUMMARY AND CONCLUSIONS

Assessments of sediment and contaminant migration out of the upper New Bedford Harbor for proposed dredging and disposal were made from information and analyses developed by field, laboratory, and various model studies. Upper New Bedford Harbor is a sheltered area with low current speeds, typical of many areas where contaminated sediments reside. Paradoxical tidal fluxes were found for suspended sediments and sediment-associated contaminants, implying that depositional sites do not retain all particle-associated contaminants. Average PCB flux from the upper estuary was seaward about 1.55 kg Aroclor per tidal cycle. However, the Acushnet estuary was found to be depositional, and a reasonably efficient sediment trap. TSM was imported from the coastal areas of Buzzards Bay, pumped upstream by tidal action, and formed turbidity maximums in the upper harbor.

The sediment fraction slowest to settle and deposit comprised 28 percent of the composite sediment and represents by far the greatest potential for sediment and contaminant migration. Other fractions will

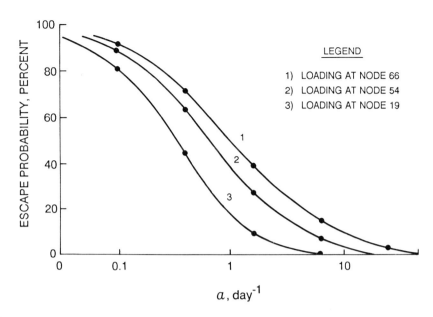

FIGURE 8 Escape probabilities for sediments released at three points along the upper harbor (see Figure 6 for locations).

not be highly mobile in the upper harbor.

Experimentally determined erosion thresholds were used with numerical hydrodynamics results to identify areas where CAD cells should not be sited. Based on these results, the area of the channel upstream from the bridge were not recommended for CAD sitting. See Figure 6.

Results from the dredge plume model indicated that an average, weighted by occurrence frequencies, of about 33 percent of the resuspended material will escape from a 100-m radius of the dredging site. Results from the CAD cell model indicated that all of the fine fraction sediment expelled with slurry pore water will escape from CAD cells.

Modeling results indicated that the CAD may cause the greatest release of sediments and contaminant materials. The escape probability for bulk sediments resuspended at the point of dredging will average 27 percent but will depend on the highly variable bed sediment composition. General information on dredge resuspension rates and suspended sediment releases during CAD filling is scarce. Pilot dredging and disposal, planned for the fall of 1988, will provide direct measurements for this site.

The methods employed in this study could be applied to similar sites. However, studies of high-current systems will have to give greater attention to the difficult experimental and theoretical aspects of erosion processes.

ACKNOWLEDGMENTS

This work was sponsored by EPA Region I, and monitored by Mr. Frank

Ciavettieri. The EPA has not reviewed this paper, and the views expressed herein are those of the author and not necessarily those of the EPA. Permission to publish was granted by the Office, Chief of Engineers.

REFERENCES

Ariathuria, R., R. C. MacArthur, and R. B. Krone. 1977. Mathematical Model of Estuarine Sediment Transport. Technical Report D-77-12. Vicksburg, Miss.: U.S. Army Engineer Waterways Experiment Station.

Averett, D. E. 1988. Study Overview. Report 1 of 12, New Bedford Superfund Project: Acushnet River Estuary Engineering Feasibility Study Series. Technical Report EL-88-15. Vicksburg, Miss.: U.S. Army Engineer Waterways Experiment Station. In preparation.

Summerhayes, C. P., et al. 1977. Fine-Grained Sediment and Industrial Waste Distribution and Dispersal in New Bedford Harbor and Western Buzzards Bay, MA. WHOI-76-115. Woods Hole, Mass.: Woods Hole Oceanographic Institution.

Teeter, A. M. 1988. Sediment and Contaminant Hydraulic Transport Investigations. Report 2 of 12, New Bedford Superfund Project: Acushnet River Estuary Engineering Feasibility Study Series, Technical Report EL-88-15. Vicksburg, Miss.: U.S. Army Engineer Waterways Experiment Station. In preparation.

Thomas, W. A. and W. H. McAnally. 1985. User's Manual for the Generalized Computer Program System: Open-Channel Flow and Sedimentation: TABS-2. Instr. Rpt HL-85-1. U. S. Army Engineer Waterways Experiment Station, Vicksburg, Miss.

PCB POLLUTION IN THE UPPER HUDSON RIVER

John E. Sanders
Barnard College, Columbia University

ABSTRACT

The upper Hudson River is one of the nation's most extensively PCB-polluted waterways. Wastewater discharge from two General Electric Company (GE) plants, and erosion of wood-laden, PCB-soaked deposits contributed to downriver supply of PCBs. As a result of a 1976 settlement between the state and GE, PCB discharges were stopped and the state has proposed to rehabilite the upper river by dredging of the PCB hot spots and encapsulation of the contaminated dredged material in a secure facility. Since that time, numerous legal and institutional obstacles--primarily funding and permitting--have delayed rehabilitation dredging to 1993 or 1994. The amount of PCBs entering the lower river has dropped from about 2 tonnes per year in the late 1970s to 1 tonne and less in the 1980s. Yet despite this drop, since 1983 the PCB content of striped bass caught in the Hudson estuary has averaged about 4 ppm. Although this amount is less than the pre-1984 Food and Drug Administration action limit of 5 ppm, it is still double the current action limit of 2 ppm.

OVERVIEW OF THE HUDSON RIVER

The Hudson River is divided into upper river and lower river where it is joined by the Mohawk River, south of Waterford. The combined river doubles the flow of water and triples the quantity of suspended sediment carried into the estuary over the Federal Dam at Green Island compared with that at Waterford (U.S. Geological Survey [USGS], 1977). The upstream limit of the estuary (the lower river and estuary are nearly synonymous) is the Green Island dam at the city of Troy, at the head of tidewater and about two miles south of the confluence of the Mohawk and Hudson rivers (Figure 1).

An additional factor related to the upper river is the Hudson-Champlain barge canal, a division of the New York State barge canal consisting of 6 dams and 7 locks. The canal enables small boats and barges to use the upper Hudson River between the Federal dam at Green Island and Fort Edward, where it cuts through the landscape in a northeast direction, away from the river, which swings west, then north.

A typical profile section across the upper Hudson River shows marginal flats underlain by silt and clay sediments up to 3 m thick and a wide channel floor underlain by a thin (1 m or less) carpet of coarse sand and gravel resting on deformed Ordovician bedrock (Sanders, 1982).

In the upper Hudson River, PCB-contaminated sediments attain their highest concentrations in the reach between the cities of Hudson Falls and Troy (Hetling and Horn, 1977; Hetling et al., 1978; Tofflemire and Quinn, 1979; Tofflemire et al., 1979; Brown and Werner, 1985; Brown et al., 1988). Just south of Bakers Falls, the river flows past two General Electric (GE) capacitor-manufacturing plants, one at Hudson Falls, which began using PCBs in 1947, and one at Fort Edward, which began using them in 1952 (Figure 2).

Hydraulic Influences

Downriver movement of PCBs in the upper Hudson is a function of natural sediment transport governed by water discharge (Turk, 1980; Turk and Troutman, 1981a, 1981b; Schroeder and Barnes, 1983a, 1983b; Barnes, 1987). A network of gauging stations maintained by the Water Resources Division of the USGS (Figure 2) monitors variations in discharge and extremes associated with floods (Figure 3). A compilation of maximum known discharge and stage for 326 localities within the Hudson River basin has been made by Robideau et al. (1984). Empirical studies of the relationship between water discharge and PCB transport into the Hudson estuary, have shown that what might be termed the "high-water mode" starts when the daily discharge at Waterford exceeds about 19,800 ft^3/sec (Schroeder and Barnes [1983b] and Barnes [1987] use a value of 650 m^3/sec). The 100-year flood flow of the Hudson River is 50,000 ft^3/sec at Fort Edward; 110,000 at Waterford; and 220,000 at Green Island (Darmer, 1987).

PCB Transport

Starting in the Water Year 1977 (October 1, 1976 to September 30, 1977), the USGS intensified monitoring of the upper Hudson River. Daily samples were collected to determine suspended sediment in the upper Hudson and Mohawk rivers (winter sampling was discontinued during winter months as of 1980) and intermittent samples were measured for for PCBs to show the range of variations of water discharge.

In the USGS laboratory, PCBs may be extracted from the total sample, or only after samples are passed through a 0.45-micron silver-oxide filter. What passes through the filter is defined as "dissolved load"; what remains is the "suspended load." PCB analyses indicated that two contrasting regimes operate in the river as a function of water discharge. At high flows, PCBs are found almost entirely in the fraction that remains on the 0.45-micron filter, and thus is attached to the sediment (Schroeder and Barnes, 1983b). In general, the more the water discharge, the more the suspended sediment, and thus, the higher the concentration of PCBs (Figure 4). But additional gauging

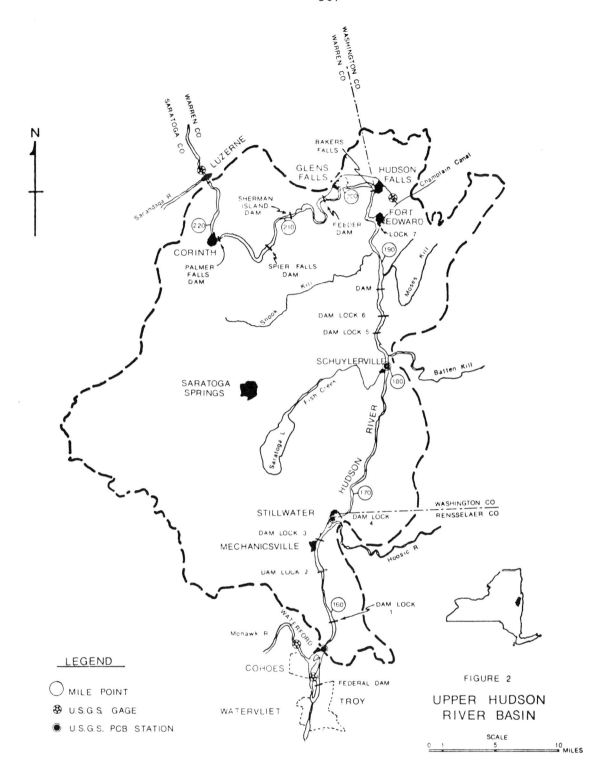

FIGURE 1 The Hudson River Basin; the dashed line marks the limit of subbasin drainage area. SOURCE: Hetling et al., 1978.

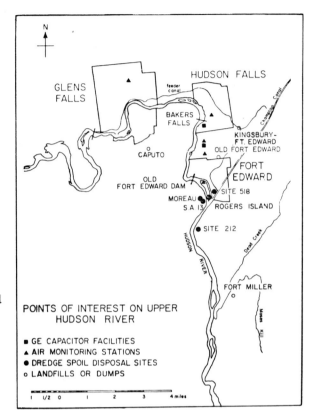

FIGURE 2 Glen Falls-Fort Edward area showing PCB-related sites. SOURCE: After Hetling et al., 1978.

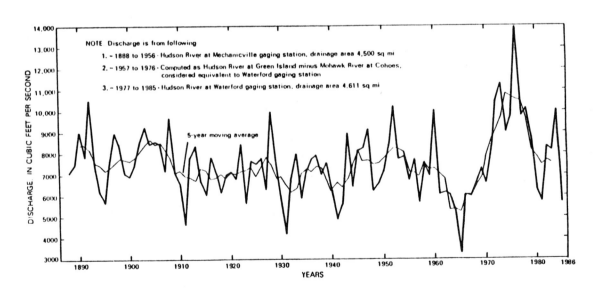

FIGURE 3 Water discharge of upper Hudson River (expressed as mean daily discharge computed on an annual basis at Mechanicville (1888 to 1956) and Waterford (1957 to 1985). SOURCE: After Darmer, 1987.

stations between Waterford and Glen Falls demonstrated that water discharge at Waterford is not a single reliable variable for estimating the quantity of PCBs transported into the estuary. Two floods in 1977 show the possible contrasts. During the March flood, in the lower drainage basin, much of the water entered the Hudson from the Hoosick River, south of the most heavily contaminated area. The ratio of PCBs to suspended sediment at Waterford for the April flood, in the upper basin, was seven times that of the March flood, indicating that the PCB source was bottom scour north of Stillwater. Closer inspection shows that it came from between Schuylerville and the Thompson Island pool (Turk and Troutman, 1981a; Schroeder and Barnes, 1983b).

At low flows, PCBs are found in the dissolved load. However, it is possible that PCBs are attached to colloidal particles, which are small enough to pass through the filter (Schroeder and Barnes, 1983a). PCB concentrations tend to increase as water discharge decreases (Turk, 1980; Turk and Troutman, 1981b; Schroeder and Barnes, 1983b; Figure 5). Because it was first thought that PCB concentrations would increase with water discharge (Figure 4, right side), this relationship was referred to as the "low-flow anomaly." The inverse relationship between water flow and PCB concentrations implies, however, that PCBs are entering the water column at a constant rate (migrating out of contaminated bottom sediments, for example) so that as the amount of water decreases, PCB concentrations increase (same amount of PCBs mixed with less water).

The downriver changes in PCB concentrations indicate that the low-flow PCBs are also derived from the reach of the river between Rogers Island (Fort Edward) and Schuylerville (Figure 6). The low-flow PCBs

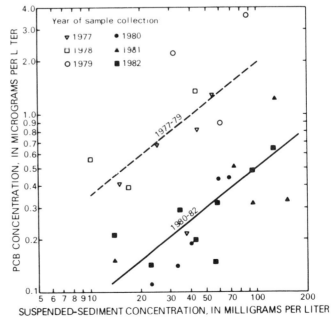

FIGURE 4 PCBs and concentration of suspended sediment in the Hudson River at Schuylerville, New York, Water Years 1977 through 1982. Lines are best-fit regressions for years indicated. SOURCE: Schroeder and Barnes, 1983b.

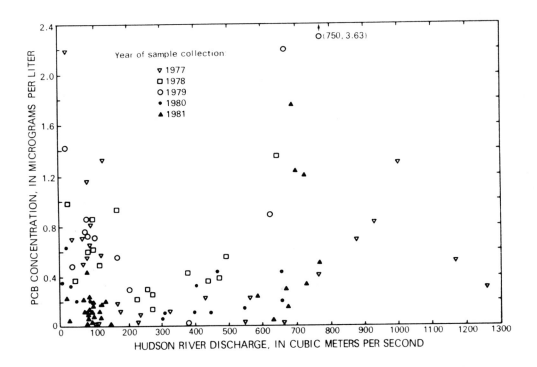

FIGURE 5   PCB concentration and water discharge, upper Hudson River at Schuylerville, New York, Water Years 1977 through 1982. SOURCE: Schroeder and Barnes, 1983b.

move with the water. Although PCB concentrations during high and flows and generally decreased after 1977, maximum concentration of total PCBs during the high-flow events did not decrease (Schroeder and Barnes, 1983b).

The foregoing discussion of sediment transport is important to understanding why computations of future PCB transport into the estuary prepared by Lawler, Matusky, and Skelly Engineers (LMS, 1978, 1979) and based on the U.S. Army Corps of Engineers (COE) HEC-6 riverbed scour model, have been so much higher than observed values. The HEC-6 model is predicated on a stepwise transport downriver from pool to pool. According to the LMS model, PCBs that wash over the Green Island Dam should come from the pool backed up behind the dam. These PCBs would have reached the pool only after having traversed all the other pools between Fort Edward and Green Island. For whatever reason, information from the USGS indicates that on the upper Hudson River since 1980, a pass-through type mechanism has been dominant (NUS Corp., 1983). The PCB load transported into the estuary is acquired not from the Green Island Pool but rather from north of the Thompson Island Dam.

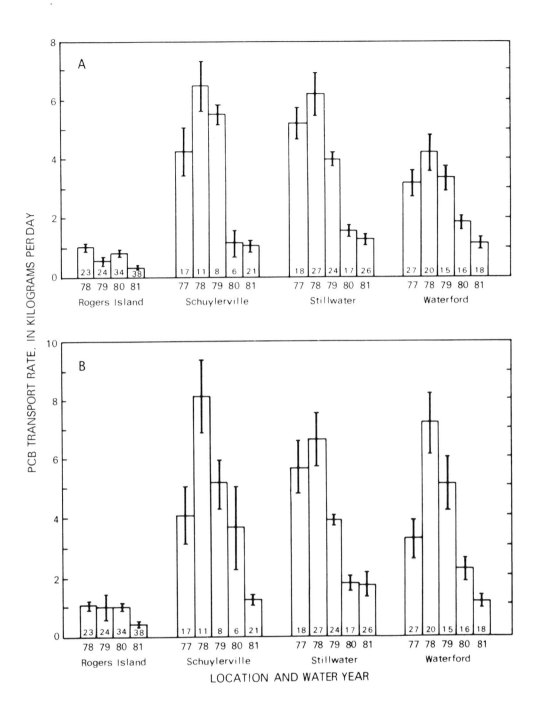

FIGURE 6 Transport rates of PCBs in upper Hudson River during nonscouring discharges, Water Years 1978 through 1981, calculated by multiplying PCB concentration by river discharge at station indicated. Standard-error bars at tops of rectangles; numbers of samples shown within and at bases of rectangles. SOURCE: Schroeder and Barnes, 1983b.

Discharge Cycles

Some investigations present evidence that the Hudson River and estuary may be subject to cyclic variables. Mathematical analyses of monthly mean flows of the upper Hudson River at Green Island (based on daily readings by the USGS), suggest that several cycles may be present. Starting with a table showing average monthly flows from October 1947 through September 1975, Texas Instruments Incorporated Ecological Services (TI) summarized various physical factors affecting the estuary, with particular attention to deriving a mathematical expression of the varying locations of the landward edge of the saltwater wedge. One factor recognized was changing freshwater discharge. TI found that discharge data could be reconstructed using five major cyclic components --105, 21, 10.5, 4.2, and 1.9 years--and that

> All except the last cycle have periods which are multiples of the value 2.1; this suggests an outside controlling influence. There is some similarity to recurring cyles of solar activity, but the relationship remains to be defined. (TI, 1976, p. IV-12)

In analyzing the so-called "no-action" alternative as part of the management alternatives explored by NYS DEC for dealing with the problem of PCB-contaminated sediments in the upper Hudson, LMS (1978) followed the TI cyclic approach. The LMS forecasts of future river discharge (the critical variable in trying to predict future PCB transport into the estuary) were made by analyzing the monthly mean flows at Spier Falls (1178N-653E, Corinth quadrangle) for the period 1930-1977 (computed by the Hudson River-Black River Regulating District). These values were then related to the flow of the combined Hudson-Mohawk rivers at Green Island, as recorded daily by the USGS. In their projections, LMS presumed that the flows from 1957 to 1976 would be repeated during the forecast period of 1977 to 1996.

In a summary of the hydrology of the Hudson River, Darmer commented that

> Extreme periods of precipitation, either high or low, are of concern because of their effect upon the environment. The extreme drought of the 1960's, followed by a series of wet years in the 1970's, imply that precipitation may follow some cyclic pattern rather than being entirely random. (Darmer, 1987)

If the flow of the Hudson River is cyclic, it must be a complex function of several interacting cycles. Cycles whose effects seem to be present include the lunar perigee-syzygy cycle of 14 months (Fergus Wood, 1978) and the 19.8-yr Saturn-Jupiter lap cycle (Fairbridge and Sanders, 1987), both of which seem to be reflected in the cyclic orbit of the Sun around the center of mass of the solar system, and thus possibly also in solar output (Landscheidt, 1987).

Other Significant Environmental Factors

Remnant Deposits

Removal of the Fort Edward Dam in 1973, exposed what are called remnant deposits, debris washed downriver from lumbering sites in the Adirondacks that accumulated behind the dam (Malcolm Pirnie, Inc., 1975; 1977a, 1977b; 1978c). Wood is the characteristic component of these deposits, which possess a strong affinity for PCBs. Considering their location just downstream from the GE wastewater discharge pipes, it is not surprising that some of the highest concentrations measured in the upper river have come from the remnant deposits.

Heavy Metals

Sediments in the upper Hudson River contain elevated levels of Pb, Hg, Zn, Cu, Cr, Cd, and Ni (Matusik, 1978; Malcom Pirnie, Inc. [MPI], 1975, 1978a; Tofflemire and Quinn, 1978; Tofflemire, 1984; Brown et al., 1988). These heavy metals likely came from the Marathon Battery plant, the Hercules Chemical (now CIBA-Geigy) chemical plant, or other sources in the Hudson Falls-Glens Falls area (Tofflemire and Quinn, 1978; Tofflemire, 1984). In general, large lead discharges from the Hercules plant occurred at the same time as PCB discharges from the GE plants. Thus sediments containing elevated PCB levels also tend to be high in lead. The details of the lead pollution of the upper Hudson River are not known and have not been carefully investigated.

Measurements of heavy-metal content have been made in samples collected near Fort Edward Dam and in the remnant deposits (Table 1).

TABLE 1 Heavy-Metal Content of Selected Upriver Sediments

Sample	Lead	Cadmium	Metal Copper	Mercury	Arsenic	Zinc
Fort Edward Dam (brown fibrous sludge and (8) black silt)	234 to 3630(8)	14 to 138(8)	27 to 159(8)	0.28 to 1.28(4)	3.2 to 22 (8)	74 to 2950
Remnant deposits	(ug/g)	(ug/g)				
Area 3A	< 3 to 5600	6 to 110				
Area 4	20 to 480	< 4 to 12				
Area 5	40 to 1100	< 4 to 93				

SOURCE: MPI, 1975, 1978a

Cesium-137 Fallout

Cesium-137 fallout from nuclear weapons tests carried out in the atmosphere during the 1950s has been used to indicate ages of sediment layers in core samples. A large network of cores in which cesium-137 has been used in this way has been established in the Hudson River by investigators from Lamont-Doherty Geological Observatory of Columbia University (Bopp, 1979; Bopp et al., 1978, 1981, 1982, 1984; Simpson et al., 1976, 1984).

ANTHROPOGENIC HISTORY

The large-scale PCB pollution of the upper Hudson River can be resolved into two components:

1. introduction of PCBs into the river starting about 1950, and until 1973, the temporary storage of most of them in the first sediments they encountered, in the pool behind the Fort Edward Dam; and
2. wholesale spreading throughout the entire system as a result of two large floods in April 1974 and April 1976, after the dam had been removed in 1973 without any acknowledgment that the sediments stored behind the dam (now known as the remnant deposits) might contain elevated levels of PCBs nor of any significant consideration of the possible environmental consequences of post-dam-removal floods in eroding and spreading of these highly contaminated sediments downriver (MPI, 1975; 1977a, b; 1978b).

PCBs were introduced into the upper Hudson River via daily discharges of plant cleanup water from two capacitor-manufacturing facilities of the General Electric Company (GE). GE began using PCBs at Hudson Falls in 1947 and at Fort Edward in 1952 (Hetling and Horn, 1977; Hetling et al., 1978). In late 1972, the U.S. Congress passed the Water Pollution Control Act, which assigned responsibility for regulating the discharges of industrial wastes into waterways to the newly formed U.S. Environmental Protection Agency (EPA) via a program of permits. In December 1972, GE applied to EPA for a permit to discharge 30 to 47.6 pounds per day of PCBs into the upper Hudson River. In January 1975, EPA granted GE a permit to discharge 30 pounds per day of PCBs into the upper Hudson River and assigned monitoring of the permit to the New York State Department of Environmental Conservation (NYS DEC).

The first public announcement of high levels of PCBs in fish from the Hudson River came from concerned private citizens. Robert Boyle, of the Hudson River Fishermen's Association, persuaded editors at *Sports Illustrated* magazine to support a program of catching and sampling coastal game fish for pesticide residues, mercury, and PCBs. The results of the analyses (carried out by the WARF Laboratories, Madison, Wisconsin) were published in October, 1970 (Boyle, 1970).

In 1975, nearly five years later NYS DEC announced that fish containing levels of PCBs well above the FDA action level of five parts per million (ppm) were being caught in the Hudson River (Boyle, 1975). The entire upper river fishery and the Hudson estuary commercial striped bass fishery were closed. An administrative proceeding was initiated against GE that sought

- cessation of PCB discharges,
- penalties from GE for having polluted the river, and
- rehabilitation of the upper river to mitigate the effects of the PCB contamination.

A settlement was negotiated between NYS DEC and GE in which GE agreed to build wastewater-treatment facilities at its two capacitor-manufacturing plants, cease PCB discharges by July 1977, make a cash payment of $3 million to the state to study the extent of PCB pollution and/or carry out rehabilitation measures, and carry out $1 million worth of environmentally oriented in-house research. For its part, NYS DEC accepted the principle of joint culpability; agreed to put up $3 million in cash or in kind for studies and/or rehabilitation; to establish an Advisory Committee of independent experts and representatives of several governmental agencies and the general public; and, should comprehensive study recommend large-scale rehabilitation, to use its best efforts to seek funds from sources "other than GE" to assist in rehabilitating the river (e.g., the federal government) (Sofaer, 1976a, b).

The Hudson River PCB Settlement Advisory Committee established by the agreement assisted NYS DEC in all phases of the comprehensive studies. Members of its remnant deposits subcommittee brought remnant deposits to the forefront of the thinking about the river by NYS DEC staff. Prior to this time, NYS DEC's view regarding the remnant deposits was to let them be eroded from their riverbank locations in a steep-walled, inaccessible bedrock gorge and be redeposited at Fort Edward, where they became more accessible and thus could be removed at least cost (MPI, 1975). The initial version of the 1976 contractor report that recommended strategies to be followed in the second cleanup of Fort Edward did not mention the PCB-pollution problem (MPI, 1977b). It even recommended disposing of the dredge spoil as usual by dumping it without treatment on Rogers Island.

The PCB Settlement Advisory Committee rejected the contractor recommendation and insisted on encapsulation of the proposed dredge spoil and

- construction of a haul road down the east wall of the gorge containing the remnant deposits to give access to heavy construction vehicles;
- removal and encapsulation of the most highly contaminated sediments in Area 3A, an area so highly polluted with PCBs that no plants were growing;
- transport of quarried stone blocks to the site for riprap to prevent further bank erosion in the Area 3.

The committee also recommended removal and encapsulation and/or final treatment of the remaining remnant deposits 3A be given the highest priority. The proposed Hudson River PCB reclamation-demonstration project had to be scaled down to fit with budget constraints, however, and the remnant deposits were dropped from the work plan, which concentrated exclusively on removal of PCB-contaminated "hot spots" in the Thompson Island Pool. The committee unanimously recommended to Commissioner in June 1978 that the only feasible means for rehabilitating the upper Hudson River was dredging and securely encapsulating the contaminated sediments. The committee further found that not only was such action environmentally sound but that no other method could be considered as having reached the stage of engineering applicability.

From 1977 to 1983, PCB values in the water, suspended sediments, and fish (both upper river and estuary) declined (Armstrong and Sloan, 1980, 1981; Sloan et al., 1984, 1983, 1988). This decline was popularly ascribed to natural cleansing of the river, however, a contrasting argument could be made that the decline resulted from two actions taken by the state and from one natural cause. The actions taken by the state were

1. forcing GE to stop its PCB discharges (ended July 1, 1977), and
2. carrying out two large-scale cleanup operations at Fort Edward and taking care of the most contaminated of the remnant deposits.

The natural action was lower rainfall, which resulted in less erosion of PCB-contaminated sediments (Barnes, 1987). Since 1983, PCB values in striped bass in the estuary have varied with discharge. When discharge increases, PCB values in striped bass increase, and vice versa (Sloan et al., 1984, 1988).

Three major unresolved issues related to the resolution of PCB pollution in the upper river are

1. depth of scour of bed sediments during floods;
2. significance of selective dechlorination of PCB congeners by anaerobic bacteria; and
3. whether the Hudson River flow varies cyclically over 20 years.

The depth of scour during floods determines what thickness of bed sediments will be mobilized into the water column, each of them only temporarily, during the flood. The stirred-up sediments move some distance downriver, and then, as the flood wanes, are placed back on the bed of the river again. Proof that hot-spot boundaries in the Thompson Island Pool have not shifted since the 1974 and 1976 floods (Brown and Werner, 1985; Brown et al., 1988) is consistent with the expectation that no such shifting should have taken place because the post-1976 floods have not attained the flow levels of the pre-1976 floods.

Congener-specific analyses of PCBs in selected hot-spot sediments (Bopp et al., 1984; Brown et al., 1984, 1987, 1988; Bush et al., 1986; Simpson et al., 1984) and followup bacteriological studies have further proved that the low-chlorine PCB congeners found only in these hot-spot sediments beneath their surficial cover have resulted from the actions of anaerobic bacteria. Such low-Cl congeners have not been found downriver; they offer additional proof that locations of the boundaries of the hot spots defined in 1978 have not changed. What is yet to be determined is the minimum PCB concentration at which the anaerobic bacteria are capable of selectively dechlorinating the high-Cl PCB congeners. Such selective dechlorinization has been reported to take place at PCB concentrations of 700 ppm in experiments carried out by James Tiedje at Michigan State University (supported by GE). But, no such selective dechlorinization has been found at concentrations of 50 ppm in experiments carried out by Dr. Rhee (Chen et al., 1988) in the New York State Department of Health (supported by NYS DEC and the Hudson River Foundation).

The possibility that a 20-yr cycle exists in discharge variation in the upper Hudson River has been reported by mathematical analyses of monthly discharge means. EPA staff rejected this possibility, arguing that the statistical base is inadequate (NUS Corp., 1983). However, if such a flow cycle does exist, the remedial dredging planned for the early 1990s will be done against a background of rising river discharges and thus rising ambient PCB values in the water, suspended sediments, and fish. Because of the delays over finding funding from "sources other than GE," and permit delays, the opportunity to carry out a PCB hot spot reclamation demonstration project against a natural background of declining PCB values has been squandered. Rather than being able to claim credit for the results of a dredging project liberally assisted by natural causes, NYS DEC may have to explain why PCB values were higher after the dredging project than they were before.

Since 1978, when it recommended dredging and secure encapsulation, the Advisory Committee has kept itself informed on the latest methods of treating PCB-contaminated sediments. The point of this effort has been to be able to recommend final treatment of encapsulated sediments as treatment technology develops (Carpenter, 1987). The committee recommended spending parts of the money remaining in the settlement fund to pay for contractor experiments with PCB-contaminated sediments from the upper Hudson River. Three successful or partly successful sets of experiments have thus been made.

A further delay in starting the final phase of the rehabilitation of the upper Hudson River has arisen from the second Siting Board's rejection of the NYS DEC Project-Sponsor Group's (PSG) application for use of a parcel of land next the the old Fort Edward dump as the proposed encapsulation site. NYS DEC ruled that PSG must re-apply for a parcel close to the river in southern Fort Edward and the site for which the certificate and DEC permits granted by the first Siting Board in 1982 were voided by a 1984 State Appellate Court decision. Moreover, NYS DEC ruled that the application for use of Site 10 must request not just secure encapsulation, as in the previous two requests, but also must include available processes for stripping PCBs from contaminated sediments and/or destroying the PCBs, and it must deal with not merely the sediments from the 20 hot spots in the Thompson

Island Pool but also from the remnant deposits, from old NYS DOT dredge-spoil sites, and from the 20 hot spots between Thompson Island Dam and Troy.

This ruling is a return to the scope of the cleanup recommended in 1978 by the Advisory Committee, but which was rescoped downward to fit available funds. Still undetermined is when EPA will decided that its "interim period of evaluation" under Superfund I is over and that its Record of Decision (ROD) of July 1984 be revisited. In July 1984, the Food and Drug Administration (FDA) action level for PCBs in fish was 5 ppm. In August 1984, this level was lowered to 2 ppm. As of 1988, PCB values in fish have not declined much below 5 ppm. Accordingly, the 1984 ROD's projection about fish has not materialized. Recently issued report by New York State on the necessity to add water from the Hudson River to the drinking water supply for the New York City metropolitan area represents a further public-health issue not considered in EPA's 1984 ROD. Accordingly, EPA's Superfund evaluation that recommended no action with respect to removing PCB-contaminated sediments in the upper Hudson River needs to be revised.

A final factor in the delay over taking any remedial action with the PCB-contaminated sediments in the upper Hudson River has been the ambivalent attitudes of the citizens of Fort Edward. They favored what they considered to be beneficial dredging operations, but opposed any dredging operations whose objective was rehabilitation of the upper Hudson River. In 1974-75 and again in 1977-78, the state's cleanup operations

1. repaired and protected their water-supply pipes where they cross the Hudson River at old Fort Edward Dam site;
2. reopened their oil terminal for barge traffic; and
3. unplugged blocked dewage outfall pipes to restore water flow in the east channel past Rogers Island, thus enabling the raw sewage they were dumping into the river to be carried away downstream once again.

Fort Edward citizens clamored for these operations and did not oppose dredging nor the spreading of the dredge spoil on Rogers Island. By contrast, starting in 1982, they opposed every proposal to rehabilitate the upper Hudson River by dredging and encapsulating the dredge spoil in a specially engineered facility to be located in Fort Edward. Moreover, they expressed satisfaction with the idea that no action at all would result in the river eventually eroding the polluted sediments and redepositing them downriver.

ASSESSMENT OF CONTAMINATION

Assessment of contamination of the Hudson River began when high levels of PCBs were found in striped bass caught in the lower river (Boyle, 1975). Once the problem had become public and the GE plants at Hudson Falls and Fort Edward were identified as the point sources, attempts were made to assess the extent of contamination. One approach was to compile GE's records of purchases of PCBs from the sole American supplier, Monsanto Industrial Chemicals Company; another approach was to collect core samples of sediments, analyze their PCB contents, and compute the quantities of PCBs discharged into the upper Hudson River. The following sections summarize these endeavors.

PCB Levels in Fish

As early as 1969, samples of biota from the upper Hudson River were found to contain high levels of PCBs. Robert Boyle's program of collecting specimens of coastal game fish and analyzing them for pesticide residues, mercury, and PCBs brought the problem to the public's attention. The Wisconsin Alumni Research Foundation (WARF) Laboratory in Madison, Wisconsin, analyzed the samples. The specimens were shipped in dry ice without ever being wrapped in plastic. High PCB values were found in most of the fish. The highest values were in the Hudson River striped bass, 4.5 to 5 ppm in the fish flesh and 11 to 12 ppm in the eggs.

Boyle communicated his results to NYS DEC, but got no response. (He learned later that NY State's response was to begin to test fish. They did not publicize their results until August 1975.) The significance of the Hudson River striped bass specimens was that they showed elevated levels of PCBs even before the Fort Edward Dam had been removed in 1973 and thus prior to the great downriver surges of PCB-contaminated sediments in 1974 and 1976.

EPA Region II staff collected fish from the upper Hudson River, downstream and upstream from the GE discharge pipes. They found clean fish upstream, but heavily contaminated fish downstream from the two GE plants (Nadeau and Davies, 1974, 1976).

In September 1975, NYS DEC announced that PCB concentrations in a bass caught near the GE plants were 350 ppm. The averages of several catches in August 1975 at Waterford were 41.5 and 53.5 ppm in smallmouth bass, 28.2 and 48.9 ppm in white suckers, 32.4 ppm in walleyed pike. Table 2 shows PCB values in fish caught during 1975-76 in various parts of the Hudson River.

Analyses of Sediment Samples

Sediment samples from the upper Hudson River have been collected and analyzed for PCBs by NYS DEC, EPA, and contractors. A systematic collection was made as part of the comprehensive study made possible by the GE settlement fund. NYS DEC contracted with Normandeau Associates, Inc. (1977) of Bedford, New Hampshire, for detailed mapping and sediment collecting in the upper river. Normandeau collected 312 short cores from channel-margin flats in the upper Hudson River, and 600 grab samples on 40 transects at approximately one-mile intervals between Fort Edward and Waterford. The 672 analyses for total PCB by O'Brien & Gere (1978) formed the basis for the NYS DEC map of 40 hot spots (Figure 7) and computations of sediment volumes and PCBs concentrations

TABLE 2 PCB Values in Fish Caught in the Hudson River during 1975-76

Location	Species	Number of fish	Total PCB (ppm) Average	Low	High
Upstream of Hudson Falls	Smallmouth bass	26	Trace		
	Yellow perch	15	Trace		
Fort Edward to Federal Dam (Troy)	Smallmouth bass	11	72.6	41.5	122.9
	Yellow perch	10	134.6	79.3	299.3
	White sucker	37	68.2	28.2	131.4
	Largemouth bass	37	61.7	12.5	164.4
Federal Dam to Battery	Yellow perch	5	5.28		
	Largemouth bass	10	10.05	1.73	23.74
	White perch	23	10.08	5.28	19.88

SOURCE: Horn et al., 1978.

in the sediments (Figure 8). The PCB content of the Thompson Island pool was calculated to be 61 tonnes (MPI, 1977c; 1978a, 1978b, 1979, 1980a, 1980b, 1980c, 1981; Tofflemire et al., 1979a, 1979b).

Figure 9 summarizes the aggregate distribution of PCBs with depth in the core samples collected in the Normeandeau I operation.

In order to obtain a clearer understanding of the nature of the channel-floor coarse sediments that had proved so difficult for Normandeau's coring attempts, Steve Selwyn and I used a box corer that he had designed and built to collect four samples from the Thompson Island pool between Normandeau transects 7-4 and 7-6, at Hot Spot No. 5. Dr. James Tofflemire, of NYS DEC participated in the operation by arranging for NYS DOT assistance, selecting the coring sites, determining positions with an optical range finder, and arranging for analyses.

The box corer measured 0.5 x 0.5 x 1.5 m. The box was rigged so that it would fall freely to the bottom from just beneath the water surface after a trigger weight attached to the end of a release lever arm had touched bottom, thus allowing the release lever to swing upward and open the release clamp. After the box penetrated the sediment, the upward pull on the wire rope would activate the jaws, drawing them shut and preventing the sediment from escaping out the bottom. In their open position, these jaws fit closely along the outside of the box.

The box had been constructed so that the vertical plate on one side could be unbolted. Therefore, after a box had been retrieved from a drop, we would lay it on the side opposite the removable plate, unbolt and remove the plate, and thus reveal a view of the sediment on a plane that was perpendicular to the water/sediment interface.

At site BC-11, the box corer was attached to a pile driver and pounded into the bottom sediments by repeated blows. This pile driver coring was done by NYS DOT work crews downstream from where they had been installing sheet piling. They drove the box corer in to a point of refusal, which was reached after the box had penetrated nearly its

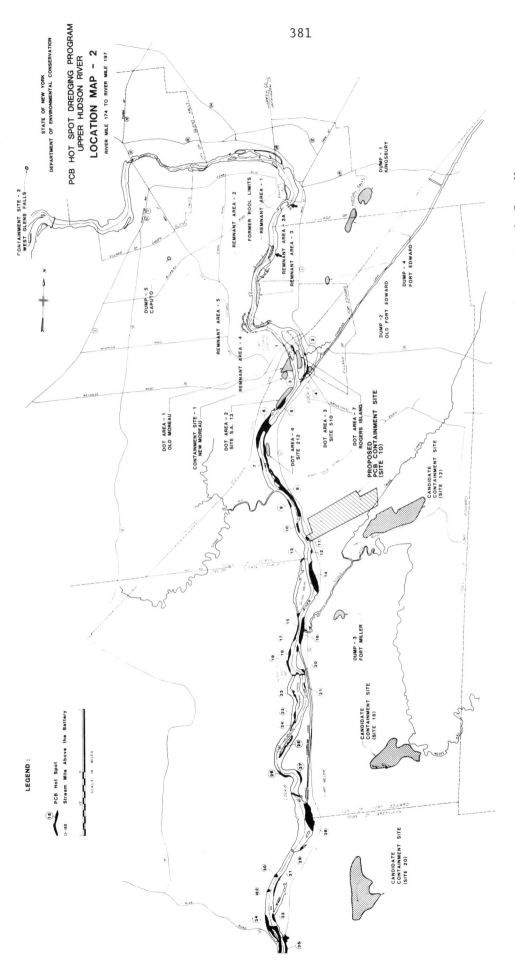

FIGURE 7 Map of PCB hot spots in the upper Hudson River north of the dam at Lock 5. Hot spots 1-20 are located in the Thompson Island pool. The map includes the remnant deposits, along with various dumps and NYS DOT dredge-spoil areas, and the Sites G and 10, subjects of the Siting Board hearings.

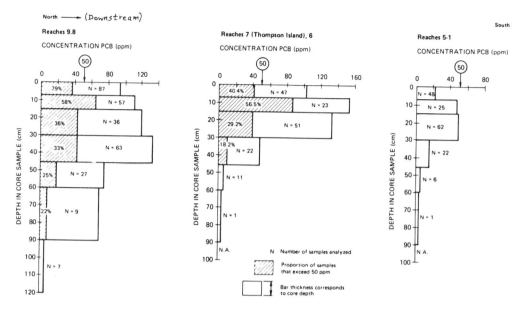

FIGURE 8 PCB analyses sediment samples selected from 312 short cores collected from channel margin flats along the upper Hudson River, grouped according to reaches (1) 9 and 8 north of Thompson Island Dam, (2) 7 and 6 between Thompson Island Dam and Lock 5, and (3) 5 to 1 between Lock 5 and Troy. High PCB values above the 90-cm level in reaches 8 and 9, and above the 45-cm level in the other reaches, mark the flow of April 1974, when debris from the then-newly exposed remnant deposits surged downriver after removal of the Fort Edward Dam. SOURCE: Hetling et al., 1978.

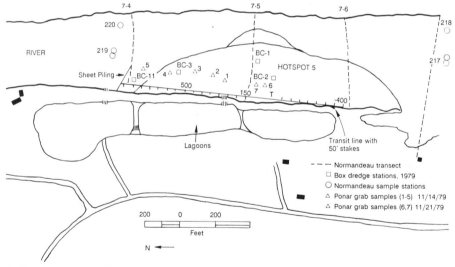

FIGURE 9 The northern part of Thompson Island pool near hot spot 5, showing Normandeau transects (dashed lines) and sampling stations (circles), NYS DEC ponar grab samples (triangles), and box cores (squares) collected in August (BC1, BC2, and BC3) and September (BC11). Areas marked lagoons are part of Special Area 13 of NYS DOT dredge-spoil sites.

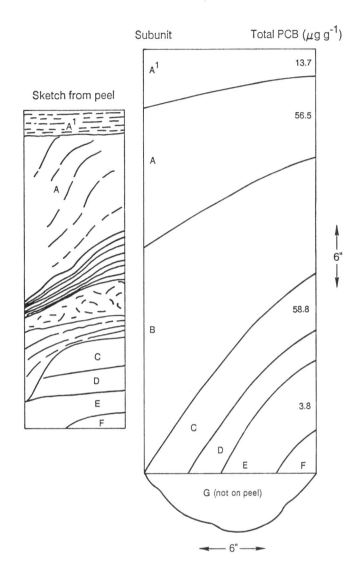

FIGURE 10 Sketch of relief peel made when the box corer was first opened and samples collected for PCB analyses.

entire length into the sediment.

When the box was opened, we found 42 inches of sediment within the rectangular part and another six inches in the closing jaws. We sketched the relationships visible on the vertical face of the contents of the box and collected samples for PCB analysis. Then, after several attempts, we finally made a successful relief peel using the techniques described in Burger et al. (1969). Figure 10 shows the relief peel with lettered subdivisions that selected for PCB analyses. The numbers are total PCBs expressed as parts per million on a dry-weight basis.

The large relief peel from station BC-11 gives a dramatic indication of the kinds of debris that washed into the northern end of the Thompson Island pool as a result of the 1974 and 1976 post-dam-removal floods. Two remarkable features stand out:

1. the steep inclination of the layers in the lower part, and
2. the distribution of PCB values.

At first sight, the steeply inclined layers might be considered to be an artifact of the coring operation. However, the fact that the topmost layer is horizontal proves that the inclination is natural. I interpret the steep dip as oversteepened cross strata that resulted from flow across a shallow area into a deeper area. I infer that the sediment being swept across the shoal avalanched into the deeper water but that the dip was steepened well beyond its normal angle of repose because of flow up the slope that was associated with a separation eddy (Friedman and Sanders, 1978). The fact that the high concentrations of PCBs are found only in subunits C through A' indicates that these layers were deposited since the 1974 flood.

As part of their assignment in preparing the Superfund I Remedial Action Master Plan (RAMP), NUS Corp. (1983) re-examined hot spots that NYS DEC had mapped in the Thompson Island pool based on the Normandeau I cores (1977-78). NUS found that no hot spots had disappeared, but the mapped values needed revision. Accordingly, NUS recommended to EPA that all the hot spots in the Thompson Island pool be remapped.

A detailed re-survey of Thompson Island pool hot spots was mandated by EPA as part of its funding eligibility requirements, and 400 sediment cores and 600 grab samples were collected and analyzed (1985; Normandeau Associates II). After all the PCB results had been submitted to NYS DEC, the results were summarized using NYS DEC's newly organized computer programs. The concentrations of PCBs were integrated at depth intervals of 0.5, 1.0, and 1.5 m. Approximately 95 percent (21.9 tonnes) of PCBs were found to be in the top 0.5 m, and 99.91 percent (23 tonnes), in the top 1 m. The total value of 24 tonnes for the PCBs in the Thompson Island pool hot spots is significantly smaller than the 61 tonnes calculated in 1978 (Brown and Werner, 1985; Brown et al., 1988). These results served as a basis for recalculating all the previous estimates of the PCB budget in the Hudson River (Figure 11).

Other cores have been collected in the upper river by John Brown of GE (Brown et al., 1984, 1987) and Richard Bopp of Lamont Doherty Geological Obervatory (Bopp and Simpson, this volume). Both have analyzed for individual PCB congeners, instead of for comparison with a given standard or total PCBs. The have found evidence that the lower chlorinated congeners are being enriched relative to the higher chlorinated congeners. According to the experimental evidence later presented to Siting Board II by Dr. James Tiedje, these changes have resulted from the effects of anaerobic bacteria. The congeners being formed by bacterial action are unlike those thought to have been washed into the river from the GE plants, are unlike those being transported downriver sampled by Bopp and Simpson, and are unlike those found by Bopp and Simpson in the lower river. These findings supply further proof that the hot spots are not being eroded and are not contributing to the downriver migration of PCBs as sampled by the USGS.

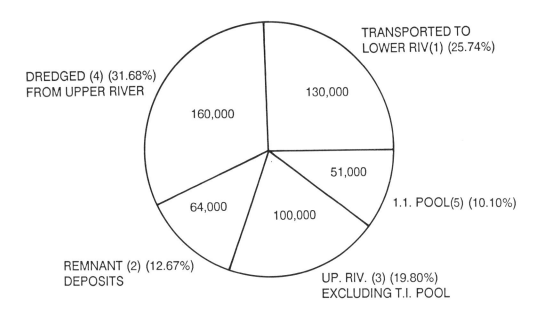

FIGURE 11 Revised budget for PCB discharge 504,000 pounds. Sources of estimates: Brown using data from Bopp and USGS, Tofflemire and Quinn (1979), Brown roughly reducing 1978 estimate in light of errors, Horn et al. (1979), Brown et al. (1985).

Organisms Other Than Fish

Although most of the early attention to PCBs in organisms concentrated on fish, because of human health concerns, many other parts of the biota proved to contain PCBs: for example, snapping turtles (Stone et al., 1980), aquatic intertebrates (Werner, 1981), caddisflies (Simpson et al., 1986), and plants (Weston, 1978; Buckley, 1982, 1983, 1987). The results of PCB measurements in plants (Table 3) support the conclusion that the PCBs found in plant samples come from the atmosphere.

Evaluation of Threat to Food Chain and Human Health

The main basis for evaluating activities with respect to Superfund sites is impact on human health. At least five threats of PCBs to human health have been evaluated:

1. from contaminated sediments and water into various organisms and up the food chain to fish and from contaminated fish to

TABLE 3 PCB Concentrations in Plants from Selected Localities in Upper Hudson Valley Region

Location and species	PCB concentration (Total PCBs; milligrams per gram, dry-weight (basis))				
	Lower leaves (0-2 m)	Upper leaves (ca 5 m)	Lower twigs (0-2 m)	Upper twigs (ca 5 m)	Cores from trunk, 1 to 1.5 m; bark discarded
Caputo dump:					
Pitch pine	317	81.2	87.1	17.7	1.2
White pine	215		120		1.16
Quaking aspen	89.6	28			

SOURCE: Hetling et al., 1977

      people who eat fish;
2. from drinking water (either ground water from wells in the upper Hudson area generally but near old PCB dump sites in particular or from public water supplies drawn from the Hudson River;
3. from forage crops and vegetables grown near the upper Hudson River;
4. from direct contact with highly contaminated sediments; and
5. from the atmosphere.

   The most direct and substantial source of PCBs into people is via ingestion of contaminated fish. Accordingly, many fish have been caught and analyzed to determine both the geographic extent of such pollution and any changes through time. The publicity over the discovery of high levels of PCBs in 1975 prompted NYS DEC to close the entire upper-river sport fishery and the commercial fishery for striped bass and eels in the Hudson estuary (a ban extended in 1984 to include the striped bass from Long Island Sound and the New York Bight after the FDA lowered its action level for banning fish from the market place from 5 to 2 ppm).
   As noted earlier, EPA's Superfund 1984 ROD for the upper Hudson River concluded that human health was sufficiently protected in New York state by enforcement of the fishing bans and reliance on "nature's remedy," which they took to be operating in the river to bring the average level of contamination of Hudson River fish below the FDA 5-ppm action level. With the change to 2 ppm, any new evaluation will find that PCB contamination in fish exceeds the FDA action level.
   EPA's 1983 Superfund RAMP concluded that Waterford's public water supply should be evaluated from the point of view of the possible threat to human health from drinking water from the Hudson River. Over

TABLE 4 PCB Concentrations in Air Samples from Selected Localities Compared with Upper Hudson River Area

Location (date)	Number of samples	Range (nanograms per m^3)	Comment on Aroclor or other information
SELECTED LOCALITIES			
Vineyard Sound, MA (1973)	02	04-05	Calc. as 1254
University of RI (1973)		2.1-5.8	do
Providence, RI (1973)		9.4	do
Chicago, IL (1975-76)	04	3.6-11	4% as 1242; 97% as "vapor"
Lake Michigan (1976-78)	06	0.57-1.6	74% as 1242; 88% as "vapor"
Milwaukee, WI (1978)	02	2.7	59% as 1254; 27% as 1260; 84% as "vapor"
UPPER HUDSON RIVER AREA		Average	
Washington Co. offices			
(Nov 76 to Jun 77)	31	990	Air pumped for 24 hr through a fluorisil column
(Jul 77 to Nov 77)	15	305	
Fort Hudson Nursing Home			
(Nov 76 to Jun 77)	23	108	
(Jul 77 to Nov 77)	10	25	
Main Street, Fort Edward			
(Nov 76 to Jun 77)	24	102	
(Jul 77 to Nov 77)	12	67	
Glens Falls			
(Nov 76 to Jun 77)	32	< 20	
(Jul 77 to Nov 77)	09	< 20	
Warrensburg			
(Nov 76 to Jun 77)	25	< 20	
(Jul 77 to Nov 77)	07	< 20	
Caputo dump			
(Nov 77)	05	3,240	3 2.5-hr samples; 2 08-hr samples
Fort Miller dump			
(Nov 77)	03	2,160	08-hr samples
New Moreau facility			
(Nov 77)	03	107	08-hr samples

SOURCE: Beeton, 1979; Hetling et al., 1978.

the years, NYS DEC has repeatedly tested Waterford's water supply, most notably during 1977 and 1978, when the much-monitored NYS DOT channel maintenance dredging of Fort Edward terminal took place. During these operations, no changes in PCB level in Waterford's intake could be detected. NYS DEC also monitored the intake of Poughkeepsie's water treatment plant, but gave it up in 1979 when the levels persistently dropped below detection level.

The chief places where people might come into direct contact with contaminated sediments are the remnant deposits. The significance of this route was emphasized by EPA's Superfund RAMP. As a result, EPA recommended that the remnant deposits be contained by capping them with 18 inches of soil. However, EPA's July 1984 ROD, indicates that

> The appropriateness of further remedial action for these sites will be reexamined if EPA decides at a later date to take additional action with respect to sediments in the river.

Nationwide, the atmosphere is the largest natural source of PCB transport (Beeton, 1979). Table 4 compares PCB concentrations in air samples nationwide with those found in the upper Hudson valley.

REMEDIAL/ALTERNATIVE TECHNOLOGIES

Evaluation Methodology

Initally, NYS DEC and the Advisory Committee evaluated claims of new schemes for PCB destruction and evaluated them against the known engineering realities of dredging and encapsulation. In addition to these efforts, EPA's contractor, NUS Corp. (1983), presented a list of known PCB-destruction processes, but nearly all of them were for PCBs in oils. Only a few dealt with destruction of PCBs in sediments, and they all presupposed dry sediments. Although EPA's 1983 Superfund I evaluation of PCBs in the upper Hudson River rejected dredging, none of the alternative technologies reviewed can in any way be considered as a realistic alternative to dredging and encapsulation.

NYS DEC supported contractor experiments with Hudson River PCB-contaminated sediments. So far, successful thermal stripping of PCBs has been achieved by American Toxics Disposal, Inc., of Waukegan, Illinois, and PCB reduction has been achieved by the use of triethylamine, taking advantage of its reverse immiscibility with water/oil at $70^{\circ}F$, a process developed by Resources Conservation Corp. of Bellevue, Washington. PCBs have been both removed from sediments and destroyed in the Wright Malta steam-gasification process.

Technologies Considered and Their Costs

In 1978, when the Advisory Committee recommended to NYS DEC that dredging was the best way to begin rehabilitating sites with PCBs in excess of 50 ppm (the hot spots) and that secure encapsulation facility should be constructed close to but out of the river, the only available process for destroying PCBs in contaminated sediments was incineration. Cost estimates for incineration (estimated at $130/m^3; $100/yd^3) were based on the GE-Nichols Engineering tests with Hudson River hot spot sediments (Nichols Engineering Research Corp., 1978). These tests yielded estimates based on energy costs calculated when crude oil was $14 per barrel. It was widely believed that the price of oil would continue to rise, increasing incineration costs beyond what was affordable. Incineration was thus rejected as a final solution. Since there was no other tested PCB-destruction process capable of use on sediments, NYS DEC and the Advisory Committee agreed upon that safe encapsulation, perhaps for many years or even tens of years. Meanwhile, they continued to search for reasonably priced methods for stripping PCBs from the sediments of PCBs or destroying them.

From 1978 to 1988, many methods for destroying PCBs in contaminated sediments have been proposed. In 1986, EPA Research Laboratories in Cincinnati commissioned engineering analyses of PCB-destruction processes. A final report (Carpenter, 1987) evaluated how various methods might apply to the upper Hudson River hot spots. The Carpenter report found three methods other than incineration that could be used to treat PCB-contaminated sediments once they were dredged from the river:

1. Basic Extraction Sludge Treatment (B. E. S. T.) process of the Resources Conservation Co. of Bellevue, Washington;
2. ozone-ultraviolet exposure in an ultrasonic bath developed by Ozonic Technology of Closter, New Jersey; and
3. microbial scheme of Bio-Clean, Inc. of Burnsville, Minnesota.

Two other processes have been or are being investigated by NYS DEC and the Advisory Committee that were not considered in Carpenter, 1987: PCB stripping by a rising current of heated air (American Toxics Disposal, Incorporated, Waukegan, Illinois), and steam-gasification (Wright Malta, Inc., Ballston Spa, New York).

Remedial Action Selection Criteria

No process has been proposed that constitutes a viable alternative to physically removing PCB-contaminated sediments from the river as a first step in final rehabilitation. The Advisory Committee believes that the proposed dredging will not involve large-scale, long-distance downriver migration of PCB-contaminated sediments as a result of the disturbance of bottom sediments that inevitably accompanies dredging. Actual dredging operations monitored in 1977 demonstrated that at low-flow stages of the river, when dredging takes place, gravity pulls the stirred-up sediment back to the river bed in the vicinity of the dredge. Indeed, no increased levels of PCB concentrations were detected one mile downriver from the dredge. That highly contaminated sediments can be encapsulated securely was demonstrated when the highly contaminated sediments from the second cleanup of Fort Edward terminal in 1978 and from area 3A of the remnant deposits were placed in the new

Moreau facility.

Accordingly, the only alternative other than dredging that has been considered is the no-action alternative. Doing nothing had been rejected as a responsible way in which to deal with a toxic-waste problem. However, the delays in obtaining permits for constructing the proposed containment site have allowed 10 years of no action to elapse.

Basis for Rejecting Alternatives

In its applications to Siting Boards I and II, the state of New York sought permission only for dredging and secure encapsulation. As noted, however, NYS DEC was directed to expand its permit application to include various PCB stripping and/or destruction technologies. For future reference, the Advisory Committee is evaluating the merits of the top three alternatives listed in the Carpenter report, and the Wright Malta process.

The basis for making a final decision on a method of treating PCB-contaminated sediments to be dredged and placed in a secure encapsulation site using Sec. 116 funds has not been determined. As with many other projects, the decision probably will depend on financial considerations.

If and when EPA re-evaluates its 1984 ROD in light of the terms of Superfund II, it will be obliged to reexamine previous decisions under Superfund I and to prefer destruction methods to encapsulation. EPA has not scheduled any activities under Superfund II. EPA Region II has raised the possibility of entering the upper Hudson River into the SITE program under SARA. Should that happen, the final decision about treating the contaminated sediments will be based on field trials on the scene. So far, among the candidate processes, only the Ozonic Technology and Wright Malta (including Zurn et al.) processes are designed to destroy PCBs while or after stripping them from contaminated sediments, and of these, only the Wright Malta process renders the heavy metals in the residue in nonleachable form. Of the stripping-only processes, only Resource Recovery Corporation's B.E.S.T. scheme using triethylamine deals with both PCBs and heavy metals.

Basis for Choosing a Remedial Action

The position reached by NYS DEC and the Advisory Committee is that any rehabilitation of the upper Hudson River has to begin with dredging. No in-river process is viable. Moreover, all available PCB recovery and/or destruction processes require that the sediment first be removed from the river. And in conformance with U.S. law, any sediment containing more than 50 ppm of PCBs that is removed from the river must be placed in a secure encapsulation facility. Thus, while both EPA and NYS DEC are moving away from so-called landfills as ways to deal with solid wastes, the law requires that a secure encapsulation facility be constructed, even if the site is to be used only for a work space for stripping PCBs from the sediments and/or destroying them.

All of the final-treatment processes mentioned previously are available only for processing contaminated sediments that have been dredged from the river. If that is to be done, a secure encapsulation facility must first be constructed.

Anticipated Benefits

A significant anticipated benefit of the proposed remedial dredging is to forestall further spread of PCBs into the lower reaches of the Hudson River. The proposed hot spot dredging in the Thompson Island pool would remove about 10 years' worth of PCB contamination at existing rates of PCB flux over the Federal dam at Green Island. However, the main benefit of carrying out proposed hot spot dredging and secure encapsulation (and/or final cleanup) is that it may trigger EPA to re-examine its "interim measures" adopted in the July 1984 ROD under Superfund I. Under Superfund II, EPA is obligated to re-examine its previous determination about public-health effects.

EPA has named GE as a "responsible and liable party" for the PCB pollution of the upper Hudson River. Although New York State has "signed off" with GE with respect to obtaining further funds to deal with the pollution, EPA has refused GE's offer to "cash out" with respect to Superfund by paying for the recommended interim treatment of the remnant deposits. Therefore, if NYS DEC succeeds in obtaining permits for the requested PCB-encapsulation site, EPA may re-examine its interim recommendation about disposition of the remnant deposits. Under Superfund I, EPA recommended only a temporary measure: covering the remnant deposits with 6 inches of clay. The Advisory Committee believes that removal and treatment of these deposits are the keys to rehabilitating the upper Hudson River. Currently, the proposed hot spot dredging is the key to the future ultimate removal and/or PCB destruction of the remnant deposits.

Costs

Since 1977, nearly $10 million has been spent on field work (including coring), mapping, PCB analyses, fish monitoring, and partial rehabilitation of the upper Hudson River. This compares with a 1977 estimate prepared for GE's attorneys of $15 million just to prove the extent of PCB contamination in the sediments and as much as $250 million to clean up all contaminated sediments by dredging.

IMPLEMENTATION/MONITORING

In the upper Hudson River, many of what might be referred to as "remedial actions" were taken before any toxic waste problems had been identified, indeed, before any toxic waste legislation had been passed. Moreover, to maintain a navigation channel to Fort Edward terminal, PCB-contaminated sediments have been dredged repeatedly out of the upper Hudson River near Fort Edward. Accordingly, records are available to show what has been dredged both before and after the public awareness of PCB contamination. In addition, NYS DEC's action against GE, which led to the 1976 Settlement Agreement, forced GE to cease PCB discharges and to take certain other steps.

Remedial Action Taken

Pursuant to the 1976 Settlement Agreement, GE took three significant actions in connection with PCBs in the upper Hudson River:

1. it stopped discharging PCBs into the river on July 1, 1977;
2. it constructed wastewater treatment facilities at its capacitor manufacturing plants at Fort Edward and Hudson Falls; and
3. it is now using alternative compounds (alkyl pthalates) in its capacitors.

NYS DOT Dredging

DOT dredging operations included routine channel maintenance and two massive clean-up operations at Fort Edward as a result of surges of remnant deposits eroded by floods in the Hudson River in 1974 and 1976.

NYS DEC Remnant Deposit Actions (1975-1978)

NYS DEC erosion control measures (I). As armor against bank scour in Area 5 (Figure 11), 4,700 yd^3 of stone purchased from a nearby quarry were used to construct riprap for 1,100 feet of riverbank at a cost of $75,000. In Area 2, the slope leading to the river along 2,800 ft of bank was graded and planted at a cost of $72,000. The 94 now exposed but former in-river cribs were dismantled and the rocks filling them placed along the riverbank for about 2,000 of the 3,100 ft of Area 3 shoreline and all along the shore of Area 4.

NYS DEC erosion control measures (II). The April 1976 flood constituted a severe test of NYS DEC's erosion control measures (I). The rock riprap of Area 4 and 5 withstood the flood waters, but the slope grading and planting and partial rock treatment did not. After the recommendations from the Advisory Committee, NYS DOT built a haul road down the steep east valley wall enabling more stone to be hauled in. To prevent further scour, a complete rock riprap was built along the eastern shore of the river at Area 3.

Area 3A sediments encapsulated at new Moreau facility. The most highly contaminated remnant deposits were found in Area 3A. As part of the rehabilitation program recommended to NYS DEC, 14,000 yd^3 of debris were scraped from the barren flats in Area 3A and trucked to the new Moreau encapsulation facility.

Monitoring of Remedial Actions

All actions dealing with PCB pollution have been extensively monitored. The results of the USGS water-monitoring program are shown in Figure 12. Monitoring of fish has shown that the PCB values in fish caught has declined from its peak in the 1970s (Figure 13). By 1980, values of PCBs in striped bass in the Hudson estuary fluctuated according to river discharge. When discharge increased, PCB values in striped bass increased, and vice versa. Other biomonitoring results are contained in Simpson et al. (1986).

INSTITUTIONAL/MANAGEMENT CONSIDERATIONS

New York State Constitutional Mandate re: Barge Canal

The New York State Constitution, Article 15, Canals, prohibits the State from disposing of the canal system, in effect a constitutional mandate to maintain the barge-canal system. This article obliges individual legislatures to appropriate funds needed to keep the canal system operative, including maintenance dredging as required. In terms of PCB pollution, NYS DOT has in the past removed an estimated 160,000 lbs of PCBs in the sediments dredged (Tofflemire and Quinn, 1979), and will have to continue to dredge to stay ahead of the accumulating sediment. Accordingly, the state will need to acquire one or more sites for the upland deposition of dredge spoils that will contain large concentrations of PCBs for the foreseeable future. The so-called no action alternative, therefore, only applies to dredging unrelated to channel maintenance.

Miscellaneous Political Considerations

No history of Hudson River PCB pollution would be complete without some mention of several political considerations--changing governors and NYS DEC commissioners, the relationships between New York State and GE, the opposition to the proposed encapsulation sites by nearby residents, Congressman Gerald B. Solomon's opposition, differences between upstate and downstate residents, organizational problems in state government, and the ambivalent attitudes of the citizens of Fort Edward, who favored beneficial dredging operations while opposing those related to rehabilitation.

CONCLUDING REMARKS

Although government action has been slowly moving toward rehabilitation of the upper Hudson River, the Hudson River has continued to transport PCBs into the estuary. As a result of the cessation in 1977 of GE discharges of PCBs, of remedial action taken with respect to the remnant deposits, and of less water flowing in the river, the amounts of

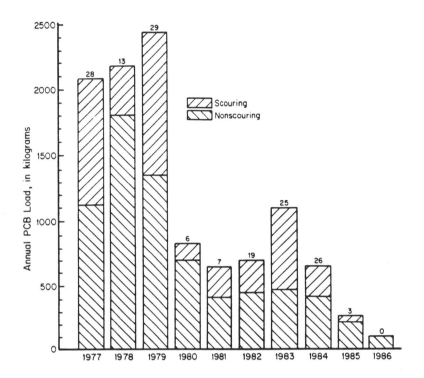

FIGURE 12 Annual transport of PCBs in the Hudson River at Waterford. Numbers above the bars indicate the number of days with flow above the estimated scour threshold of 600 m^3/sec (Barnes, 1987).

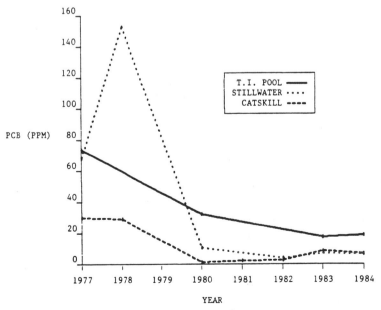

FIGURE 13 PCBs in largemouth bass, 1977 to 1984.

PCBs per year has dropped from about 2 tonnes in the late 1970s to 1 tonne and less in the 1980s. PCB values in fish showed a comparable decline until 1983. Since 1983, the PCB content of striped bass caught in the Hudson estuary has averaged about 4 ppm, but has fluctuated with river discharge. The 1980s values of PCBs in striped bass, are less than the pre-1984 FDA action limit of 5 ppm but more than the current action limit of 2 ppm.

Upstate opponents of the proposed dredging project are content with the no-action alternative. They consider that time is on their side. Moreover, if no remedial action is taken, the possibility exists that the PCB-contaminated sediments will wash away from their existing up-state locations and be transported downstate.

If NYS DEC is able to carry out its proposed hot-spot dredging project, the earliest date for beginning work is probably 1993 or 1994. This is about 20 years after the high-water flows of the early and mid-1970s. According to the disputed concept that a 20-year flow cycle exists, dredging done in the early 1990s will be done against a background of flows much larger than those of the 1980s. The possibility of doing the dredging project during the low-flow decade of the 1980s has been squandered.

I consider it urgent to re-evaluate the EPA's ROD of July 1984. NYS DEC's attempt to establish an intellectual basis for the upriver PCB pollution situation does not include any effort to pressure EPA to carry out the terms of SARA and re-visit its Superfund I conclusions. It is unlikely that NYS DEC can carry out any significant rehabilitation of the upper Hudson River unless EPA reverses its previous ROD and finds that the continuing downriver transport of PCBs constitutes a threat to human health.

NYS DEC should develop a public-relations campaign setting forth its arguments in favor of the proposed rehabilitation measures that would compare favorably with the one of December 1984 that was orchestrated by GE on the subject of "biodegradation" of PCBs. If New York City continues to press for permission to augment its drinking water supply by tapping into the Hudson River, the human health impacts of PCBs in the Hudson River will be magnified many times. Only aroused public demand for ridding the Hudson River of PCBs is likely to stimulate public officials into taking significant actions.

REFERENCES

Armstrong, R. W. and R. J. Sloan. 1980. Trends in Levels of Several Known Chemical Contaminants in Fish from New York Waters. Technical Report 80-2. Albany, New York: Department of Environmental Conservation. 77p.

Armstrong, R. W. and R. J. Sloan. 1981. PCB patterns in Hudson River fish. I. Resident/freshwater species. Proceedings, Hudson River Environmental Society, Hyde Park, New York.

Barnes, C. R. 1987. Polychlorinated biphenyl--transport in the upper Hudson River, New York, 1977-83. Northeastern Environ. Sci. 6(1).

Beeton, A. M., Chairman. 1979. Polychlorinated Biphenyls. Washington, D.C.: National Research Council. 182 p.

Bopp, R. F. 1979. The Geochemistry of Polychlorinated Biphenyls in the Hudson River. New York: Columbia University, Department of Geological Sciences. Ph.D. dissertation. 191 p.

Bopp, R. F., H. J. Simpson, B. L. Deck, and N. Kostyk. 1984, Persistence of PCB components in sediments of the lower Hudson. Northeastern Environ. Sci. 3(3/4):180-184.

Bopp, R. F., H. J. Simpson, and C. R. Olsen. 1978. PCB Analysis in the Sediments of the Lower Hudson. Palisades, New York: Lamont-Doherty Geological Observatory. 35 p.

Bopp, R. F., H. J. Simpson, C. R. Olson, and N. Kostyk. 1981. Polychlorinated biphenyls in sediments of the tidal Hudson River, New York. Environ. Sci. Technol. 15(2):210-216.

Bopp, R. F., H. J. Simpson, C. R. Olsen, and N. Kostyk. 1982. Chlorinated hydrocarbons and radionuclide chronologies in sediments of the Hudson River and Estuary, New York. Environ. Sci. Technol. 16(10):666-676.

Boyle, R. H. 1970. Poison roams our coastal seas. Sports Illustrated 33: 70-74. Oct. 16, 1970.

Boyle, R. H. 1975. Of PCB ppms from GE and a SNAFU from EPA and DEC. Audubon 77(6):127-133.

Brown, J. F., R. E. Wagner, D. L. Bedard, M. M. Brennan, J. C. Carnahan, H. Feng, and R. E. Wagner. 1987. Polychlorinated biphenyl dechlorination in aquatic sediments. Science 236:709-712.

Brown, J. F. Jr., R. E. Wagner, D. L. Bedard, M. J. Brennan, J. C. Carnahan, R. J. Mayh, and T. J. Tofflemire. 1984. PCB transformations in upper Hudson sediments. Northeastern Environ. Sci. 3(3/4):167-179.

Brown, M. P. and M. B. Werner. 1985. Distribution of PCBs in the Thompson Island Pool of the Hudson River, PCB Hot Spot Confirmation Report. Albany, New York: Department of Environmental Conservation. 34 p.

Brown, M. P., B. Bush, G-Y. Rhee, and L. Shane. 1988. PCB dechlorination in Hudson River sediments. Science 240: 1674-1675.

Brown, M. P., M. B. Werner, C. R. Carusone, and M. Klein. 1988. Distribution of PCBs in the Thompson Island Pool of the Hudson River. Albany, New York: Department of Environmental Conservation. 94 p.

Buckley, E. H. 1982. Accumulation of airborne polychlorinated biphenyls in foliage. Science 216: 520-522.

Buckley, E. H. 1983. Decline of background PCB contamination in vegetation in New York State. Northeastern Environ. Sci. 2:181-187.

Buckley, E. H. 1987. PCBs in the atmosphere and their accumulation in foliage and crops. In Phytochemical Effects of Environmental Comkpounds, J. A. Saunders, L. Kosak-Channing, and E. E. Conn, eds. New York: Plenum Press. Pp. 175-201.

Bush, B., L. A. Shane, M. Wahlen, and M. P. Brown. 1986. Sedimentation of 74 PCB cogeners in the upper Hudson River. Chemosphere 16:733-744.

Carpenter, B. H. 1987. PCB Sediment Decontamination Processes--Selection for Test and Evaluation. Research Triangle Park, North Carolina: Research Triangle Institute. 173 p.

Chen, M., C. S. Hong, B. Bush, and G-Y. Rhee. 1988. Anaerobic biodegradation of polychlorinated biphenyls by bacteria from Hudson River sediments. Ecotoxicol. and Environ. Safety 16:915-105.

Darmer, K. I. 1987. Overview of Hudson River Hydrology. New York: Hudson River Foundation for Science and Environmental Research, Inc. 174 p.

Fairbridge, R. W. and J. E. Sanders. 1987. The Sun's orbit, A. D. 750-2050: Basis for new perspectives on planetary dynamics and Earth-Moon linkage. Pp. 446-471.

Friedman, G. M. and J. E. Sanders. 1978. Principles of sedimentology. New York: John Wiley and Sons. 792 p.

Hetling, L. J. and E. G. Horn. 1977. Summary of Hudson River PCB Study Results. Albany, New York: Department of Environmental Conservation. 62 p.

Hetling, L. J., E. G. Horn, and T. J. Tofflemire. 1978. Summary of Hudson River PCB Study Results. Technical Paper 51. Albany, New York: Department of Environmental Conservation. 88 p.

Landscheidt, T. 1987. Long-range forecasts of solar cycles and climate change. In Climate: History, Periodicity, and Predictability, M. R. Rampino, J. E. Sanders, W. S. Newman, and L. K. Konigsson, eds. New York City: Van Nostrand Reinhold. Pp. 421-445.

Lawler, Matusky and Skelly, Engineers. 1978. Upper Hudson River No Action Alternative Study. Pearl River, New York: Department of Environmental Conservation. 190 p.

Lawler, Matusky and Skelly, Engineers. 1979. Upper Hudson River PCB Transport Modeling Study. Pearl River, New York: Department of Environmental Conservation.

Malcolm Pirnie, Inc. 1975. Investigation of Conditions Associated with the Removal of Fort Edward Dam, Fort Edward, New York. White Plains, New York: MPI. 118 p.

Malcolm Pirnie, Inc. 1977a. Environmental Assessment of Maintenance Dredging at Fort Edward Terminal Channel, Champlain Canal. White Plains, New York: MPI. 271 p.

Malcolm Pirnie, Inc. 1977b. Engineering Report, Investigation of Conditions Associated with the Removal of the Fort Edward dam, Fort Edward, New York. Review of 1975 Report. White Plains, New York: MPI. 141 p.

Malcolm Pirnie, Inc. 1978a. Phase I Engineering Report--Dredging of PCB Contaminated Hot Spots, Upper Hudson River, New York. Albany, New York: Department of Environmental Conservation. 134 p.

Malcolm Pirnie, Inc. 1978b. Feasibility Report, Dredging of PCB-Contaminated River Bed Materials from the Upper Hudson River, New York. White Plains, New York: MPI.

Malcolm Pirnie, Inc. 1978c. Environmental Assessment of Remedial Measures at the Remnant Deposits of the Former Fort Edward Pool, Fort Edward, New York. White Plains, New York: MPI. 173 p.

Malcolm Pirnie, Inc. 1979. Removal and Encapsulation of PCB-Contaminated Hudson River Bed Materials. White Plains, New York: MPI.

Malcolm Pirnie, Inc. 1980a. Engineering report, PCB Hot Spot Dredging Program, Upper Hudson River: Containment Site Investigations. Program Report No. 1. White Plains, New York: MPI.

Malcolm Pirnie, Inc. 1980b. PCB Hot Spot Dredging Program, Upper Hudson River: Dredging System Report. Program Report No. 2. White Plains, New York: MPI.

Malcolm Pirnie, Inc. 1980c. Draft Environmental Impact Statement. New York State Environmental Quality Review: PCB hot spot Dredging Program, Upper Hudson River, New York. White Plains, New York: MPI.

Malcolm Pirnie, Inc. 1981. Draft PCB Hot Spot Dredging Program, Upper Hudson River, New York: Rescoping Report. White Plains, New York: MPI.

Matusik, J. J. 1978. Data on Heavy Metals in Hudson River Sediments. Albany, New York: Department of Health, Radiological Science Laboratory.

Nadeau, R. J. and R. A. Davies. 1974. Investigation of Polychlorinated Biphenyls in the Hudson River: Hudson Falls-Fort Edward Area. New York City: EPA Region II.

Nadeau, R. J. and R. A. Davies. 1976. Polychlorinated biphenyls in the Hudson River (Hudson Falls-Fort Edward, New York State). Bull. Environ. Contam. Toxicol. 16(4):436-444.

Nichols Engineering and Research Corporation. 1978. Decontamination of PCB-Laden Hudson River Bottom Sediment for General Electric in the 36 Inch Nichols/Herreshoff furnace. New Jersey.

Normandeau Associates, Inc. 1977. Hudson River Survey 1976-1977 with Cross-Section and Planimetric Maps. Bedford, New Hampshire: Normandeau Associates. 351 p.

NUS Corporation. 1983. Feasibility Study: Hudson River PCBs Site, New York. Pittsburgh, Pennsylvania: NUS Corporation.

NUS Corporation. 1984. Draft Feasibility Study of Remedial Action Alternatives. Acushnet River Estuary above Coggeshall Street Bridge, New Bedford site, Bristol County, Massachusetts. Pittsburgh, Pennsylvania: NUS Corporation.

O'Brien and Gere, Engineers. 1978. PCB Analysis of Hudson River Samples: Final Report. Syracuse, New York: O'Brien and Gere. 150 p.

Robideau, J. A., P. M. Burke, and R. Lumia. 1984. Maximum Known Stages and Discharges of New York Streams through September 1983. Open-File Report 83-927. Reston, Virginia: U.S. Geological Survey. 83 p.

Schroeder, R. A. and C. R. Barnes. 1983a. Polychlorinated Biphenyl Concentrations in Hudson River Water and Treated Drinking Water at Waterford, New York. USGS Water Resources Investigations, Report 83-4188. Reston, Virginia: U.S. Geological Survey. 13 p.

Schroeder, R. A. and C. R. Barnes. 1983b. Trends in Polychlorinated Biphenyl Concentrations in Hudson River Water Five Years after Elimination of Point Sources. USGS Water Resources Investigations, Report 83-4206. Reston, Virginia: U.S. Geological Survey. 28 p.

Simpson, H.J., R. F. Bopp, B. L. Deck, S. Warren, and N. Kostyk. 1984. Polychlorinated biphenyls in the Hudson River: The value of individual packed-column peak analysis. Northeastern Environ. Sci. 3(3/4):159-165.

Simpson, H. J., C. R. Olsen, R. M. Trier, and S. C. Williams. 1976. Man-made radionuclides and sedimentation in the Hudson River Estuary. Science 194:179-183.

Simpson, K. W. 1986. Biomonitoring of PCBs in the Hudson River. I. Results of long-term monitoring using caddisfly (insecta: Trichoptera: Hydropsychidae) larvae and multiplate residues, p. 1-68. II. Development of field protocol for monitoring PCB uptake by caged live *Chironomus tentans* (Insecta: Diptera: Chironomidae) larvae during dredging operations. Pp. 69-99.

Sloan, R. J., M. P. Brown, and C. R. Barnes. 1984. Hudson River PCB relationships between resident fish, water and sediments. Northeastern Environ. Sci. 3(3/4):148-152.

Sloan, R.J., K. Simpson, R. A. Schroeder, and C. R. Barnes. 1983. Temporal trends toward stability of Hudson River PCB contamination. Bull. Environ. Contam. Toxicol. 31:377-385.

Sloan, R. J., B. Young, V. Vecchio, K. McKown, and E. O'Connell. 1988. PCB Concentrations in the Striped Bass from the Marine District of New York State. Technical Report 88-1. Albany, New York: Department of Environmental Conservation. 23 p.

Sofaer, A. D. 1976a. Interim opinion and order. Opinion in the matter of violations of the Environmental Conservation Law of the State of New York by General Electric Company. File No. 2822, 9 February 1976, Department of Environmental Conservation, Albany, New York.

Sofaer, A. D. 1976b. Recommendation of settlement. Opinion in the matter of violations of ECL by General Electric Company, New York. File No. 2833, Department of Environmental Conservation, Albany, New York.

Stone, W. B., E. Kiviat, and S. A. Butkas. 1980. Toxicants in snapping turtles. NY Fish and Game J. 27(1):39-50.

Texas Instruments Incorporated Ecological Services. 1976. A synthesis of available data pertaining to major physiochemical variables within the Hudson River Estuary emphasizing the period from 1972 through 1975. Prepared for Consolidated Edison Company of New York, Inc.

Tofflemire, T. J. and S. O. Quinn. 1979. PCB in the upper Hudson River: Mapping and sediment relationships. Technical Paper No. 56. Albany, New York: Department of Environmental Conservation. 144 p.

Tofflemire, T. J., L. J. Hetline, and S. O. Quinn. 1979a. PCB in the Upper Hudson River: Sediment Distributions, Water Interactions and Dredging. Technical Paper No. 55. Albany, New York: Department of Environmental Conservation. 68 p.

Tofflemire, T. J., S. O. Quinn, and P. R. Hague. 1979b. PCB in the Hudson River: Mapping, Sediment Sampling and Data Analysis. Technical Paper No. 57. Albany, New York: Department of Environmental Conservation.

Turk, J. T. 1980. Applications of Hudson River basin PCB-transport studies. In Contaminants and Sediments, R. A. Baker, ed. Ann Arbor, Michigan: Ann Arbor Science Publications. Pp. 171-183.

Turk, J. T. and D. W. Troutman. 1981a. Relationship of water quality of Hudson River, New York, during peak discharges to geologic characteristics of contributing subbasins. USGS Water-Resources Investigations 80-108. Reston, Virginia: U.S. Geological Survey. 15 p.

Turk, J. T. and D. W. Troutman. 1981b. Polychlorinated Biphenyl Transport in the Hudson River, New York. USGS Water-Resources Investigations 81-9. Reston, Virginia: U.S. Geological Survey. 11 p.

U.S. Geological Survey. 1974-1987. Water Resources Data for New York. Albany, New York, Water-Data Reports. Volume 1, Eastern New York Excluding Long Island. Issued annually. Reston, Virgina: U.S. Geological Survey.

Werner, M. B. 1981. The Use of a Freshwater Mollusc (*Elliptio complaplanatus*) in Biological Monitoring Programs. B. The Freshwater Mussel as a Biological Monitor of PCB Concentrations in the Hudson River. Albany, New York: Department of Environmental Conservation.

Weston Environmental Consultants. 1978. Migration of PCBs from Landfills and Dredge Spoil Sites in the Hudson River Valley, New York. West Chester, Pennsylvania: Weston Environmental Consultants.

Wood, F. J. 1978. The Strategic Role of Perigean Spring T6des in Nautical History and North American Coastal Flooding, 1635-1976. Washington, D.C.: NOAA. 538 p.

CONTAMINATION OF THE HUDSON RIVER
The Sediment Record

Richard F. Bopp and H. James Simpson
Lamont-Doherty Geological Obervatory of Columbia University

ABSTRACT

Measurements of natural and man-made radionuclides have been used to trace fine-grained sediment accumulation throughout the Hudson River system. The results, when combined with measurements of particle-associated pollutants, such as PCBs, chlorinated hydrocarbon pesticides, and trace metals, provide information on the sources, transport, distribution, history, and fate of these contaminants. This technique has proven quite useful for monitoring contaminant levels in natural water systems and assessing the effect of various remedial actions, particularly the "no-action" alternative.

INTRODUCTION

For over a decade, geochemists have studied the history of contamination of natural water systems with particle-reactive pollutants by using radioactive tracers to establish a time scale of sediment accumulation. Early practitioners of this technique have investigated polychlorinated biphenyl (PCB) and DDE accumulation in the Santa Barbara Basin (Hom et al., 1974); the accumulation of fallout radionuclides in Lake Michigan sediments (Robbins and Edgington, 1975); trace metal pollution in Narragansett Bay, Chesapeake Bay, and the Savannah River Estuary (Goldberg et al., 1977, 1978, 1979), and kepone contamination of the James River (Cutshall et al., 1981).

Our work on contaminated Hudson River sediments has relied primarily on measurements of a few naturally occurring and man-made radionuclides that have a high affinity for fine-grained particles and thus serve as tracers for both recent sediment and sediment-associated pollutant accumulation. Cesium-137 (Cs-137) and Plutonium-239,240 (Pu-239,240) are both derived from global fallout resulting from atmospheric testing of nuclear weapons. Measurable fallout began in about 1954 and peaked in 1963 (Hardy et al., 1973). An additional source of Cs-137 to the lower 60 mi of the Hudson system is effluent from the Indian Point nuclear reactors (Figure 1), which began operation in the mid 1960s and had a maximum release in 1971 (Booth, 1975). This source also contributes measurable amounts of Cobalt-60 (Co-60) to the system. A particularly useful natural radionuclide is Beryllium-7 (Be-7),

FIGURE 1 A map of the Hudson River with sediment core locations designated by mile point (mp).

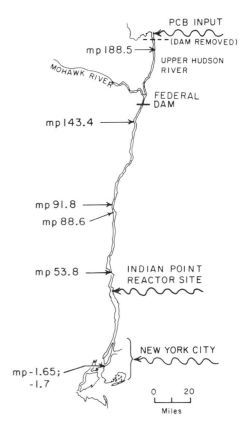

which is produced in the atmosphere via cosmic ray-induced spallation of oxygen and nitrogen. It reaches the surface of the earth primarily via precipitation and has a half-life of about 53 days. Because of this relatively short half-life, measurable Be-7 is generally confined to near-surface sediments deposited within about a year prior to analysis.

The particle-associated pollutants we have analyzed in Hudson sediment samples include PCBs, chlorinated hydrocarbon pesticides, and trace metals. A major source of PCBs to the system was discharges from two General Electric (G.E.) company capacitor manufacturing facilities in the upper part of the drainage basin (Figure 1) over the period between about 1950 and 1976. Another significant source of PCBs to the system is wastewater discharges, which are dominated by inputs from the New York metropolitan area (Mueller et al., 1982; Figure 1). Chlorinated hydrocarbons analyzed include DDT-derived compounds and chlordane, while trace metal analyses focused on copper (Cu), lead (Pb), and zinc (Zn). For both classes of compounds, a strong New York metropolitan area source is indicated by the sediment data.

Detailed descriptions of the sampling and analytical procedures used in the work described below can be found in Bopp (1979), Olsen (1979), and Williams et al. (1978). Sediment cores were sectioned at 2- to 4-cm intervals with 0 cm defined at the sediment-water interface. Control Number (CN) designations unambiguously identify a particular core and are reported in all publications from this laboratory. Locations on the Hudson River are given in mile points that correspond to the number of statute miles upstream of the southern tip of Manhattan measured along the axis of the channel (Figure 1). All contaminant concentrations and radionuclide activities in sediment samples are reported on a dry weight basis.

RADIONUCLIDES IN HUDSON RIVER SEDIMENTS

The advantage of combining sediment contaminant analyses with measurements of independent indicators of the time of deposition results from the heterogeneity of net sediment accumulation rates observed over short distances in natural water systems. Net sedimentation rates in the Hudson River range from less than 1 mm per year in much of the natural channel to more than 10 cm per year in some dredged areas of New York Harbor (Olsen, 1979; Bopp, 1979). Coarse resolution "dating," where the presence of Cs-137 in a sediment sample indicates a significant component of post-1954 deposition, has proven quite useful for establishing first order budgets for sediments and associated contaminants in the Hudson. It has been applied to fine particles (Olsen et al., 1984-85), trace metals (Williams et al., 1978; Bower et al., 1978), and PCBs (Bopp et al., 1981). Occasionally, cores were collected that could be dated in more detail. Most often, such cores exhibited interpretable Cs-137 profiles, penetrating to the first appearance of that radionuclide (1954) and reaching a midcore maximum Cs-137 level (1963) which then decreases toward the sediment-water interface (the date of coring).

The interpretation of such sediment profiles can often be supported by data on other radionuclides, such as measurements of fallout Pu-239,240 that yield a similar distribution with depth in the core, Be-7 determinations to provide a constraint on very recent rates of sediment accumulation, and, when downstream of mp 60, detection of a second Cs-137 maximum that can be associated with the 1971 release from the Indian Point nuclear reactors (Figure 1, Table 1). This second maximum can be unambiguously identified by the presence of reactor-derived Co-60, an excellent marker for post-1971 sediment deposition in the lower Hudson (Simpson et al., 1976; Olsen et al., 1978; Bopp et al., 1982). Ideal cores with continuous records of sediment deposition over the past few decades are relatively rare as a result of both natural disturbances, including large storms and other resuspension events, and human intervention, particularly dredging. Fortunately, when such cores are collected they have large enough net sediment accumulation rates (on the order of 1 cm per year or more) to prevent biological or tidal current mixing in the Hudson from significantly altering the sediment-associated radionuclide and contaminant profiles (Olsen et al., 1981).

From our collection of over 200 Hudson River sediment cores, we have found about 20, spanning the system, with radionuclide profiles that indicate a continuous record of sediment accumulation. Several examples are given in Table 1, and a few others were discussed by Bopp et al. (1982). More recent sediment core data is shown in Figure 2. The core at mp 188.5 (CN 1852) was taken about 10 mi downstream of the

TABLE 1 The Bases for Establishing the Timescale of Sediment
Accumulation in Selected Hudson River Cores

Location (mp)	Date of collection	Principal time indicators	Reference
188.5 CN 1852	July 1983	Fallout Cs-137 (1954, 1963)	Figure 2
143.4 CN 1298	July 1977	dredge boundary (1972)	Bopp et al., 1982
91.8 CN 1329	July 1977	Fallout Cs-137 and Pu-239,240 (1954, 1963)	Bopp et al., 1982
88.6 CN 1984	July 1986	Fallout Cs-137 (1954, 1963); Be-7	Figure 2
53.8 CN 1240	January 1977	Fallout Cs-137 and Pu-239,240 (1954-1963) and reactor Cs-137 and Co-60 (post-1971)	Bopp et al., 1982
3.0 CN 1380	September 1975	Fallout Cs-137 and Pu-239,240 (1954, 1963); reactor Cs-137 and Co-60 (post-1971)	Bopp et al., 1982
-1.7 CN 1472	September 1979	Fallout Cs-137 and Pu-239,240 (1954, 1963); reactor Cs-137 and Co-60 (post-1971)	Figure 2
-1.65 CN 1923	July 1984	Be-7; reactor Co-60 (post-1971)	Figure 2

PCB inputs from the G.E. capacitor plants. This reach of the Hudson is characterized by a series of dams, the southernmost being the Federal Dam at mp 154. Downstream of this dam, the Hudson is a tidal system. The core at mp 88.6 (CN 1984) was collected in 1986 and was used to complement contaminant chronologies developed from a mp 91.8 core (CN 1329) collected in 1977 (Bopp et al., 1982). The core at mp -1.7 contained measurable Co-60 in the top five samples, confirming the assignment of the upper Cs-137 maximum to releases from the Indian Point nuclear reactors. The 8-cm penetration of Be-7 in the mp -1.65 core indicates very rapid sediment accumulation (several cm per year) confirmed by the detection of reactor-derived Co-60 to a depth of at least 50 cm. These two cores were used to develop contaminant chronologies

FIGURE 2 Profiles of radionuclide activity versus depth in some Hudson River sediment cores.

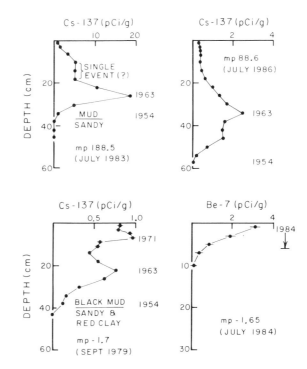

for New York Harbor sediments from the mid 1950s to 1984.

The radionuclide tracers in such cores can be used to determine levels of sediment contamination as a function of time at various locations on the river (Bopp et al., 1982). We estimate the uncertainty on the time of deposition for any given sample at approximately +2 years to account for possible changes in sediment accumulation rates in short time scales and the possibility of gaps in the depositional record between the individual radionuclide-based stratigraphic markers.

POLYCHLORINATED BIPHENYLS

Commercially used PCBs are mixtures of up to several dozen distinct congeners that vary in degree of chlorination and the arrangement of chlorine atoms on the molecules. An individual PCB congener may contain between one and ten chlorine atoms per molecule, but the particular mixtures most used at the G.E. capacitor plants on the upper Hudson River (NYSDEC, 1975), designated Aroclor 1242 and Aroclor 1016, are dominated by di, tri, and tetrachlorobiphenyls with Aroclor 1242 also containing about 10 percent penta and hexachlorobiphenyls (Webb and McCall, 1973).

Throughout the Hudson River system, dated sediment cores show a maximum in total PCB concentration in the early to mid 1970s. This feature, shown in Figure 3 and for some additional cores in Bopp et al. (1982) has been attributed to the removal of a dam in 1973 that had provided the first impoundment of water downstream of the G.E. discharges (Figure 1). The dam removal destabilized large amounts of highly contaminated sediments that were transported downstream in the fall of 1973 and with the unusually high runoff the following spring (Hetling,

FIGURE 3 Total PCB concentrations in Hudson River sediment cores versus time of deposition. For all cores except mp 188.5, PCB concentrations were based on measurements of 22 components resolved by packed-column gas chromatography (Webb and McCall, 1973). For samples from mp 188.5, quantification was based on three PCB congeners that were observed to be highly persistent (i.e., resistant to dechlorination) in another core from the same cove (core 18-6, Brown et al., 1984). This second core had a similar profile of total PCB concentrations as determined by quantification of all major congeners present (Brown et al., 1984; Bopp et al., 1985).

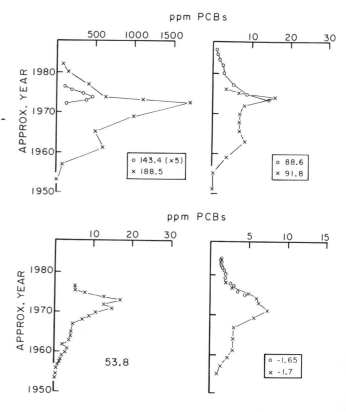

et al., 1978). The maximum observed PCB concentration in sediments decreases with distance downstream from the former dam, reaching over 1,000 ppm in the upper Hudson. In the tidal Hudson, maximum observed concentrations range from about 100 ppm a few miles downstream of the Federal Dam to about 8 ppm in New York Harbor sediments.

PCB Composition

With the exception of the core at mp 188.5, the composition of sediment-associated PCBs observed near this maximum closely resembles that of Aroclors 1242 and 1016. The composition of PCBs in sediments at mp 188.5 is shifted dramatically toward lower chlorinated biphenyls. This has been attributed to bacterially mediated anaerobic dechlorination of PCBs (Brown et al., 1984) and has been observed in other highly contaminated sediments upstream of the Federal Dam (Brown et al., 1984; Bopp et al., 1984). While anaerobic dechlorination may occur in sediments of the tidal Hudson, it is certainly much less significant than in the more highly contaminated sediments of the upper Hudson. Based on PCB component analysis in dated sediment cores of the tidal Hudson, Bopp et al. (1984) found no evidence of significant compositional changes in PCBs during 20 years of anaerobic burial. Such observations could significantly influence management decisions related to systems with PCB contaminated sediments. Since dechlorination generally lowers the persistence of PCBs in organisms and thus decreases their chronic toxicity

(Hansen, 1979), the occurrence of significant in situ dechlorination would be one factor supporting a no-action alternative, while a lack of significant dechlorination would argue in favor of dredging or other remedial action. Furthermore, bacterially mediated dechlorination shows considerable promise as a hazardous waste treatment technology (Roberts, 1987).

In the upper Hudson River, between the former dam site and the Federal Dam, several surface sediment samples and each of the 10 suspended matter samples collected in 1983 and 1984 had PCB compositions that closely resembled Aroclors 1242 and 1016 (Bopp et al., 1985). This maintains the connection between G.E. discharges and PCB contamination throughout the system, but implies significant isolation of the highly contaminated sediments of the upper Hudson that exhibit gross dechlorination of PCBs.

In the tidal Hudson, the composition of PCBs observed in the sediments is most affected by suspended matter-water partitioning, water to air transport, and inputs from the New York metropolitan area (Bopp, 1979, 1983; Bopp et al., 1981). All of these factors tend to increase the average degree of chlorination of sediment-associated PCB components; however, even in New York Harbor sediments, the PCB composition closely resembles Aroclor 1242 (Bopp et al., 1981).

Sources of PCBs to New York Harbor Sediments

Figure 4 shows the importance of New York metropolitan area inputs of the highly chlorinated PCB congeners. Since about 1970, levels of hepta, octa, and decachlorobiphenyls have been much higher in New York Harbor sediments (mp -1.65 and -1.7) than in upstream tidal Hudson sediments (mp 88.6 and 91.8), despite the fact that during the 1970s the upstream cores had total PCB levels about twice those of the harbor cores. Although these highly chlorinated congeners generally comprise less than 10 percent of the total PCBs in Hudson River sediment samples, their toxicological significance is magnified by the fact that they are most persistent in higher organisms and tend to increase in relative abundance along a food chain (Hansen, 1979).

The major problem in quantifying New York metropolitan area inputs of PCBs to the Hudson River lies in the limited number of analyses performed on sewage effluent and urban runoff. Simple mass balances and PCB component ratio analysis (Bopp et al., 1981) indicates that between 1971 and 1976 approximately 75 percent of the total PCBs deposited in New York Harbor sediments were derived from downstream transport and about 25 percent from local metropolitan area inputs. Detailed analysis of the cores at mp -1.65 and 88.6 indicates that by the mid 1980s, the relative importance of local sources had increased significantly. In 1984, both cores recorded total PCB levels of about 1.3 ppm and by 1986, the level in the mp 88.6 core had dropped to 0.8 ppm (Figure 5).

Since at peak levels the harbor core had about half the total PCB concentration of the upstream core, local New York metropolitan area inputs now appear to at least equal the downstream supply of total PCBs to New York Harbor sediments.

FIGURE 4 Hepta + octachlorbiphenyl and decachlorobiphenyl concentrations in upstream tidal Hudson (mp 88.6 and 91.8) and New York Harbor (mp -1.65 and -1.7) sediment cores versus time of deposition. Quantification of hepta + octachlorobiphenyls are based on seven PCB components resolved by packed-column gas chromatography and the composition of Aroclor 1260 (Webb and McCall, 1973). Decachlorobiphenyl quantifications are based on standards prepared from the pure compound.

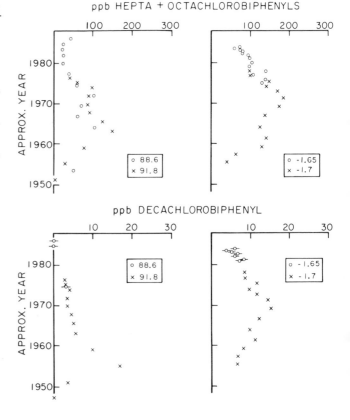

Response Time To Pollution Events

We believe that this situation is due primarily to the recovery of the system from the major pulse of PCBs associated with the dam removal in 1973 discussed above.

Figure 5 shows details of the decrease in total PCB levels. At mp 88.6 concentrations over the last decade can be modelled rather well by a simple exponential decrease toward zero with a half-time of 3.5 years (Figure 4). At mp -1.65, applying a similar time constant produces a curve that decreases asymptotically toward a value of about 0.5 to 0.7 ppm total PCBs. We interpret this "residual level" as resulting from local New York metropolitan area PCB inputs. This type of analysis was first employed by Bopp et al. (1982), who described the response of the system to two distinct types of pollution events involving sediment-associated contaminants. For pollutant inputs to the drainage basin, such as with fallout radionuclides or local DDT applications, a half-response time for Hudson sediments of about six to eight years was found. As would be expected, for pollutant inputs directly to the river, such as the pulse of PCBs associated with the 1973 dam removal, a much shorter half-response time was determined. Analysis of five sediment cores throughout the system gave half-response times of 1.3 to 3.8 years for PCB concentrations. This is in good agreement with the data presented in Figure 5, which is much more detailed than the earlier analysis and tracks the recovery for several additional years.

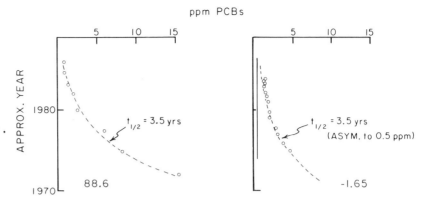

FIGURE 5 Detailed recent chronologies of total PCB concenations in sediments from an upstream tidal Hudson core (mp 88.6) and a New York Harbor core (mp -1.65).

The natural cleansing of the river as indicated by decreasing sediment contaminant levels results both from burial of the most contaminated sediments and the removal of pollutants from the system with river discharge and associated suspended particle transport.

PCB Budgets

The PCB burden in sediments upstream of the Federal Dam and possible remedial action in this reach of the river has been discussed in detail elsewhere (Sanders, this volume; Carcich and Tofflemire, 1982). Although no PCB-directed remedial action has been planned for sediments of the tidal Hudson, we have identified two extensive depositional areas where such action is feasible. The first is New York Harbor. Based on average PCB concentrations and Cs-137 penetration depths in 16 sediment cores, Bopp (1979) estimated that about 23,000 kg of PCBs were associated with in situ sediments of New York Harbor. It was also estimated that an additional 37,000 kg of PCBs had been removed from the harbor as part of normal maintenance dredging and deposited on the shelf at the dredge spoil dump site about 11 mi from the mouth of the Hudson River. These estimates were considered accurate to about a factor of two. The only other reach of the river identified as having significant recent sediment accumulation in the channel was near mp 90, where the river both widens and turns. From the only six cores available at the time, an estimate of 12,000 kg of sediment-associated PCBs was obtained for the area from mp 85 to 93. Additional core collection in 1986 produced the coverage shown in Figure 6. Significant spatial heterogeneity of net sediment accumulation rates is indicated by the Cs-137 penetration depths given in centimeters by the numbers in parentheses. From these data and the average recent sediment PCB concentration of about 7 ppm observed in this region, our best (factor of two) present estimate is that about 21,000 kg of PCBs are associated with sediments of this reach.

CHLORINATED HYDROCARBON PESTICIDES

Chlorinated hydrocarbons, including DDT and chlordane, formed the

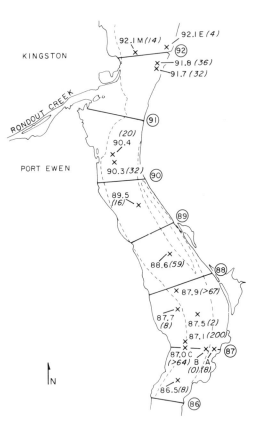

FIGURE 6 A map of the Hudson River from mp 86 to 93. Sediment core locations are marked with x's and labelled by mile point. Numbers in parentheses indicate the depth of penetration of Cs-137 in centimeters.

basis of our insect control strategy in the 1950s and 1960s. They are now well-known for their persistence and ubiquity in the environment. A DDT-derived compound found in recent sediments throughout the Hudson is pp'-DDD produced via anaerobic dechlorination of pp'-DDT carried out by bacteria. Pollution chronologies for this compound in Hudson sediments (Bopp et al., 1982; Figure 7) are characterized by maximum values in the 1960s and early 1970s that decline significantly toward the present. This indicates the effectiveness of the ban on DDT use in the United States imposed by the U.S. Environmental Protection Agency (EPA) in 1972. Sediments from New York Harbor (mp -1.65 and -1.7) show much higher levels of pp'-DDD than sediments from mp 91.8 at comparable time horizons (Figure 7), however, other upstream cores (e.g., mp 53.8) have been found that reach peak levels of almost 100 ppb pp'-DDD (Bopp et al., 1982). This suggests that both downstream transport, resulting from DDT applications in the drainage basin, and local New York metropolitan area inputs are significant contributors to the pp'-DDD contamination observed in New York Harbor sediments.

Chlordane in New York Harbor sediments shows a similar profile. Peak levels of a major chlordane component, γ-chlordane, were found in the 1960s and early 1970s. The mp -1.65 and -1.7 cores peak at about 48 ppb (Figure 7). By the mid 1980s levels had decreased by more than 50 percent, to about 16 ppb in the above mentioned cores. This decrease is consistent with the banning of chlordane for most uses by the EPA in 1975.

For chlordane, the dominance of New York metropolitan area inputs is most pronounced. Tidal Hudson sediments upstream of New York Harbor were always found to contain less than 5 ppb γ-chlordane with mid 1980s levels typically less than 2 ppb. Further discussion of chlorinated hydrocarbon pesticide chronologies in the Hudson River can be found in Bopp et al. (1982).

FIGURE 7 Chronologies of pp'-DDD concentrations in sediments from an upstream tidal Hudson core (mp 91.8) and pp'-DDD and γ-chlordane concentration in sediments from New York Harbor cores (mp -1.65 and -1.7). The most recent data point on the upstream (mp 91.8) pp'-DDD chronology is from the 0-2 cm sample of a core at mp 88.6 collected in 1986. This sample contained significant Be-7, indicating that it was deposited within about a year prior to coring.

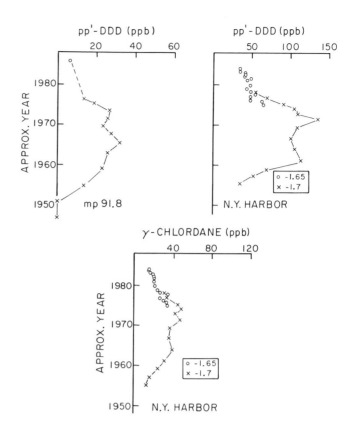

TRACE METALS

A characteristic of our industrial society is elevated levels of trace metals such as copper, lead, and zinc in sediments of natural water systems. Williams et al. (1978) described tidal Hudson sediment levels of these trace metals in terms of three end members: "old" (pre-industrial) sediment with average shale levels, recent (i.e., Cs-137 bearing) sediments upstream of New York Harbor, and recent New York Harbor sediments. Over the past two years, several of the cores in Table 1 have been analyzed to provide chronological data on the trace metal content of recent Hudson sediments. The results, shown for Cu and Pb in Figure 8, confirm the observations of Williams et al. (1978), that New York metropolitan area inputs dominate the sources of these metals to New York Harbor sediments. Average harbor sediment levels of Cu and Pb are more than twice as large as levels observed in upstream sediments of the tidal Hudson at comparable time horizons (Figure 8).

The other outstanding feature of this data is the recent decline in trace metal levels seen in sediments of the upstream tidal Hudson core (mp 88.6). The drop of about 50 percent in Cu and Pb levels over the past decade could indicate recent decreases in the substantial industrial discharges of trace metals to the upper Hudson (Rohmann et al., 1985) or it could be related to the recent implementation of secondary treatment for sewage of the city of Albany, whose discharge is to the

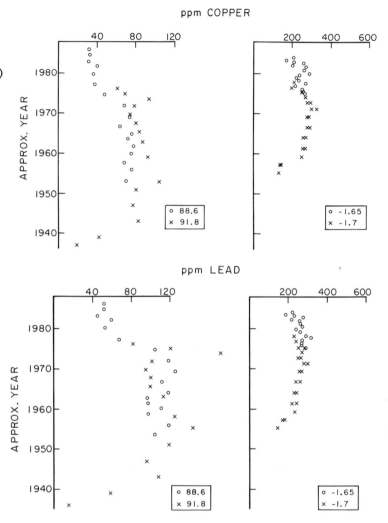

FIGURE 8 Chronologies of copper and lead concentrations in sediments from upstream tidal Hudson cores (mp 88.6 and 91.8) and New York Harbor cores (mp -1.65 and -1.7). Results of replicate analyses of separate aliquots of sediment are shown for several sections of the mp -1.7 core.

Hudson a few miles downstream of the Federal Dam. Similar trace metal results have been reported for a core at mp 46 (Peller and Bopp, 1985). This core also penetrated to pre-industrial sediment with Cu and Pb levels of about 25 ppm.

Most curious is the lack of any substantial improvement in Cu and Pb levels in New York Harbor sediments over the past decade, as indicated by the data from cores at mp -1.65 and 1.7. This is particularly puzzling in the case of Pb. Mueller et al. (1982) report that wastewater, urban runoff, and downstream transport are the dominant sources of this metal to the lower Hudson. Improvements in sewage treatment between the early 1970s and 1982 (Mueller et al., 1982) should have decreased wastewater loading of Pb. The switch from leaded to unleaded gasoline, which produced a two-thirds decrease in atmospheric Pb deposition in New York City between 1970 and 1980 (Freely et al., 1976; Toonkel et al., 1980) should also have significantly decreased the urban runoff of Pb, and--as discussed above--our best estimate for the downstream transport of Pb would be a decrease of 50 percent over the past decade. Despite these indications of decreases in loading, New York Harbor sediments show little or no recent improvement in Pb levels (Figure 8). This apparent contradiction indicates the need for further research into the sources and behavior of trace metals in the lower Hudson and suggests that additional attention be paid to regulation of these important environmental contaminants.

CONCLUSIONS AND RECOMMENDATIONS

The case study of sediment contamination in the Hudson River demonstrates that the measurement of radionuclide time indicators is crucial to the interpretation of pollutant levels in sediments. We recommend that this technique be universally applied to particle-associated contaminant monitoring and the assessment of related problems. We are continuing this practice in detailed studies of adjacent systems, including Raritan Bay, Newark Bay, Jamaica Bay, and the nearshore coastal environment.

Pollutant chronologies that can be developed from this technique are useful indicators of contaminant sources and can provide a detailed assessment of what is commonly called the "no-action" alternative. In the case of PCBs in the Hudson River, levels on particles transported downstream in the mid 1980s are several times lower than in the mid 1970s, despite continued postponement of the removal of the most highly contaminated sediments from the upper Hudson (i.e., no action). Assessment of our general nationwide efforts to limit pollution is also possible. There is evidence that New York Metropolitan area inputs of PCBs to the Hudson have decreased recently, probably in response to EPA restrictions on the manufacture and use of PCBs in the United States in the late 1970s. Hepta, octa, and decachlorobiphenyl levels in New York Harbor sediments have declined by about a factor of two over the past decade (Figure 4), while the New York metropolitan area contribution to total PCBs in New York Harbor sediments was estimated at 0.8 ppm in the mid 1970s (Bopp, 1979) and 0.5 to 0.7 ppm in the mid 1980s (Figure 5). Restrictions on chlorinated hydrocarbon pesticide use in the United States promulgated by the EPA in the 1970s are most likely responsible for the recent decline in levels of pp'-DDD and chlordane in Hudson sediments (Figure 7). Finally, with respect to trace metals, New York Harbor sediment chronologies indicate little or no recent improvement (Figure 8) despite recent upgrading of sewage treatment and restrictions on lead in gasoline. This is a most direct recommendation for further study of the sources and fate of these contaminants, not only in the Hudson River, but in other major natural water systems as well.

ACKNOWLEDGMENTS

We would first like to thank our scientific and technical collaborators, Bruce Deck, Curt Olsen, Nadia Kostyk, Dave Robinson, Kathleen Ledyard, Charles Lester, Ellen Kalb, Yu-Pin Chin, Robert Trier, Sue

Williams, Peter Kay and Linda Hubbard. Financial support for our present work on the Hudson is provided by the Hudson River Foundation. Past research efforts have been funded by the New York State Department of Environmental Conservation, the National Science Foundation, the National Oceanic and Atmospheric Administration, and the U.S. Environmental Protection Agency. This is LDGO contribution number 4342.

REFERENCES

Booth, R. S. 1975. A compendium of radionuclides found in liquid effluents of nuclear power stations. ORNL-TM-3801. Oak Ridge National Laboratory, Oak Ridge, Tenn.

Bopp, R. F. 1979. The Geochemistry of Polychlorinated Biphenyls in the Hudson River. Ph.D. Thesis, Columbia University, New York.

Bopp, R. F., H. J. Simpson, C. R. Olsen, and N. Kostyk. 1981. Polychlorinated biphenyls in sediments of the tidal Hudson River, New York. Environ. Sci. Technol. 15:210-216.

Bopp, R. F., H. J. Simpson, C. R. Olsen, R. M. Trier, and N. Kostyk. 1982. Chlorinated hydrocarbons and radionuclide chronologies in sediments of the Hudson River and estuary, New York. Environ. Sci. Technol. 16:666-676.

Bopp, R. F. 1983. Revised parameters for modeling the transport of PCB components across an air water interface. J. Geophys. Res. 88:2521-2529.

Bopp, R. F., H. J. Simpson, B. L. Deck, and N. Kostyk. 1984. The persistence of PCB components in sediments of the lower Hudson. Northeastern Env. Sci. 3:180-184.

Bopp, R. F., H. J. Simpson, and B. L. Deck. 1985. Release of Polychlorinated Biphenyls from Contaminated Hudson River Sediments, NYS C00708 Final Report. New York State Department of Environmental Conservation.

Bower, P. M., H. J. Simpson, S. C. Williams, and Y.-H. Li. 1978. Heavy metals in the sediments of Foundry Cove, Cold Springs, New York. Environ. Sci. Technol. 12:683-687.

Brown, J. F., R. E. Wagner, D. L. Bedard, M. J. Brennan, J. C. Carnahan, R. J. May, and T. J. Tofflemire. 1984. PCB transformations in upper Hudson sediments. Northeastern Env. Sci. 3: 166-178.

Carcich, I. G. and T. J. Tofflemire. 1982. Distribution and concentration of PCB in the Hudson River and associated management problems. Environ. Internat. 7:73-85.

Cutshall, N. H., I. L. Larsen, and M. M. Nichols. 1981. Man-made radionuclides confirm rapid burial of kepone in James River sediments. Science 213:440-442.

Feely, H. W., H. L. Volchok, and T. N. Toonkel. 1976. Trace Metals in Atmospheric Deposition. EML-308. Washington, D.C: Department of Energy.

Goldberg, E. D., E. Gamble, J. J. Griffin, and M. Kiode. 1977. Pollution history of Narragansett Bay as recorded in its sediments. Estuar. Coast. Mar. Sci. 5:549-561.

Goldberg, E. D., V. Hodge, M. Kiode, J. Griffin, E. Gamble, O. P.

Bricker, G. Matisoff, G. R. Holdren, and R. Braun. 1978. A pollution history of Chesapeake Bay. Geochim. Cosmochim. Acta 42:1413-1425.

Goldberg, E. D., J. J. Griffin, V. Hodge, M. Kiode, and H. Windom. 1979. Pollution history of the Savannah River estuary. Environ. Sci. Technol. 13:588-594.

Hansen, L. G. 1979. Selective accumulation and depletion of polychlorinated biphenyl components: Food animal implications. Ann. N.Y. Acad. Sci. 320:238-246.

Hardy, E. P., P. W. Krey, and H. L. Volchok. 1973. Global inventory and distribution of fallout plutonium. Nature 241:444-445.

Hetling, L., E. Horn, and T. J. Tofflemire. 1978. Summary of Hudson River PCB Results. Technical Paper No. 51. New York State Department of Environmental Conservation.

Hom, W., R. W. Risebrough, A. Soutar, and D. R. Young. 1974. Deposition of DDE and PCB in dated sediments of the Santa Barbara Basin. Science 184:1197-1199.

Mueller, J. A., T. A. Gerrish, and M. C. Casey. 1982. Contaminant inputs to the Hudson-Raritan estuary. Chapter VI, Wastewater inputs. Technical Memorandum OMPA-21. Rockville, Md.: NOAA.

NYSDEC. 1975. NYS Department of Environmental Conservation, in the Matter of Alleged Violations of Sections 17-0501, 17-0511, and 11-0503 of the Environmental Conservation Law of the State of New York by General Electric Co., Interim Opinion and Order, File No. 2833.

Olsen, C. R., H. J. Simpson, R. F. Bopp, S. C. Williams, T.-H. Peng, and B. L. Deck. 1978. Geochemical analysis of the sediments and sedimentation in the Hudson estuary. J. Sediment. Petrol. 48: 401-418.

Olsen, C. R. 1979. Radionuclides, Sedimentation and the Accumulation of Pollutants in the Hudson Estuary. Ph.D. Thesis, Columbia University, New York.

Olsen, C. R., H. J. Simpson, T.-H. Peng, R. F. Bopp, and R. M. Trier. 1981. Sediment mixing and accumulation rate effects on radionuclide depth profiles in Hudson estuary sediments. J. Geophys. Res. 86: 11020-11028.

Olsen, C. R., N. H. Cutshall, I. L. Larsen, H. J. Simpson, R. M. Trier, and R. F. Bopp. 1984-5. An estuarine fine-particle budget determined from radionuclide tracers. Geo-Marine Letters 4:157-160.

Peller, P. and R. F. Bopp. 1985. Recent sediment and pollutant accumulation in the Hudson River National Estuarine Sanctuary. In Polgar Fellowship Reports of the Hudson River National Estuarine Sanctuary Program, 1985, Jon C. Cooper ed. New York: Hudson River Foundation.

Robbins, J. A. and D. N. Edgington. 1975. Determination of recent sedimentation rates in Lake Michigan using PB-210 and Cs-137. Geochim. Cosmochim. Acta 39:285-304.

Roberts, L. 1987. Discovering microbes with a taste for PCBs. Science 237:975-977.

Rohmann, S. O., N. Lilienthal, R. L. Miller, R. M. Szwed, and W. R. Muir. 1987. Tracing a River's Toxic Pollution, A Case Study of the Hudson, Phase II. New York: Inform.

Simpson, H. J., C. R. Olsen, S. C. Williams, and R. M. Trier. 1976.

Man-made radionuclides and sedimentation in the Hudson River estuary. Science 194:179-183.

Toonkel, L. T., H. W. Feely, and R. J. Larsen. 1980. The Chemical Composition of Atmospheric Deposition. EML-381. Washington, D.C.: Department of Energy.

Webb, R. G. and A. C. McCall. 1973. Quantitative PCB standards for electron capture gas chromatogspahy. J. Chromatogr. Sci. 11:366-373.

Williams, S. C., H. J. Simpson, C. R. Olsen, and R. F. Bopp. 1978. Sources of heavy metals in sediments of the Hudson River estuary. Mar. Chem. 6:195-213.

KEPONE AND THE JAMES RIVER

Robert J. Huggett
College of William and Mary

ABSTRACT

The James River in Virginia was contaminated by the pesticide kepone when the material entered the river as early as 1968 and continued until its discovery in 1975. The river became so contaminated that commercial fisheries were closed. In 1988, 13 years after closure, all fishing restrictions were lifted. The contaminated sediments have been diluted and covered enough by uncontaminated material that the kepone flux back into the water column has diminished. Kepone concentrations in organisms inhabiting the river are finally below the U.S. Environmental Protection Agency and Food and Drug Administration action levels. Biological, chemical, physical and geological aspects of the contamination indicate that remedial actions to remove kepone would be expensive and environmentally unwise.

INTRODUCTION

In 1988, there were no restrictions on commercial fishing in the James River. It has been more than a decade since workers at a kepone manufacturing facility in Hopewell, Virginia became ill from occupational exposure to the pesticide. The knowledge of their exposure, the fact that kepone is a mammalian carcinogen, and the subsequent determination that the adjacent river had become contaminated with the compound led Governor Mills Godwin to close the tidal portion of the James and its tributaries to commercial fishing.

Kepone (decachlorooctahydro-1,3,4-metheno-2H-cyclobuta[cd]-pentalen-2-one) was produced from hexachlorocyclopentadiene in the presence of sulfur trioxide. A solution of sodium hydroxide was used in the purification process (Huggett et al., 1980). The conditions used in the formulation of the compound suggest that it should be resistant to chemical degradation under natural environmental conditions, a supposition that has been verified by field observations. If kepone has degraded significantly in the river, it is not obvious even after thousands of chemical observations over 13 years.

The fishing restrictions were relaxed because the contaminated sediments were diluted and covered by uncontaminated materials. Since the kepone flux back into the water column diminished, finfish and shellfish inhabiting the river contain concentrations below action levels established by the U.S. Environmental Protection Agency (EPA) and Food and Drug Administration (FDA).

THE JAMES RIVER

The James River extends from its mouth near Norfolk and Newport News, Virginia, to West Virginia (Figure 1). It is tidal for the first 160 km with the city of Richmond located at its fall line. The drainage basin encompasses approximately 25,600 km^2 and runoff from this area results in it being the third largest tributary of the Chesapeake Bay, delivering approximately 16 percent of the fresh water entering the system. The average discharge over the fall line at Richmond is 212 m^3/sec.

The river is a coastal plain estuary for its first 60 to 80 km with the location of the freshwater-saltwater interface varying depending on rainfall in the drainage basin. Fresh water from upstream flows over more dense saline water creating a two-layer circulation pattern. As the fresh water flows to the sea, there is some mixing between layers, giving rise to a net downstream flow in the surface layer and a net upstream flow on the bottom (Pritchard, 1952). Suspended particulate matter is carried downstream in the tidal freshwater portion of the river (i.e., above the freshwater-saltwater interface) and downstream in the surface layer of the estuary. If the particles sink into the bottom layer, they are transported upstream toward the interface (Figure 2). This phenomenon is mainly responsible for the higher sedimentation rate and more turbid water in the interface region of the river, which is appropriately called the "turbidity maximum zone."

The circulation pattern and its influence on the movement of particulate matter controls the transport of kepone in the James River. The pesticide entered the river at Hopewell, associated with particulate material, and was transported downstream. Most of the kepone deposited in the turbidity maximum zone.

THE KEPONE SOURCE

Allied Chemical Corporation began producing kepone in 1966 and intermittently continued until 1974. At this time, Life Science Products, Inc., began production and continued until July, 1975 (Huggett et al., 1980; Huggett and Bender, 1980). During this period over 1.5 x 10^6 kg of the substance were produced (Batelle Memorial Institute, 1978). It is likely that kepone entered the James River throughout the period of production. Analyses of oysters (*Crassostrea virginica*) and bottom sediments collected as early as 1967, but analyzed in 1976, revealed that the James River was contaminated in the 1960s (Huggett et al., 1980; Huggett and Bender, 1980).

Kepone entered the river at Hopewell via a number of routes. The most significant was the discharge of the local municipal sewage system. Kepone-laden industrial waste entered the sewage treatment plant and the pesticide exited with little or no degradation. Other sources

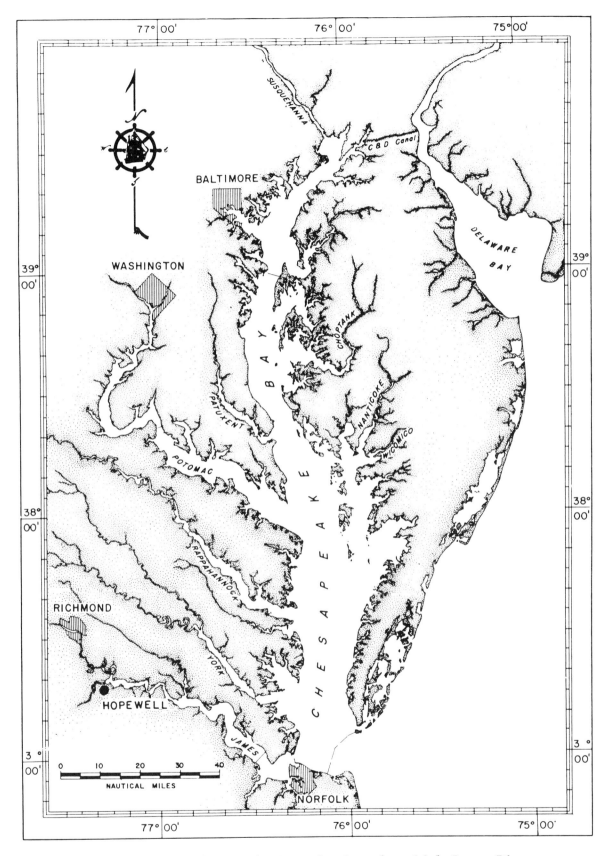

FIGURE 1 Map of the Chesapeake Bay showing the tidal James River.

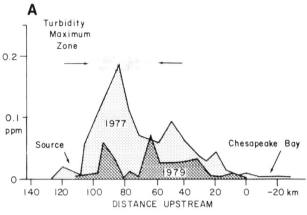

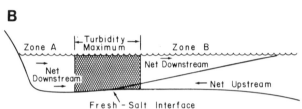

FIGURE 2 A. Kepone in the top 2 cm of channel bottom sediments from the James River. B. Hypothetical coastal plain estuary with two-layered circulation and turbidity maximum. SOURCE: Huggett and Bender (1980), reprinted with permission from the American Chemical Society.

included runoff from contaminated soils near the manufacturing facilities and solid waste dumped into a freshwater marsh on a small tributary of the James River (Huggett et al., 1980). The material entered the river either as particulates or in solution. In the latter case, it rapidly sorbed to bottom and suspended solids to be transported by the river's currents.

CONTAMINATED SEDIMENTS

Kepone readily partitions from solution to solids. Dawson (Batelle Memorial Institute, 1978) suggested that a sediment-water partitioning coefficient of 10^4 to 10^5 be used. Other laboratory experiments as well as measurements of kepone in suspended sediments and associated waters from the James indicate that the value is between 1.6×10^3 and 7.7×10^3 (Huggett et al., 1980; Strobel et al., 1981). The partitioning coefficient, as derived in the laboratory, does not appear to be affected by changes in salinity from 0 to 20 $^o/oo$ or by pH values from 6 to 9 (Huggett et al., 1980). These span both the salinity and pH ranges normally found in the contaminated portion of the river. Field investigations verify these findings (Strobel et al., 1981).

The bottom sediments of the James River are contaminated with kepone to varying degrees. The main factors governing the concentrations appear to be the makeup of the sediments and the currents of the overlying water. These two factors, in combination, distribute kepone in a nonuniform pattern over an area of approximately 500 km^2 (Huggett and Bender, 1980, 1982).

Kepone associates more with the organic portion of the bottom sediments (Huggett et al., 1980). Sandy or coarse-grained sediments

generally contain less kepone than fine-grained sediments. This is due
to the ordinarily high organic content of the latter. The organic
content of the sediments can have a dramatic influence on kepone
distribution. For instance, the highest sediment concentrations found
(except within several kilometers of the Hopewell source) were near the
outfall of a sewage treatment facility 75 km downstream (Huggett and
Bender, 1982). The organic content of this sediment was approximately
20 percent. There was no indication that kepone had ever been disposed
of by this treatment plant.

The distributions of the pesticide in the top 2 cm of bottom sediments in the channel of the river in 1977 and 1979 are given in Figure
2. The highest concentrations in 1977 were found in the vicinity of
the turbidity maximum zone. The mass of kepone in the sediments at
that time was estimated to be between 1×10^4 kg and 3×10^4 kg
(Huggett and Bender, 1980). The range was due to the large area contaminated (500 km^2) and the relatively few samples analyzed at the time.

By 1979, surface sediment concentrations were greatly diminished.
Analyses of sediment cores with depth showed that kepone was becoming
diluted and buried by newly deposited material rather than being transported away or decomposing. This trend has continued, but in areas
where the sedimentation rate is low, kepone is most concentrated near
the surface. Where the sedimentation rates are high, concentrations
increase with depth (Figure 3) (Helz and Huggett, 1987).

As mentioned previously, most of the kepone is deposited in the
James' turbidity maximum zone, which has a high sedimentation rate.
This has resulted in a continual reduction in the pesticide's concentration in surface sediment (Figure 3). This reduction is reflected in
the residue concentrations in edible tissues of male blue crabs
(*Callinectes sapidus*) and oysters (*Crassostrea virginica*)

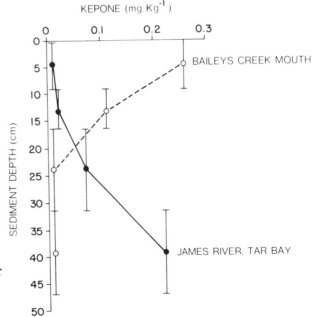

FIGURE 3 Kepone concentrations in
sediment cores from the James River.
Bars indicate the depth interval of
the sediments analyzed. SOURCE: Reprinted with permission from Majumdar
et al., 1987.

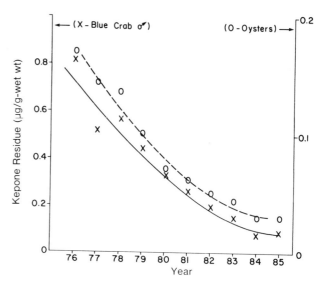

FIGURE 4 Kepone concentrations in blue crabs and oysters.
SOURCE: Majumdar et al., 1987.

collected from 1976 to 1985 (Figure 4). The data are interesting in that they show similar rates of concentration decrease for both species although crabs obtain most of their kepone from food, while oysters obtain kepone both from solution and from suspended particles (Schimmel and Wilson, 1977; Morales-Alamo and Haven, 1983; Bender and Huggett, 1987). Apparently the equilibration times between sediments and water, sediments and food are relatively short.

DISCUSSION AND CONCLUSION

Kepone concentrations in the James River are much lower now than in the past, therefore the biota are at less risk from the toxicant now than during the period of production. A comparison of existing toxicity data and kepone concentrations in solution or in tissues of the biota indicates that there has been little or no biological impact due to the contamination (Bender and Huggett, 1984). The impact has been economic; commercial fishermen couldn't harvest the seafood and consumers couldn't buy it.

Any consideration of mitigation must balance the benefits of cleanup, which would be solely economic, with the costs, which are not only economic (e.g., the cost of dredging) but also ecological. *Any cleanup effort* will have detrimental biological impact relative to doing nothing. Natural forces, such as sedimentation, are cleansing the river and the time frame for this cleanup is on the order of decades.

Studies have been conducted, however, to assess the feasibility of mitigating the kepone contamination of the James (Batelle Memorial Institute, 1978). Options ranged from dredging, at an estimated cost of 3×10^9 not including the cost of disposal, to stabilizing the sediments with molten sulfur (often called "the Yellow Brick Road theory"). None of these options were feasible, either economically or environmentally; therefore, nothing has been done.

Kepone concentrations in finfish and shellfish are now low enough to again allow commercial harvesting in the river. The pesticide is buried under a veneer of clean sediments. A major hurricane could stir up these sediments and recontaminate the river (Huggett and Bender, 1980). Such a storm has not occurred in the area since the 1950s. Another complicating factor is that the channels of the James River will need to be dredged in the near future. In the past, dredged material was disposed by placing it on the flanks of the river, adjacent to the channel being dredged. Such a practice now would place buried kepone-contaminated sediments back on the surface. The biota would again be exposed to the pesticide. The resulting body burdens could result in fisheries closures.

Given the uncertainties involved in predicting the transport and fate of kepone under the conditions mentioned above, deciding whether or not to dredge the James River will be difficult. The benefits of continued shipping on the James River by allowing dredging will have to be compared to the potential costs of fisheries closures due to kepone contamination. One solution to the dilemma may be to bear the expense of upland disposal and containment of the dredged materials rather than pumping them back overboard. (VIMS Publication Number 1502.)

REFERENCES

Battelle Memorial Institute. 1978. The Feasibility of Mitigating Kepone Contamination of the James River Basin. Final Report to the U.S. Environmental Protection Agency. Washington, D.C.: Battelle.

Bender, M. E. and R. J. Huggett. 1987. Contaminant effects on Chesapeake Bay shellfish. In Contaminant Problems and Management of Living Chesapeake Bay Resources, S. K. Majumdar, L. W. Hall, and H. M. Austin, eds. Penn. Acad. of Sci. Pub. Pp. 373-393.

Bender, M. A. and R. J. Huggett. 1984. Fate and effects of kepone in the James River estuary. In Reviews in Environmental Toxicology, E. Hodgson, ed. New York: Elsevier Science Publishers. Pp. 5-50.

Helz, G. R. and R. J. Huggett. 1987. Contaminants in Chesapeake Bay. In Contaminant Problems and Management of Living Chesapeake Bay Resources, S. K. Majumdar, L. W. Hall, Jr. and H. M. Austin, eds. Penn. Acad. of Sci. Pub., pp. 270-297.

Huggett, R. J., M. M. Nichols, and M. E. Bender. 1980. Kepone contamination in the James River estuary. In Contaminants and Sediments, R. A. Baker, ed. Ann Arbor, Mich.: Ann Arbor Science Publishers. 1:33-52.

Huggett, R. J. and M. E. Bender. 1982. Scientific sessions taught by Kepone. In Proceedings of a Symposium on Agrichemicals and Estuarine Productivity. Beaufort, N.C.: Duke University Marine Laboratory. Pp. 53-61.

Huggett, R. J. and M. E. Bender. 1980. Kepone in the James River. Environ. Sci. Tech. 14(8):918-923.

Morales-Alamo, R. and D. S. Haven. 1983. Uptake of kepone from sediment suspensions and subsequent loss by the oyster *Crassostrea virginica*. Mar. Biol. 74:187-201.

S. K. Majumdar, L. W. Hall, Jr., and H. M. Austin, eds. 1987. Contaminant Problems and Management of Living Chesapeake Bay Resources, Penn. Acad. of Sci. Pub.

Pritchard, D. W. 1952. Salinity distribution and circulation in the Chesapeake Bay estuarine system. J. Mar. Res. 11:106-123.

Schimmel, S. C. and A. S. Wilson. 1977. Acute toxicity of Kepone to four estuarine animals. Chesapeake Sci. 18:224-227.

Strobel, C. J., R. E. Croonenberghs, and R. J. Huggett. 1981. The suspended sediment-water partitioning coefficient for kepone in the James River, Virginia. Environ. Poll. 2:367-372.

ASSESSMENT OF CONTAMINATED SEDIMENTS
IN COMMENCEMENT BAY (PUGET SOUND, WASHINGTON)

Thomas C. Ginn
PTI Environmental Services

ABSTRACT

Sediments in Commencement Bay have been contaminated by a wide variety of inorganic and organic contaminants resulting from numerous industrial activities and pollutant discharges. Because of this contamination and associated biological effects, the area has been the subject of a Remedial Investigation/Feasibility Study to evaluate alternatives for sediment cleanup and source control. Prior to these evaluations, a decision-making framework was needed to focus the evaluation of remedial alternatives on those areas and contaminants posing the greatest hazards to the environment and to public health. The resulting assessment approach developed for the Commencement Bay investigations is described, showing how a preponderance of evidence on sediment contamination and biological effects is used in an Action Assessment Matrix to define and rank problem contaminants and problem sediments.

INTRODUCTION

Commencement Bay is an urban embayment of approximately 9 mi^2 in south-central Puget Sound, Washington (Figures 1 and 2). The bay opens to Puget Sound in the northwest, with the city of Tacoma situated on the south and southeast shores. The Commencement Bay study area consists of a series of eight waterways, the lower Puyallup River, and the Ruston shoreline.

Industrialization of Commencement Bay began in the late 1800s, at which time dredging and filling operations began in the tideflats area of the Puyallup River Delta. Numerous industrial and commercial operations were located in the filled areas of the bay, including pulp and lumber mills, shipbuilding facilities, metal smelting, oil refining, marinas, food processing, chemical manufacturing, and many other commercial operations. Much of the tideflats area was constructed on slag from a copper smelter at Ruston that was used as fill and ballast material.

Pollutant loadings in Commencement Bay result from numerous point and nonpoint sources. Recent surveys have indicated over 281 industrial activities in the nearshore/tideflats area. Comprehensive shoreline surveys have identified over 429 point- and nonpoint-source discharges in the study area, consisting primarily of seeps, storm drains, and open channels. Only 27 of the point sources were identified as NPDES-permitted discharges.

Several investigations conducted in the late 1970s and early 1980s indicated that Commencement Bay waterways were contaminated by a wide variety of metals (e.g., arsenic, copper, and mercury) and organic chemicals (e.g., PCBs, PAH, and chlorinated butadienes). The historical data suggested that sediment contamination was spatially extensive and highly heterogeneous. These studies also indicated areas of high sediment toxicity, accumulation of toxic substances in indigenous biota, and the presence of liver abnormalities and tumors in flatfish. As a result of these findings, the Commencement Bay nearshore/tideflats area was added to the U.S. Environmental Protection Agency's (EPA) National Priorities List of hazardous waste sites in 1983. Subsequently, EPA entered into a cooperative agreement with the Washington Department of Ecology to conduct a remedial investigation in Commencement Bay under the Comprehensive Environmental Response, Compensation, and Liability Act (CERCLA).

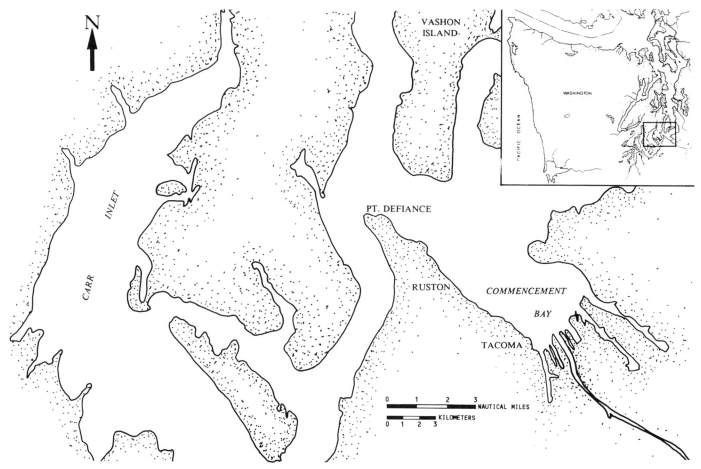

FIGURE 1 South-central Puget Sound showing locations of Commencement Bay and Carr Inlet.

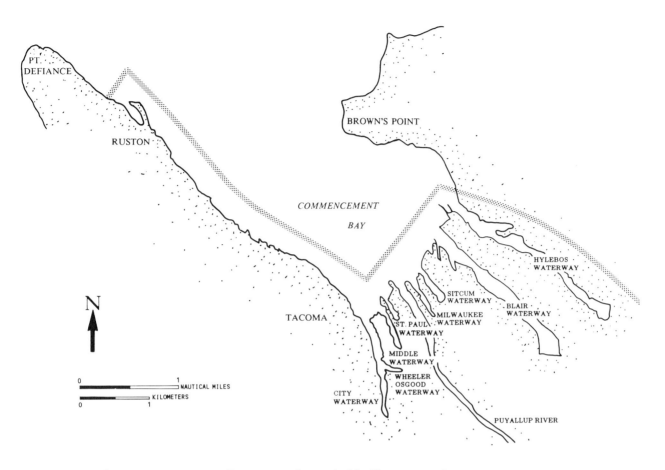

FIGURE 2 Commencement Bay nearshore/tideflats study area.

The goal of the remedial investigation was to identify potential remedial alternatives that could be used to reduce or eliminate the risks to the environment and to public health resulting from contaminated sediments in Commencement Bay. However, because of the complexity of sediment contaminants and pollutant sources in Commencement Bay and the lack of available cleanup criteria for sediment contaminants, the investigation required the development of a decision-making framework to assess and prioritize contaminated sediments prior to evaluating cleanup alternatives. The development of this framework included the specification of several objectives associated with the assessment of sediment contamination:

- characterization of sediment contamination, sediment toxicity, and biological effects;
- development of criteria to define problem sediments;
- application of the criteria to define problem areas;
- determination of problem chemicals for the problem areas; and
- prioritization of problem areas and problem chemicals relative to environmental and human health risks.

This paper describes the approach used in the Commencement Bay remedial investigation to meet these objectives and the results of the overall assessment of contaminated sediments.

GENERAL ASSESSMENT APPROACH

The decision-making framework developed for Commencement Bay incorporates a "preponderance-of-evidence" approach that is implemented in a step-wise manner to identify toxic problem areas (Figure 3). Information on the extent of sediment contamination, adverse environmental effects, and potential threats to public health form the basis for prioritization of areas for cleanup and/or source control. The decision-making framework for the Commencement Bay investigations was developed to integrate these kinds of technical information in a form that could be understood by regulatory decision makers and the public. The framework uses six steps to identify and rank problem areas and problem chemicals. Study areas that exhibit high values of indices for contamination and biological effects relative to reference areas receive a ranking of "high priority" for evaluation of pollutant sources and potential cleanup alternatives.

A review of site characteristics and historical data for Commencement Bay in conjunction with available information on the effects of contaminated sediments led to the development of three important premises as part of the decision-making process. First, it was determined that criteria to define problem sediments could not be established a priori because of limitations in the historical database and the absence of regulatory sediment criteria at the national or state levels. Therefore, site-specific criteria would be developed based on an integration of historical data and data gathered as part of this investigation. Second, it was determined that no single chemical or biological measure of environmental conditions could be used to define problem sediments. Therefore, problem areas would be defined according to the magnitude and extent of contamination and effects evidenced by several independent sediment and biological observations. These measurements would not be combined into a single index. Instead, the approach defines multiple environmental conditions to define problem areas and prompt possible remedial action. Third, it was assumed that adverse biological effects are linked to environmental conditions and that these links may be characterized empirically. Proof of cause/effect relationships would therefore not be provided by the studies. However, quantitative relationships derived from analysis of field observations would be used, where possible, to demonstrate links between sediment contamination and biological effects. In this sense, cause/effect relationships may be implied by a preponderance of field and laboratory evidence, including the correlation of specific contaminant concentrations with the occurrence of adverse biological effects.

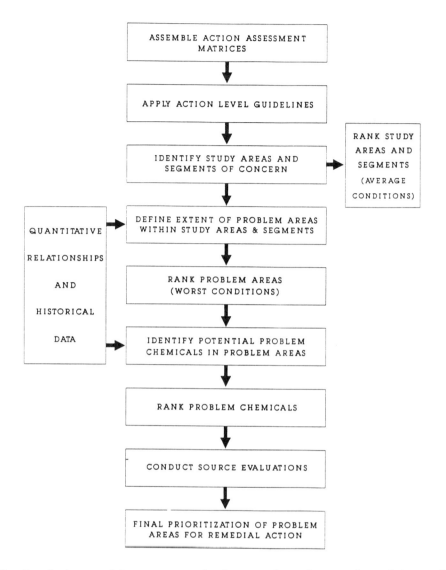

FIGURE 3 Decision-making approach for evaluation and ranking of problem areas and problem chemicals.

CHEMICAL AND BIOLOGICAL INDICATORS

The preponderance-of-evidence approach discussed above required the selection of several measurements that would serve as indicators of contamination and biological effects in Commencement Bay. To conserve costs, the objective was to select the minimum numbers of indicators that could be used to adequately characterize the contaminant situation, as well as enable a prioritization of problem sediments. The following five groups of indicator variables were selected:

1. sediment contamination--concentration of chemicals and chemical groups;

2. bioaccumulation--contaminant concentrations in English sole;
3. sediment toxicity--acute mortality of amphipods; abnormalities in oyster larvae;
4. benthic infauna--abundances of major taxa;
5. fish histopathology--prevalences of liver lesions in English sole.

Chemical contaminants of concern selected for study in the remedial investigation included many EPA priority pollutants, EPA Hazardous Substance List compounds, and several organic compounds identified in Commencement Bay samples that are not on the EPA lists. Chemicals were selected based on their historical occurrence and their documented potential for toxicity or bioaccumulation. Sediment concentrations of individual chemicals or groups of contaminants were normalized to sediment dry weight and organic carbon content of the sediments.

Bioaccumulation of contaminants in fish was selected to evaluate the bioavailability of sediment contaminants and the potential for human health effects resulting from consumption of contaminated seafood. English sole was selected as the target species because it is abundant in the study area, lives in close contact with the bottom, and has been shown to accumulate many chemicals of concern at relatively high levels.

The toxicity of sediments was based on amphipod (*Rhepoxynius abronius*) mortality as a measure of acute lethality and oyster larvae shell deformation as an indicator of sublethal effects. The amphipod bioassay was selected based on its demonstrated sensitivity in Commencement Bay, its ecological significance, and the availability of a routine protocol. The oyster larvae bioassay has also been shown to be sensitive to a wide range of contaminants and was selected to gauge potential sublethal effects. Assessment of benthic infaunal assemblages was performed because of their sensitivity to sediment contamination, their importance in local trophic relationships, and their site-specific response gradients relative to sediment contamination. English sole liver was selected for pathological analyses because it is the organ most closely associated with regulation and storage of many toxic chemicals and has been shown to be afflicted with pathological disorders in sole living in contaminated areas of Puget Sound.

FORM OF THE INDICATORS

A series of simple indices was developed for each of the five indicators to enable ranking of areas based on the relative magnitude of observed contamination and effects. These indices have the general form of a ratio between the value of a variable at a Commencement Bay site and the value of the variable at a reference site. For the Commencement Bay studies, Carr Inlet (Figure 1) was selected as a reference area based on its proximity and its documented low contamination.

The indicator ratios are structured so that the value of the index increases as the deviation from reference conditions increases. Thus, each ratio is termed an Elevation Above Reference (EAR) index. For example, the EAR for sediment toxicity is expressed as

$$EAR_T = M_{Si}/M_{Ri}$$

where

M_{Si} = mortality or abnormality rate i at a Commencement Bay study area, and

M_{Ri} = mortality or abnormality rate i at the reference area.

Benthic community structure is somewhat different from the other variables because it is not expressed as a single ratio. Instead, four separate ratios were used for the abundances of crustacea, molluscs, annelids, and total organisms. The benthic EAR was also expressed as the inverse ratio of study area and reference area because affected study areas would be expected to have lower infaunal abundances. Therefore, all of the EAR are in a form where increasing magnitudes of contamination or biological effects are expressed as increasing values of the index.

It should be noted that these indices were not used in lieu of the original data, but in addition to them. The original data were used to identify statistically detectable increases in sediment contamination, sediment toxicity, or biological effects indicators, and to determine quantitative relationships among these variables. The indices are used to reduce large, complex data sets into interpretable numbers that reflect the magnitudes of the different indicators among study areas.

ACTION ASSESSMENT MATRIX

The environmental contamination and effects indicators (EAR) were organized into an action assessment matrix used to compare the Commencement Bay study areas (Table 1). This matrix contains the EAR for each indicator as well as the reference values for that indicator. Therefore, original values for any indicator can be obtained by multiplying the EAR by the appropriate reference value. For the Commencement Bay project, such matrices were developed for the entire study area (i.e., with waterways as the study units as in Table 1) and for individual waterways with individual sampling stations or groups of stations (i.e., waterway segments) as the study units.

In assembling the matrix, each study area indicator is evaluated relative to reference conditions to determine if there is a significant difference. All biological indicators were tested using parametric or nonparametric statistical tests to determine statistical differences at $P < 0.05$. Because sediment chemistry data were not replicated at each site, chemical contamination in the study area was determined to be significant if it exceeded the upper end of the range of values from all Puget Sound reference areas.

Development of an action assessment matrix enables the decision maker to answer the following kinds of questions relative to sediment contamination:

TABLE 1 Action Assessment Matrix of Sediment Contamination, Sediment Toxicity, and Biological Effects Indices for Commencement Bay Study Areas

Variable	Study Area Elevations[a]								Reference Value
	Hylebos	Blair	Sitcum	Milwaukee	St. Paul	Middle	City	Ruston	
Sediment Chemistry									
Sb	10	4.0	8.0	3.6	4.2	9.3	7.0	510	110 ppb
As	12	7.6	11	3.6	2.2	9.6	7.5	620	3,370 ppb
Cd	2.4	1.9	2.8	1.7	1.7	2.8	5.5	27	950 ppb
Cu+Pb+Zn	10	4.8	24	7.3	5.5	18	22	120	35,000 ppb
Hg	8.1	<3.7	5.0	3.8	5.1	26	10	160	40 ppb
Ni	1.4	0.7	0.6	0.8	0.8	0.7	1.4	2.8	1,740 ppb
Phenol	<6.4	<5.2	4.3	<2.1	12	11	9.4	4.5	<33 ppb
Pentachlorophenol	1.7	<2.3	<2.1	<0.90	U1.9	5.6	<1.9	<1.0	U33 ppb
LPAH	<45	<28	<68	<60	<73	<110	<120	<87	<41 ppb
HPAH	<120	<42	<65	<68	<27	<97	<140	<85	<79 ppb
Chlor. benzenes	9.9	<4.4	2.6	<2.5	<1.8	<6.1	<9.0	<3.3	U21 ppb
Chlor. butadienes	130	<2.7	<2.4	<1.2	<1.3	<6.8	<1.9	<1.7	U62 ppb
Phthalates	4.0	<2.6	<0.58	<0.66	<0.56	<5.1	<7.1	4.5	<280 ppb
PCBs	<48	<6.3	7.6	6.6	<17	8.5	<12	19	<6.0 ppb
4-Methylphenol	<7.3	<12	10	13	1,300	<33	30	<10	<13 ppb
Benzyl alcohol	5.0	<2.2	2.4	3.4	<6.7	3.3	4.7	<1.2	U10 ppb
Benzoic acid	<0.7	<3.2	<0.5	<0.7	<1.0	U0.1	<1.6	<0.5	<140 ppb
Dibenzofuran	29	25	73	59	52	80	58	<160	U3.7 ppb
Nitrosodiphenylamine	2.1	<2.4	<7.3	U1.0	U1.2	<1.0	14	<22	U4.1 ppb
Tetrachloroethene	12	<0.6	U1.0	--	U1.0	--	U1.0	--	U10 ppb
Sediment Toxicity									
Amphipod bioassay	2.1	1.9	2.9	2.4	4.8	1.4	2.7	3.9	9.3%
Oyster bioassay	2.2	1.6	1.3	1.4	3.8	1.8	2.6	2.2	13.0%
Infauna									
Total benthos	1.2	1.0	0.7	0.8	1.9	1.5	0.7	0.6	
Polychaetes	0.6	1.0	0.4	0.7	1.5	0.7	0.8	0.5	
Molluscs	3.4	1.0	1.4	1.1	6.8	5.4	2.5	1.2	
Crustaceans	1.5	1.0	3.8	0.4	1.0	4.6	1.2	0.7	
Fish Pathology									
Lesion prevalence	3.6	2.5	3.5	3.7	2.7	5.7	1.7	2.1	6.7%
Fish Bioaccumulation									
Copper	5.6	1.0	4.0	2.3	9.1	1.0	3.8	2.5	U38 ppb
Mercury	1.5	0.93	0.80	1.6	0.76	1.3	0.82	0.96	U55 ppb
Naphthalene	0.67	0.41	0.33	24	0.19	0.19	4.1	0.19	<54 ppb
Phthalates	21	11	0.53	3.6	0.41	0.41	6.7	5.6	<74 ppb
PCBs	9.2	7.0	4.8	2.8	1.1	4.7	9.8	1.9	<36 ppb
DDE	3.8	5.1	3.3	3.4	1.7	1.7	6.2	2.9	<1.8 ppb

NOTE:

[a]Boxed numbers represent elevations of chemical concentrations that exceed all Puget Sound reference area values, and statistically significant toxicity and biological effects at the $P < 0.05$ significance level compared with reference conditions. The "U" qualifier indicates the chemical was undetected and the detection limit is shown. The "<" qualifier indicates the chemical was undetected at one or more stations. The detection limit is used in the calculations.

- Is there a significant increase in sediment contamination, sediment toxicity, or biological effects at any study site?
- What combination of indicators is significant?
- What are the relative magnitudes of the elevated indices (i.e., which represent the greatest relative hazard)?

Evaluation of the action assessment matrix for Commencement Bay waterways (Table 1) revealed many significant areas of contamination and effects. For example, one or more metals were significantly elevated in all areas except St. Paul Waterway. All areas contained at least several significant organic contaminants, with maximum levels (averaged over the waterways) over 100 times higher than reference concentrations. The matrix indicates that Blair and Milwaukee waterways had the least chemical contamination, based on the number and magnitude of significantly elevated chemical indices.

Average sediment toxicity, based on one or both of the bioassay indices, was significantly elevated in all areas except Middle Waterway. The maximum toxicity occurred in St. Paul Waterway, where amphipod toxicity was 4.8 times the reference level of 9.3 percent (i.e., the average toxicity in the waterway was about 45 percent). Significant depressions in infaunal abundances were detected in five of the eight study areas. Depressions in the abundances of molluscs were the most frequently observed effect.

Evaluation of the fish pathology and bioaccumulation indices indicated that St. Paul Waterway and the Ruston shoreline area had the least impacted fish assemblages. The value of these two indices as independent measures of contaminant effects is also indicated in the matrix, since the highest lesion prevalence occurred in an area (Middle Waterway) where there was no significant bioaccumulation in English sole. The probable explanation is that the chemicals suspected as important causative agents in the development of liver lesions (e.g., PAH) are not bioaccumulated because of rapid metabolism in fishes.

Overall, the matrix demonstrated that Hylebos Waterway had the largest number of significant indicators (significant EAR for 18 chemicals or chemical groups and eight toxicity or biological effects indicators). The lowest number of significant indicators averaged over a study area was found in St. Paul Waterway.

The chemical and biological data for Commencement Bay indicated that contamination, toxicity, and biological effects were heterogeneous within the eight primary study areas. Therefore, the waterways were subdivided into segments based on patterns of chemical contamination. Action assessment matrices were then constructed for each segment.

DEFINITION AND RANKING OF TOXIC PROBLEM AREAS

Toxic problem areas were defined as those areas with sufficient evidence of contamination and biological effects to warrant the evaluation of contaminant sources and possible remedial alternatives. The identification of these problem areas required the specification of criteria that could be applied to the action assessment matrices. Such criteria were specified as "action levels" for combinations of contamination and effects indices that would result in problem area identification. Exceedance of action levels was dependent on specific combinations of the indices being significantly elevated in the area or segment matrices. It was assumed that an area or segment requires no action unless at least one of the indicators of contamination, toxicity, or biological effects was significantly elevated above reference conditions.

In this approach, problem areas are defined according to two basic criteria: (1) the number of indicators that are significantly elevated, and (2) the magnitude of elevation of each indicator. The action level criteria are summarized as follows:

- Significant elevation above the reference for three or more indices identifies a problem area requiring evaluation of sources and potential remedial action.
- For any two indices showing significant elevations, the decision to proceed with source and remedial action evaluations depends on the actual combination of indices and the relative degree to which they are site specific.
- Even when only a single index is significantly elevated, a problem area may be defined when additional criteria are met (i.e., the magnitude of the index is sufficiently above the significance threshold to warrant further evaluation).

Specific action-level guidelines for the Commencement Bay project are presented in Table 2. Application of these guidelines resulted in classification of all of the areas and segments as problem areas. All areas would therefore require further evaluation. A ranking system was also implemented to identify the problem areas posing the greatest environmental or public health risks. The ranking system was independent of the criteria used to define the problem areas.

The criteria for ranking problem areas are based on numerical scores of 0 to 4 for each indicator (Table 3). Rank scores for each problem area or segment were based on the sum of individual scores for each indicator. All areas and segments were ranked according to average conditions within the areas and on a worst-case basis according to the maximum observed value at any station within the area. Based on this approach, eight high-priority problem areas were identified (Figure 4), including three within Hylebos Waterway, two within City Waterway, one within each of Sitcum and St. Paul Waterways and along the Ruston shoreline. These areas all exhibited significant contamination, toxicity, and benthic effects, in addition to having at least one significant indicator of fish pathology or bioaccumulation. The lowest ranking problem areas did not exhibit significant sediment toxicity or benthic effects.

The boundaries of the problem areas were established based on data collected in the Commencement Bay investigations as well as from the historical chemical and toxicity data available for the area. Application of Apparent Effects Thresholds (AET) were especially useful for defining boundaries based entirely on sediment chemistry data. The

TABLE 2 Action-Level Guidelines

	Condition Observed	Threshold Required for Action
I.	Any THREE OR MORE significantly elevated indices[a]	Threshold exceeded, continue with definition of problem area.
II.	TWO significantly elevated indices	
	1. Sediments contaminated, but below 80th percentile PLUS:	No immediate action. Recommend site for future monitoring.
	Bioaccumulation without an increased human health risk relative to that at the reference area, OR	
	Sediment toxicity with less than 50 percent mortality or abnormality, OR	
	Major benthic invertebrate taxon depressed, but by less than 95 percent	
	2. Sediments contaminated but below 80th percentile PLUS elevated fish pathology	Threshold for problem area definition exceeded if elevated contaminants are considered to be biologically available. If not, recommend site for future monitoring.
	3. Any TWO significantly elevated indices, but NO elevated sediment contamination	Conduct analysis of chemistry to distinguish site from adjacent areas. If test fails, no immediate action warranted. Otherwise, threshold exceeded for characterization of problem area. Re-evaluate significance of chemical indicators.
III.	SINGLE significantly elevated index	
	1. Sediment contamination	If magnitude of contamination exceeds the 80th percentile for all study areas, recommend area for potential source evaluation at a low priority relative to areas exhibiting contamination and effects.
	2. Bioaccumulation	Increased human health threat, defined as: prediction of ≥ 1 additional cancer cases in the exposed population for significantly elevated carcinogens, OR
		For noncarcinogens, exceedance of the acceptable daily intake value is required.
	3. Sediment toxicity	Greater than 50 percent response (mortality or abnormality).
	4. Depressed benthic abundance	95 percent depression or greater of a major taxon (equals an EAR of 20 or greater).
	5. Fish pathology	Insufficient as a single indicator. Recommend site for future monitoring. Check adjacent areas for significant contamination, toxicity, or biological effects.

NOTES:
[a] Combinations of significant indices are from independent data types (i.e., sediment chemistry, bioaccumulation, sediment toxicity, benthic infauna, fish pathology).

Significant indices are defined as follows: Sediment chemistry = chemical concentration at study site exceeds highest value observed at any Puget Sound reference area.

Sediment toxicity, benthic abundance, bioaccumulation, and pathology = statistically significant ($p < 0.05$) difference between study area and reference area.

TABLE 3 Summary of Ranking Criteria for Sediment Contamination, Toxicity, and Biological Effects Indicators

Indicator	Criteria	Score
Total Metals Contamination	Concentration not significant	0
	Significant; EAR < 10	1
	Significant; EAR 10-< 50	2
	Significant; EAR 50-< 100	3
	Significant; EAR > 100	4
Total Organic Compound Contamination	Concentration not significant	0
	Significant; EAR < 10	1
	Significant; EAR 10-< 100	2
	Significant; EAR 100-< 1,000	3
	Significant; EAR > 1,000	4
Toxicity[a]	No significant bioassay response	0
	Amphipod or oyster bioassay significant	2
	Amphipod and oyster bioassay significant	3
	$\geq$ 50 percent response in either bioassay	4
Macroinvertebrates (abundance)[b]	No significant depressions	0
	1 significant depression	1
	2 significant depressions	2
	$\geq$ 3 significant depressions	3
	$\geq$ 1 taxon with >95% depression	4
Bioaccumulation (fish muscle)	No significant chemicals	0
	Significant chemical	1
	2 significant chemicals	2
	$\geq$ 3 significant chemicals	3
	Significant bioaccumulation of $\geq$ 1 chemical posing a human health threat[c]	4
Fish Pathology (liver lesions)[d]	No significant lesion types	0
	1 significant lesion type	1
	2 significant lesion types	2
	$\geq$ 3 significant lesion types	3
	$\geq$ 5% prevalence of hepatic neoplasms	4
Maximum Possible Score		24

NOTES:
[a] Toxicity based on amphipod mortality and oyster larvae abnormality bioassays.
[b] Taxa considered were total benthic taxa, Polychaeta, Mollusca, and Crustacea.
[c] As defined in Table 2 (Action-Level Guidelines).
[d] Lesions considered were hepatic neoplasms, preoplastic nodules, megalocytic hepatosis, and nuclear pleomorphisms.

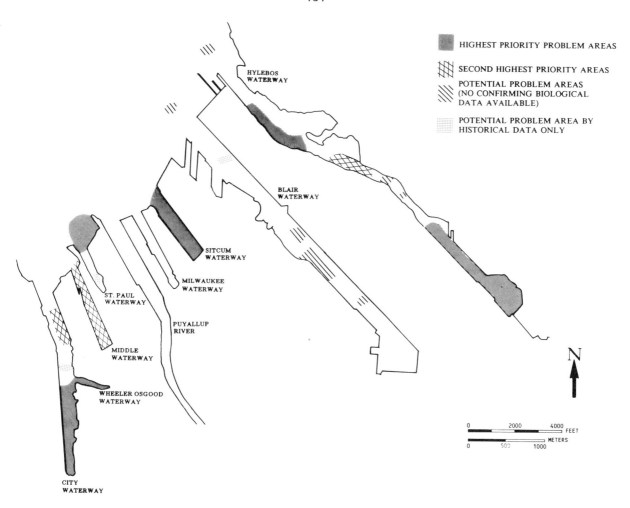

FIGURE 4 Definition and prioritization of Commencement Bay problem areas.

development and application of AET are discussed in Barrick et al. (these proceedings).

IDENTIFICATION AND RANKING OF PROBLEM CHEMICALS

The Commencement Bay investigations indicated that area sediments were contaminated by numerous inorganic and organic chemicals at levels substantially above Puget Sound reference conditions. Because of the extensive list of sediment contaminants, a procedure was developed to identify and rank problem chemicals so that source and cleanup evaluations could be focused on the chemicals posing the greatest environmental or public health risk.

The overall prioritization of sediment contaminants is described in Figure 5. Of all detected chemicals, chemicals of concern are defined as chemicals with concentrations exceeding all Puget Sound reference conditions. These chemicals are not necessarily considered problem chemicals, because sediments may be contaminated above reference conditions without exhibiting toxicity or biological effects. However, chemicals that are detected at concentrations exceeding 80 percent of the values determined for all stations within the project area were of greater concern based entirely on the magnitude of contamination. Such chemicals may be subject to source evaluations depending on available resources and the identification of problem chemicals.

Problem chemicals were defined as those chemicals whose concentration exceeded the AET in the problem area. Because the AET was defined as the contaminant concentration above which toxicity or benthic effects are always observed, chemicals present above this threshold may be contributing to observed biological effects. Problem chemicals were further ranked according to their association with toxicity or biological effects. Based on this approach, three priorities of problem chemicals were given for each problem area. The highest priority (Priority 1) chemicals are those that were present above an AET in a problem area and that also exhibited a concentration gradient corresponding to observed changes in sediment toxicity or benthic effects. For example, strong linear relationships were found between sediment toxicity and PCB concentrations in Hylebos Waterway and between sediment toxicity and 4-methylphenol concentrations in St. Paul Waterway. Other contaminants were found above AET in these problem areas, but none displayed these strong relationships with sediment toxicity. Therefore, these chemicals were given the highest priority for source evaluation and cleanup actions because of their demonstrated correspondence with the observed tox- icity. It is recognized that some unidentified contaminant(s) with similar distributions may have been the actual problem chemicals in these areas. However, source identifications for PCBs and 4-methylphenol would still be recommended based on the assumption that the problem chemical came from the same source(s), and that corrective action at the source may effectively control its release as well as the release of the identified target chemical.

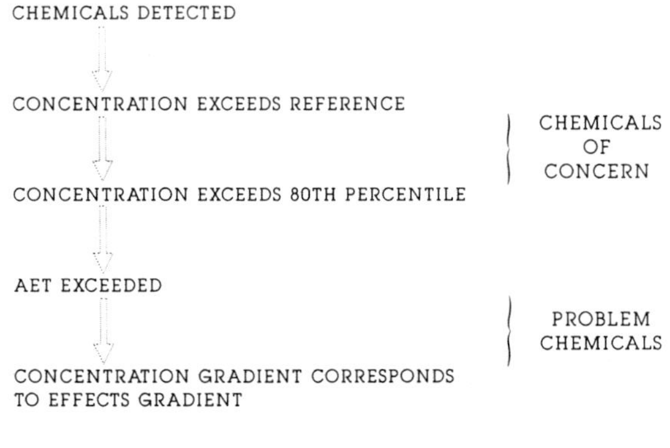

FIGURE 5 Prioritization of problem chemicals.

```
CHEMICALS DETECTED
        ↓
CONCENTRATION EXCEEDS REFERENCE           ⎫
        ↓                                 ⎬ CHEMICALS
CONCENTRATION EXCEEDS 80TH PERCENTILE     ⎭  OF CONCERN
        ↓
AET EXCEEDED                              ⎫
        ↓                                 ⎬ PROBLEM
CONCENTRATION GRADIENT CORRESPONDS        ⎭ CHEMICALS
TO EFFECTS GRADIENT
```

Priority 2 chemicals were defined as those that occur above the AET in the problem area but show no particular relationship with effects gradients (or insufficient data were available to evaluate their correspondence with gradients). Chemicals with concentrations above AET only at nonbiological stations were therefore placed no higher than Priority 2 because of the lack of biological data. Finally, chemicals with concentrations above AET at only one station within the problem area were assigned Priority 3. Problem chemicals for problem areas that were small hot spots of sediment contamination usually fell into this category.

Using this approach, Priority 1 chemicals were identified in six of the eight highest priority problem areas. These chemicals included mercury, lead, zinc, and arsenic; and PCBs, 4-methylphenol, low-MW PAH, and high-MW PAH.

Priority 1 chemicals were not identified in any of the remaining 12 problem areas with lower overall priority. For most of these areas there were insufficient sampling stations to establish correspondence between sediment contamination and biological effects.

Priority 2 chemicals were identified in all eight of the highest priority problem areas. Priority 2 chemicals were also identified in three of the lower priority problem areas. These chemicals included cadmium, nickel, and antimony; and hexachlorobutadiene, chlorinated benzenes, chlorinated ethenes, phenol, 2-methylphenol, N-nitrosodiphenylamine, dibenzofuran, selected phthalate esters, and selected tentatively identified compounds (e.g., 2-methoxyphenol).

SUMMARY AND RECOMMENDATIONS

In Commencement Bay, the complex situation of sediment contamination by multiple contaminants resulting from numerous point and nonpoint sources of pollution is typical of many urban coastal areas throughout the United States (e.g., New York Harbor, Los Angeles/Long Beach Harbor, and San Francisco Bay). Such situations require an initial assessment and ranking of contaminants and areas so that remedial resources can be directed appropriately.

A decision-making approach and associated criteria were developed for the Commencement Bay Superfund investigations to identify and rank problem areas and problem chemicals. This approach was successfully implemented in the study area. For this complex case of sediment contamination, it enabled regulatory agencies to focus source control and sediment cleanup activities on those sediments and contaminants posing the greatest environmental and public health hazards. This approach is recommended for other areas with contaminated sediments requiring similar assessments and prioritizations before initiating cleanup activities.

ST. PAUL WATERWAY REMEDIAL ACTION
AND HABITAT RESTORATION PROJECT

Jerry K. Ficklin, Simpson Tacoma Kraft
Don E. Weitkamp, Parametrix, Inc.
Ken S. Weiner, Preston, Thorgrimson, Ellis & Holman

ABSTRACT

The Simpson Tacoma Kraft Company's St. Paul Waterway Area Remedial Action and Habitat Restoration Project consists of environmental improvement actions to remedy past and present practices to protect Commencement Bay and Puget Sound Water quality and restore fish and wildlife habitat. The project will correct sediment contamination by permanently capping this shallow area with clean Puyallup River sediments; initiating chemical source control through raw material and process changes; installing a new secondary treatment plant outfall; collecting and providing secondary treatment for stormwater; containing wood chip spillage; and creating substantial new intertidal habitat for bird and marine life. The studies resulting in this action concluded that the project would create few adverse impacts and provide greater environmental benefits than other remedial action alternatives, such as dredging contaminated sediments, then landfilling or incinerating them. The project began in December 1987 and was completed in September 1988, after 18 months of planning and preliminary consultation with agencies, the Puyallup Indian Tribe, mill workers, environmental groups, citizens, and scientists.

THE TACOMA KRAFT MILL AND NEARSHORE CONTAMINATION

Simpson Paper Company Bought the Tacoma Kraft mill in late August 1985. The mill produces products common and essential to commercial and consumer use: paper (both bleached and natural), pulp, liner board, and similar materials. The mill has been operating since 1927, and was most recently owned by the Champion International and St. Regis corporations.

The mill is located on Commencement Bay in Tacoma, Washington. It is built on a 57-acre peninsula of filled tidelands between the mouths of the Puyallup River and the St. Paul Waterway. The area was originally an intertidal mudflat between two forks of the mouth of the Puyallup River (Figure 1). The original 1,750 acres of productive mudflats in Commencement Bay have been reduced to less than 100 acres through the bay's harbor development over the past 100 years. The bay is very shallow and calm near the mill, ranging in depth from a sandbar that is exposed at low tides to about a -20 ft mean sea level (MSL).

For 37 years, the mill's wastewater was discharged without treatment. Its wastewater started receiving primary treatment in the 1960s and has received secondary treatment for the past decade. Its outfall was located on the bank of the former log pond, an area immediately offshore used to store logs. Logs were stored, handled, debarked, and chipped with few controls to prevent runoff containing woody debris from entering the bay. There was also chip spillage into the bay from the unloading of barges and the storage of chips immediately adjacent to the bay (Figure 2).

The sediments of Commencement Bay next to the mill have become contaminated with chemicals and with organic debris (Figure 3). In the area near the outfall, chemicals toxic to marine life--such as phenolics, cresols, and cymenes--are the dominant contaminants. In this and other areas, logs, limbs, sawdust, wood chips, and similar organic materials blanket the bottom. Due to the organic and chemical contamination of these sediments, the general area was designated as one of the "hot spots" in Commencement Bay needing remedial action. The U.S. Environmental Protection Agency (EPA) has designated Commencement Bay's entire nearshore tideflats as a Superfund site on the National Priority List (Figure 4). The nearshore log pond area near the mill was included in the site based on results from the Commencement Bay studies conducted by Tetra Tech, Inc. (consultants for EPA) and the Washington Department of Ecology (WDOE).

PROJECT DESCRIPTION

The objectives of the St. Paul Waterway Area Remedial Action and Habitat Resotration Project are to

- install an improved outfall.
- permanently isolate the sediment in an environmentally acceptable and cost-effective manner,
- restore and preserve intertidal habitat, and
- take preventative measures against future sediment contamination from the mill.

Corollary objectives are to minimize dredging of contaminated sediment, preserve existing water dependent and harbor uses, use reliable and appropriate technology, and design the project to complement the natural forces at work in Commencement Bay.

FIGURE 1 Commencement Bay tideflats (1886) and outline of 1986 shoreline.

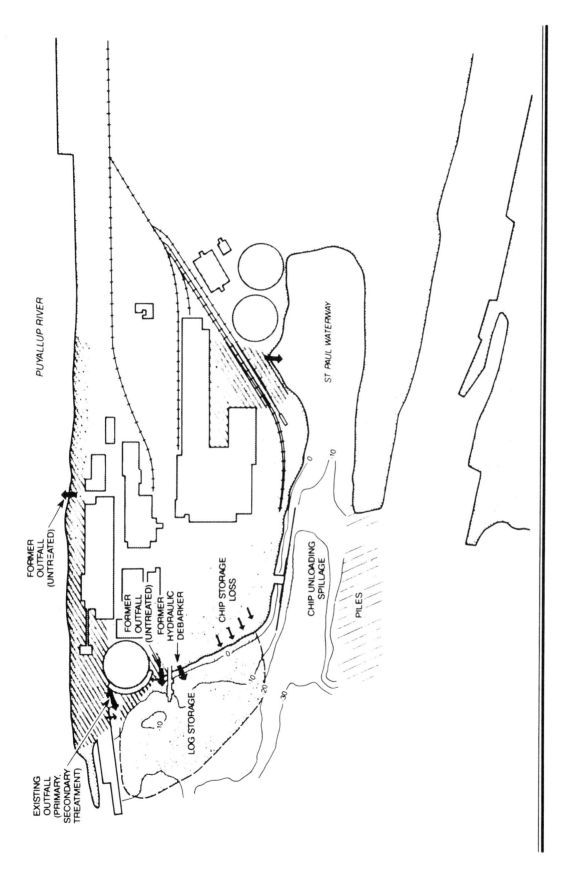

FIGURE 2 Sources of sediment contaminants from the Tacoma Kraft mill.

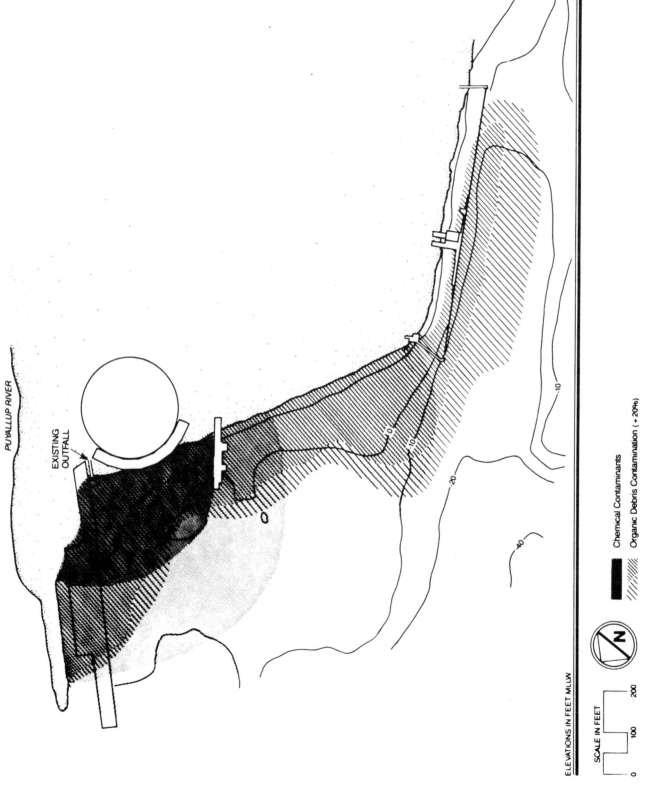

FIGURE 3 General areas of concern.

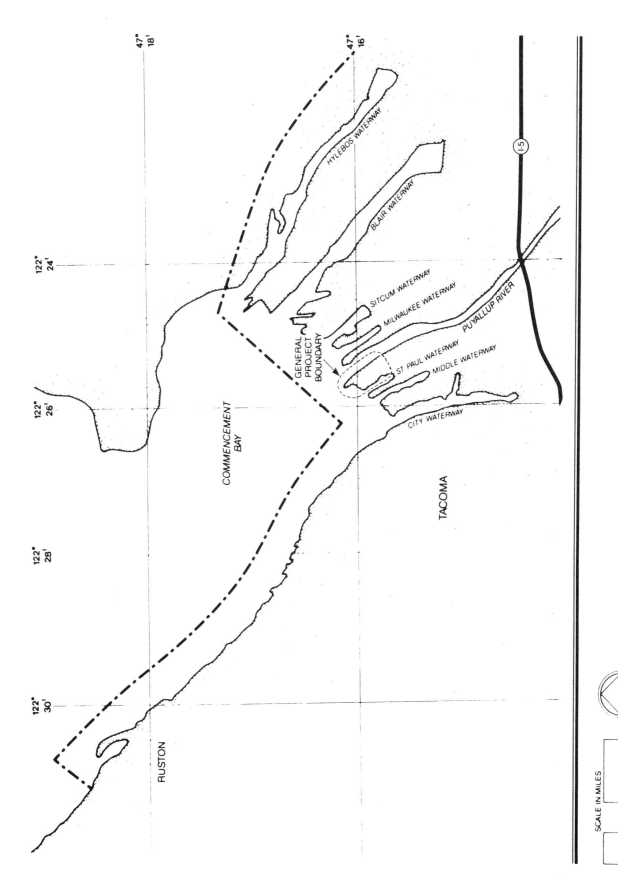

FIGURE 4 Vicinity map of Commencement Bay nearshore/tideflats Superfund study area.

The main components of the project are

1. new secondary treatment plant outfall with diffuser,
2. remedial action for contaminated sediments and habitat restoration,
3. source control by stormwater collection and treatment, chip containment, and an ongoing chemical source control program, and
4. monitoring for the physical, chemical, and biological effectiveness of the implemented actions.

These components are described below and illustrated in Figure 5.

NEW OUTFALL

The existing outfall for the mill's secondary treatment plant is permitted to discharge approximately 30 million gallons per day (MGD) through a discharge structure located immediately adjacent to the shoreline of Commencement Bay. This structure is within 0 to 12 ft of the water's surface, depending on the tide. The initial dilution from the existing outfall was estimated to be in the range of 2:1 to 5:1, the ratio of seawater to effluent.

This low level of initial dilution allowed flocculation to occur, where dissolved material and small suspended particles combine. The flocculated material tended to sink and accumulate in the vicinity of the outfall, which contributed to the contamination of sediment there. Pilot plant studies determined that the minimum dilution required to prevent flocculation was approximately 20:1. Although the wastewater that emerges from the mill's secondary treatment system is well-treated, it is important for natural decompositional processes to continue by having good dilution and dispersion after discharge.

The new outfall will take advantage of the dominant current patterns within Commencement Bay to move the diluted effluent offshore, away from the more sensitive shoreline environment. Adequate dilution, offshore movement, and monitoring will ensure that a problem is not simply being moved from one place to another.

The proposed outfall will be moved from the old log pond area and extended 920 ft offshore from the northwest corner of the mill to a depth of about -70 ft MSL. The first 220 ft will be buried below the mud line for engineering and regulatory reasons, which will require dredging about 3,000 yd^3. This new location will enable the outfall to have a minimum initial dilution of 55:1, with the initial dilution normally expected to be 70:1 or greater. In addition, at this depth, location, and initial dilution, the effluent will be trapped in a layer of water moving offshore. The outfall planning studies found that other alternatives, including deeper outfalls with higher initial dilutions, provided less shoreline protection because they trapped in water moving shoreward.

The new outfall will therefore eliminate the low dilution, thus preventing flocculation, and will ensure that the effluent is dispersed in an offshore direction that minimizes the opportunity for it to be

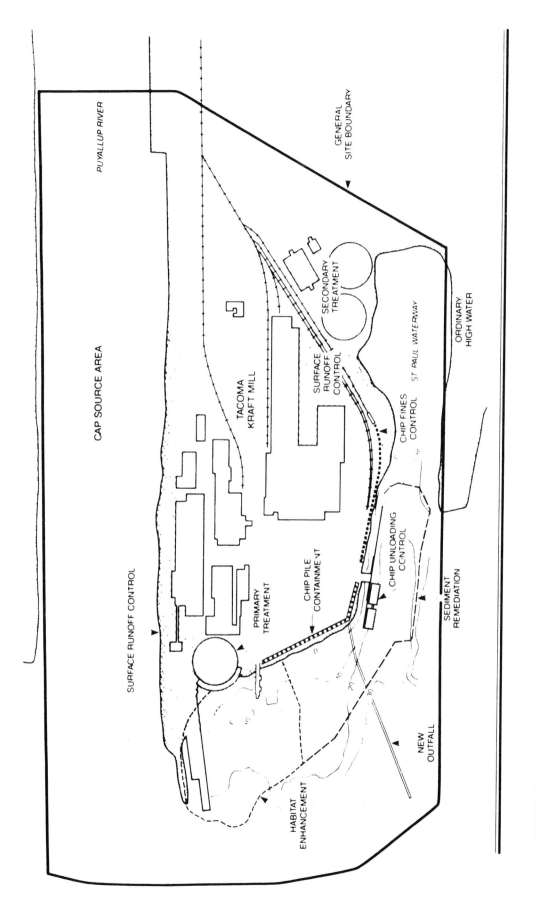

FIGURE 5 Site plan and basic elements of the project for the Tacoma Kraft mill.

carried back to shore. WDOE reviewed and approved the environmental studies and engineering plans for the outfall, which was constructed and placed into operation during February 1988.

REMEDIAL ACTION

Sediments near the mill can be viewed as having three general areas of contamination, with each area blending into the next (Figure 6).

1. Area A contains chemical contamination with some organic debris. Located near the old outfall discharge, it is about a five-acre area encompassing roughly half of the old log pond.
2. Area B contains a high level of organic debris intermixed with sandy, silty sediment, with some chemical contamination. The high level of organic material binds the sediment together, making it very cohesive. Area B is located on the southerly half of the old log pond, between area A and the St. Paul Waterway.
3. Area C is a blanket of wood chips on the bottom of the bay at the entrance to St. Paul Waterway, resulting from spillage of wood chips from the chip unloading facility.

As noted above, the contaminated sediments were brought to public attention through the Superfund designation of Commencement Bay and the subsequent Nearshore/Tideflats Remedial Investigation prepared by Tetra Tech under the direction of WDOE and EPA. Although hazardous waste conditions or public health threats do not exist, the sediments of the St. Paul Waterway are sufficiently contaminated to severely depress biological populations in the vicinity.

Many of the chemicals identified by Tetra Tech in the remedial investigation were not those typically associated with a paper mill or the pulping processes. After detailed analysis conducted during the mill's source control efforts, it was discovered that raw materials supplied to the mill contained those chemicals. Studies have established that the highest level of chemical contamination is close to the old mill outfall and that the level of contamination decreases rapidly with distance from the old discharge point (Figures 7-14). Organic debris, however, is a problem for marine life over a much greater portion of the area than are the chemical contaminants (Figure 14). Most of the chemical contamination is within the area of high organic contamination (Figure 3).

Permanent and significant reduction in volume, toxicity, or mobility of contamination is a principal goal of a remedial action under Superfund. Remedial actions must be cost-effective and consider, among other factors, long-term potential threat to human health and the environment. Other impacts associated with excavation, on-site treatment, transportation, and ultimate disposal must also be considered.

The St. Paul Waterway site is unusual compared to most areas of contamination in Puget Sound because there is no foreseeable need to dredge the site. The absence of a need to dredge the site allows the

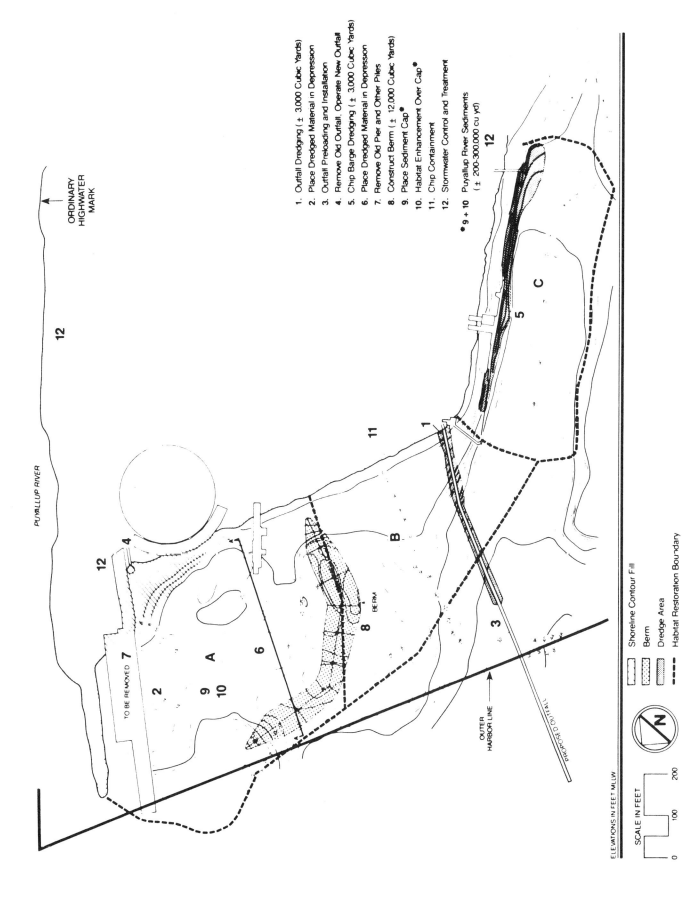

FIGURE 6 Sequence of remedial actions and source control.

use of alternatives that involve capping in place. These alternatives could not be employed at locations that require dredging for new deep water moorage or maintenance dredging for navigational purposes. The proposed action will minimize the environmental impacts of stirring up the contaminated sediments. The absence of the need for dredging also offers an opportunity for habitat enhancement.

The St. Paul Waterway site is also unusual compared with many areas of contamination in Puget Sound because of its shallow depth (less than 20 ft deep). Shallow water allows better control of dredging activities as well as more observable monitoring. The shallow water also offers an opportunity to restore intertidal habitat.

The proposed action will, without disturbing the contaminated sediments, cover them with a cap of clean Puyallup River sediment. Puyallup River sediment is currently building a bar at the northeast corner of the area. Over 30 ft of deposition has occurred in this area over the past 30 to 40 years. With the surface currents generated by the old outfall discharge (30 MGD) no longer present, Puyallup River sedimentation in the area is expected to increase. This sedimentation would naturally cap the contaminated sediments: however, such a time frame is not acceptable.

The remedial action will be tailored to the nature and location of contaminated sediment. Area A will receive an 8- to 12-ft cap of clean Puyallup River sediment. During the outfall relocation and site preparation work this past winter contaminated sediment from the chip barge unloading area and from the trench for the outfall line was placed in bottom depressions in area A and then covered with 2 ft of clean cover material to temporarily isolate the material until final cover is placed this summer [1988]. This summer Area A will be covered by a 4-ft layer of clean sediment to chemically isolate the contaminated material. Another 4- to 8-ft of sediment will be placed on top of the remedial action cap to raise the elevation to intertidal depths. This cap will be contoured as a habitat enhancing feature and to work with natural forces (tides, storms, sediment accretion)(Figure 15).

Area B will receive a 4-ft cap to isolate the organic debris, lower levels of chemical contamination from biological activity, and provide new benthic substrate. This depth will provide an additional 25 percent of clean substrate than is normally considered needed for biological protection. Area C, the area covered with wood chips, will receive a 2-ft cap of clean material to provide new substrate for benthic organisms. The purpose of the Area C cap is not to isolate the bottom, but to provide a suitable base for new marine life.

Finally, the entire area will be randomly strewn with boulders in order to provide not only a new habitat that varies with depth, but one which has a varying substrate. The habitat enhancement measures will create a new sandy, mudflat area similar to what has been lost on the outer edge of the Puyallup River delta over the past 100 years of harbor development (Figure 1). This restoration will provide valuable rearing habitat for anadromous migrants and juvenile marine species, as well as feeding habitat for shorebirds (Figure 15). Current plans do not include intervention with the natural biological repopulation of the area. The expectation is that, given the high productivity of this

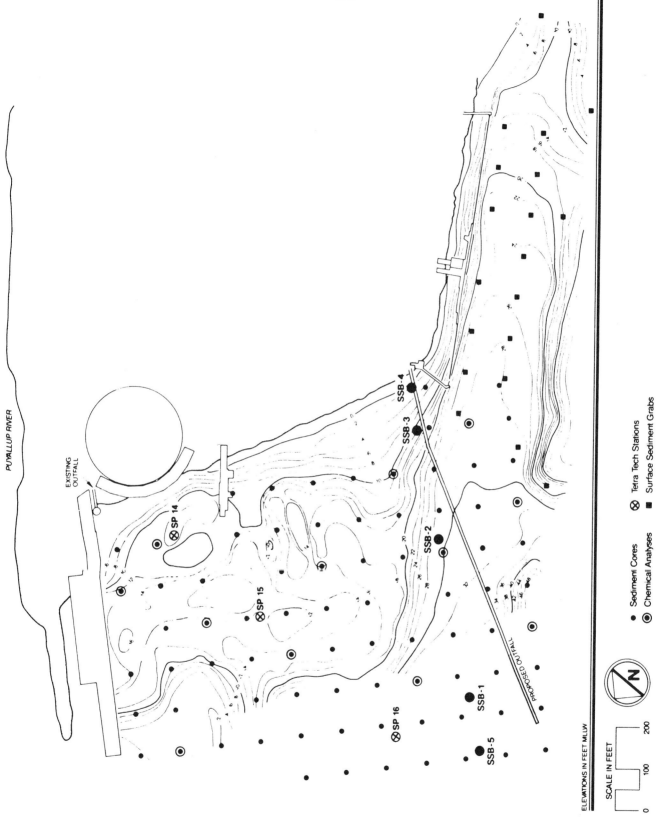

FIGURE 7 Sampling grid adjacent to the Tacoma Kraft mill.

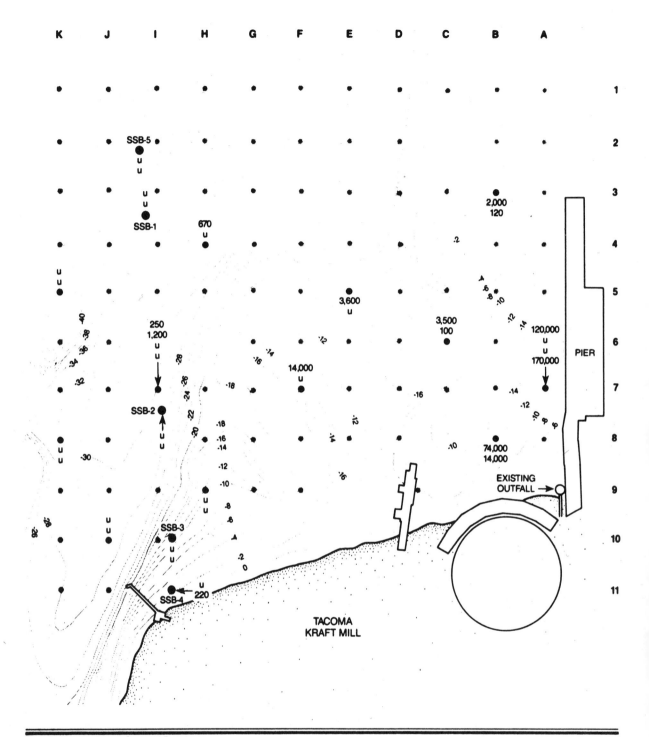

FIGURE 8 P-Cresol concentration in sediments adjacent to the Tacoma Kraft mill (amphipod AET, 1,200 µg/kg; benthic AET, 670 µg/kg).

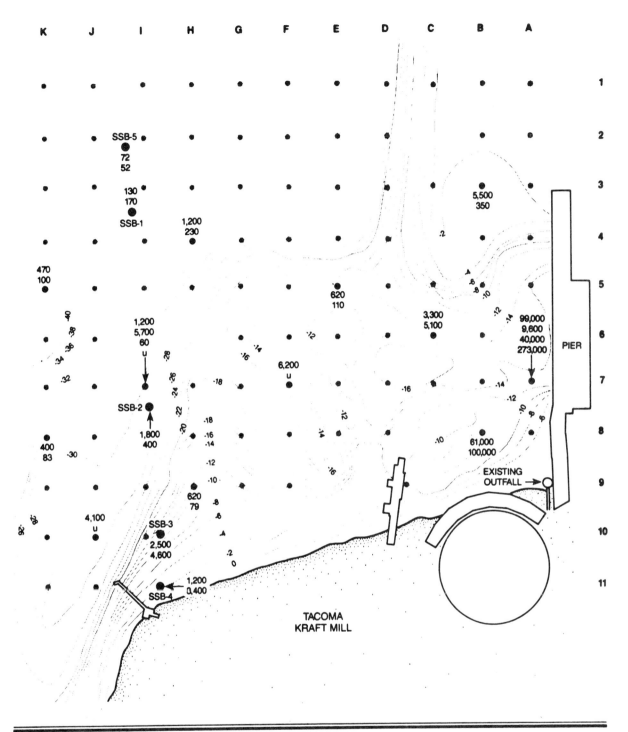

FIGURE 9 P-Cymene concentration in sediments adjacent to the Tacoma Kraft mill (amphipod AET, undetermined).

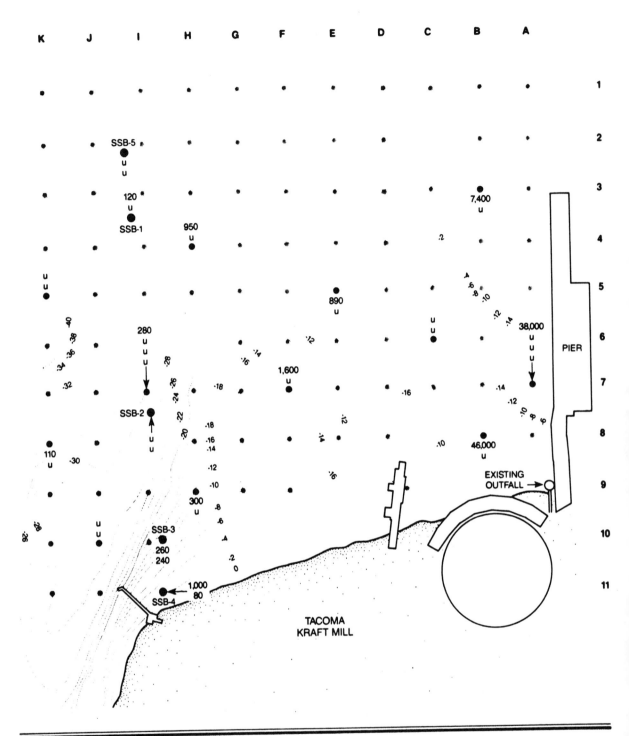

FIGURE 10 Guaicol concentration in sediments adjacent to the Tacoma Kraft mill (amphipod AET, 930 µg/kg).

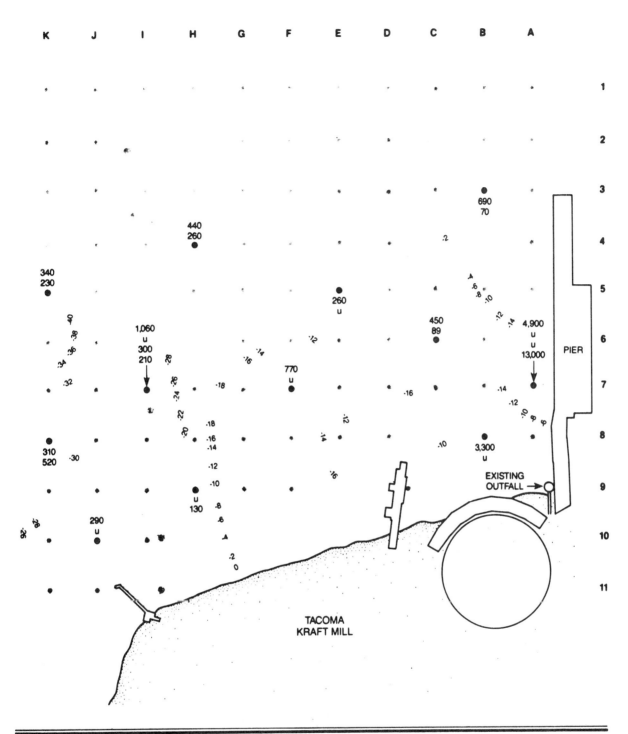

FIGURE 11 Phenol concentration in sediments adjacent to the Tacoma Kraft mill (amphipod AET, 670 µg/kg).

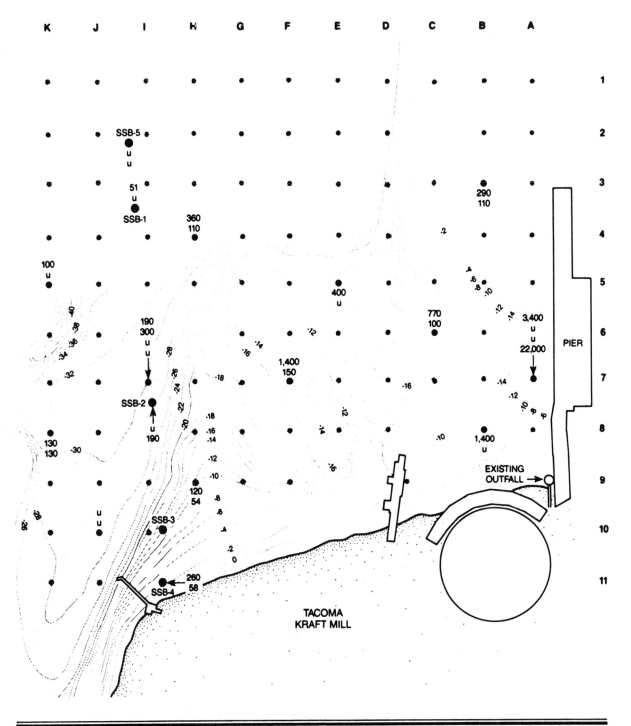

FIGURE 12 Napthalene concentration in sediments adjacent to the Tacoma Kraft mill (amphipod AET, 2,400 μg/kg).

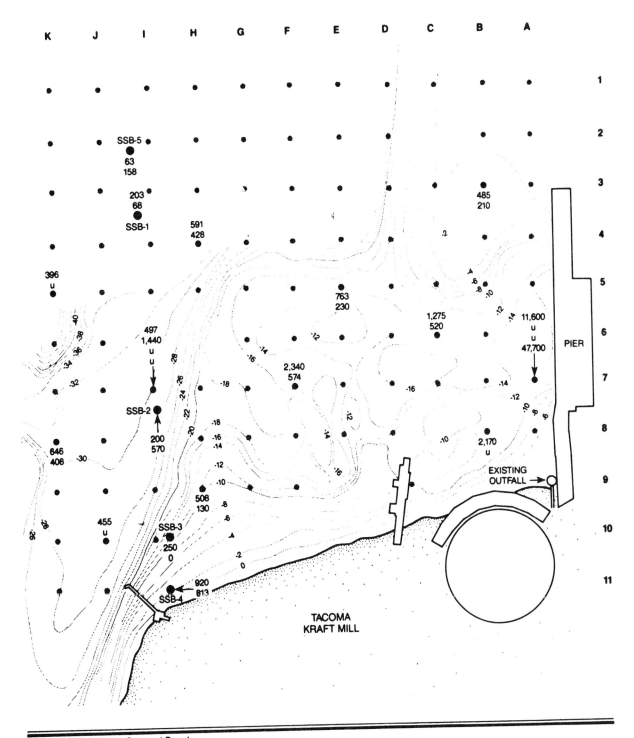

FIGURE 13 Low-molecular-weight polyaromatic hydrocarbons concentration in sediments adjacent to the Tacoma Kraft mill (amphipod AET, 5,500 µg/kg).

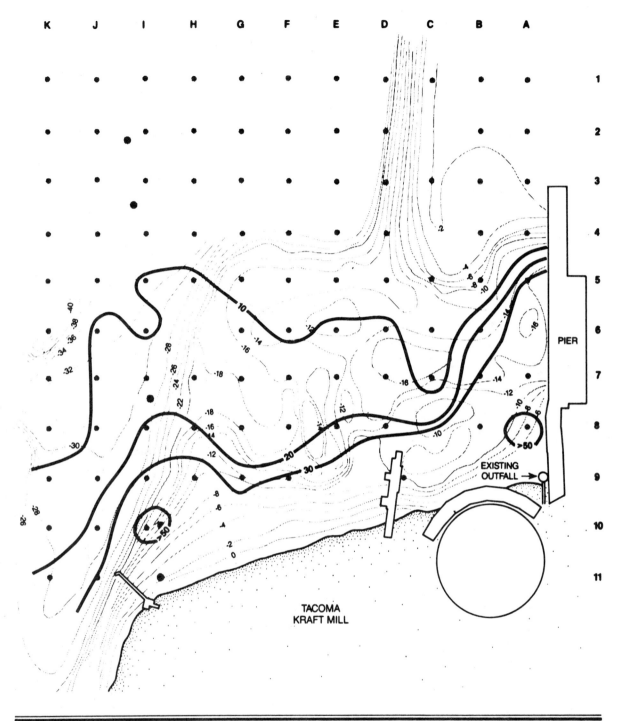

FIGURE 14 Total volatile solids (percent isopleths) observed in surface sediments adjacent to Tacoma Kraft mill.

Puget Sound estuary, marine life will rapidly reestablish itself under natural physical conditions (clean sediments, cobbles, boulders, and varying topography).

SOURCE CONTROL

The mill has undertaken a source control program on several fronts. Source control involves preventative action to reduce or eliminate chemicals or other substances that are potential pollutants. Much of the effort to date has been directed toward reducing or eliminating contaminants in raw materials brought to the plant. The other primary effort has involved changing manufacturing processes to control or reduce chemicals of concern.

Examples of source control measures implemented are as follows:

- Reducing the concentration of copper contaminating the mill's dominant source of purchased sodium sulfate (Vanillin Black Liquor), from 10,000 mg/liter to less than 10 mg/liter, represents a reduction of more than one-half million pounds of copper to the mill each year.
- Eliminating a sodium hypochlorite stage from the bleaching sequence, by adding more efficient chemical-pulp mixers, and by adding an oxygen extraction stage, resulted in a 68 percent reduction in the aqueous discharge of chloroform.
- Ongoing measures that have significant source control benefits include replacing the 1937 vintage bleach plant with a new plant which, employing chlorine dioxide substitution for chlorine, state-of-the-art mixers at each bleaching stage, and scrubbers for air emissions, will effect an estimated 50 percent further reduction in chloroform emissions and reduce the total organically bound chlorine emissions to well below standards recently established for Swedish bleached kraft mills.

The source control program has removed well over a million pounds of pollutants from the facility on an annual basis.

The source control program also includes control of woody debris. Operation of the newly installed chip unloading facility significantly reduced chip spillage during chip barge unloading operations. The containment of wood chip fines (sawdust-sized material) will be improved this summer when fencing, improved conveyor cleaners, and water sprays have been installed. By the first of October 1988, all rainfall onto the facility will be collected and secondarily treated prior to discharge to Commencement Bay.

Source control must be an intrinsic part of any remedial action effort to prevent recontamination.

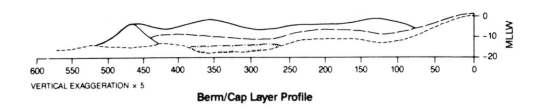

Berm/Cap Layer Profile

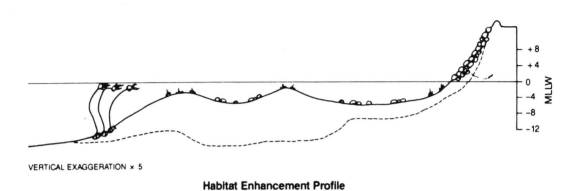

Habitat Enhancement Profile

Final Profile

——————— Habitat Enhancement Elevation
— — — — Cap Surface Elevation
—·—·—·— Organic Fill Elevation
- - - - - - - Existing Elevation

FIGURE 15 Typical profiles of perm and cap layers, habitat enhancement and final profile without vertical exaggeration.

MONITORING

A comprehensive monitoring program has been designed for the St. Paul Waterway Area Remedial Action and Habitat Restoration Project to determine whether

1. the outfall and remedial cap are working as designed,
2. the area is becoming recontaminated from the capped sediment, the mill, or other sources,
3. marine life is returning to the newly created habitat, and
4. whether any problems develop that may require corrective action.

The monitoring program will be directed by representatives from Washington State Departments of Ecology, Natural Resources, and Fisheries; the U. S. Army Corps of Engineers, Fish and Wildlife Service, and National Marine Fisheries Service; the Puyallup Indian Tribe; and Simpson Tacoma Kraft Company. The monitoring program will be most frequent during the first five years of the project, but will continue until contamination in the capped sediment has degraded to "acceptable" levels. The program will collect physical, chemical, and biological data to be compared to similar data collected from reference areas within Commencement Bay. The program can generally be divided into three stages: preconstruction, construction, and post-construction. The principal monitoring before construction will be to establish the biological and chemical baselines. Several types of construction monitoring will occur, including

- establishing sediment elevations,
- verifying proper sediment cover placement, and
- monitoring water quality to measure impacts occurring due to the remedial activities.

Post-construction monitoring will be the most extensive. In addition to detailed sampling near the new outfall diffuser, several different types of monitoring will occur within the remedial action area. That monitoring will focus on the physical integrity of the sediment cap by measuring depth of sediment cover and deposition of new sediment, chemical analysis of core samples to ensure absence of upward contaminant migration, and biological monitoring to document repopulation of the area.

DREDGING AND DISPOSAL OF CONTAMINATED MARINE SEDIMENT
FOR THE U.S. NAVY CARRIER BATTLEGROUP HOMEPORT PROJECT,
EVERETT, WASHINGTON

Edward Lukjanowicz, U. S. Navy
J. Richard Faris, U. S. Navy
Paul F. Fuglevand, Hart Crowser, Inc.
Gregory L. Hartman, Ogden Beeman & Associates, Inc.

ABSTRACT

In April 1984 the U.S. Navy selected the East Waterway of Port Gardner Bay in Puget Sound as the homeport site for a carrier battlegroup. Construction involves dredging over 3.3 million yd^3 of sediment; approximately 928,000 yd^3 of it is treated as contaminated. The Navy selected and has received permits to proceed with in-water disposal of the sediment using a method identified as confined aquatic disposal (CAD). Using the CAD technique, contaminated sediment will be deposited in depths of 400 ft and capped with uncontaminated sediment to isolate contaminants from the aquatic environment of Puget Sound. This summary case study, discusses the development of the CAD option and the stringent environmental monitoring requirements imposed on this disposal technique by federal and state regulatory agencies. These requirements are intended to safeguard ecologically sensitive Puget Sound and gather technical information on the effectiveness of CAD in deep water.

OVERVIEW

This summary case study draws from the many reports completed for this project. The primary documents are

- draft environmental impact statement (DEIS), November 1984
- final environmental impact statement (FEIS), June 1985
- sediment testing and disposal alternatives evaluation, June 1986
- public notice of Section 404 permit, October 1986
- draft supplements to FEIS (DSEIS), July 1986
- final supplements to FEIS (FSEIS), November 1986
- final dredging and disposal monitoring plan, phase I (Monitoring Plan), November 1987
- plans and specifications, environmental monitoring of dredge/disposal activities (Monitoring P&S), November 1987

Additional environmental and engineering studies have been undertaken and completed by the Navy. Some of these studies address the following topics:

- disposal site bioturbation,
- homeport master plan,
- homeport soils analysis,
- air quality modeling,
- slope enhancement for fisheries,
- seabird survey,
- crab surveys (ten seasonal trawls),
- CAD site benthic analysis,
- sediment sampling and analysis,
- characterization of East Waterway sediments,
- water column chemistry,
- physical model of East Waterway,
- confined aquatic disposal (CAD) feasibility analysis,
- Port Gardner bathymetric survey,
- Port Gardner current measurements
- leachate/sediment settlement tests,
- dump modeling,
- navigational plans for accurate sediment placement,
- preconstruction/construction/post-construction CAD site monitoring plan,
- Smith Island Upland-Dredge Disposal Feasibility Study & Evaluation,
- re-characterization of P-111 and P-905 sediment (clean/contaminated),
- geochemical evaluation of Norton Terminal,
- biological assessment for marine mammals, and
- biological assessment for bald eagles.

Project Development

The proposed project site is located in Puget Sound within the city of Everett, Washington. The site is located on the east side of Port Gardner Bay, just west of the central downtown area. Deep water for navigational purposes is available near the site, although dredging of the East Waterway would be required.

Operation of a carrier battlegroup homeport at the site would require newly constructed facilities to accommodate 13 ships including the aircraft carrier U.S.S. *Nimitz*, plus up to 10 additional small craft needed for support services. Location of a homeport facility at the 117-acre site in Everett would require construction of many support facilities.

Dredging and Disposal Preferred Alternative

Dredging of the East Waterway will be completed to depths necessary to accommodate homeported vessel drafts. All dredging will daylight to deeper depths into Port Gardner. No dredging or disposal will be done within designated "fish" windows from December 1 to June 15 of any construction year. Dredged material will be disposed of at the proposed CAD site, which is located in water depths of approximately 310 to 430 ft in Port Gardner, and which will impact an area of approximately 380 acres (Figure 1).

Total dredging volume is estimated at 3,305,000 yd^3, including 1 ft of overdepth. Of the total dredged volume, approximately 928,000 yd^3 of materials will be treated as contaminated, although only 486,900 yd^3 of in situ contaminated sediments exist. They contain organics, such as decomposed sawdust and wood chips, oils, grease, and industrial contaminants, such as polyaromatic hydrocarbons and polychlorinated biphenols. Within the total dredge volume, approximately 2,377,000 yd^3 are clean native materials and will be removed from the East Waterway to be used as cap and mound materials at the CAD site. The dredging volumes presented are estimates based on data from various studies. Actual dredging volumes may differ slightly due to minor variations encountered during construction, or any redefinition by regulatory agencies.

Standard equipment and methods will be used for dredging and disposal. Contaminated material will be dredged and disposed of using clamshell, tug, and bottom dump scow to ensure minimum induced turbidity and maximum compaction of contaminant mass on the bottom. Hydraulically dredged, native uncontaminated material will be used as capping material, the release rate and density of which will be controlled to prevent displacement of the deposited contaminated sediments through use of a floating pipeline with submerged diffuser.

The sequence and placement of the dredged materials into the CAD site are shown in Figure 2. The dredging operation will begin by removing approximately 500,000 yd^3 of uncontaminated material from the area of the carrier pier and breakwater. This material will be dredged by clamshell dredge and disposed of in such a manner as to create a mound (1). Contaminated material (approximately 97,000 yd^3) will be dredged and disposed of in a similar manner and placed (2) to the "uphill" side of the mound. Immediately thereafter, this material will be capped (3) using approximately 239,000 yd^3 of uncontaminated material. Capping will be by hydraulic pipeline dredge with disposal by submerged discharge. It is anticipated the cap thickness will be in excess of 3 ft.

The foregoing will complete the first year's dredging. Second-year dredging consists of approximately 831,000 yd^3 of contaminated material and will be dredged by clamshell and disposed of by surface bottom dump barge (4). Finally, and in sequenced manner, 1,638,000 yd^3 of uncontaminated material will be hydraulically dredged and disposed of by pipeline as represented by (5). It is anticipated the final minimum capping thickness over contaminated material will be 4.5 ft.

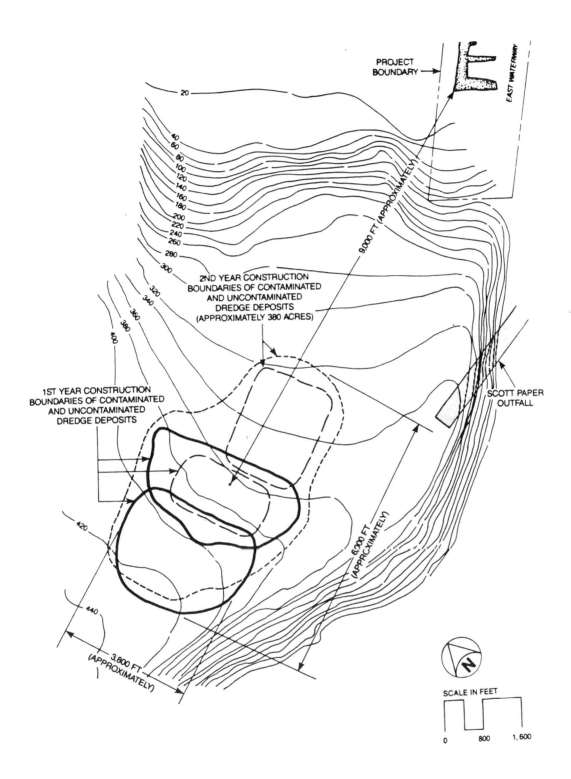

FIGURE 1 Revised application for deep confined aquatic disposal.

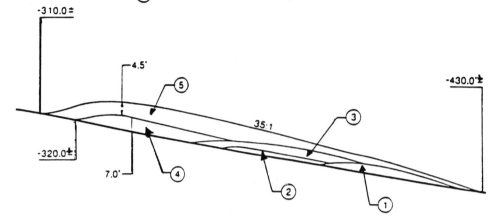

FIGURE 2 CAD site final consolidated section.

SEDIMENT ASSESSMENT AND EVALUATION

Contaminated Sediment

The contaminated sediment consists of the upper layer of sediment in most areas of the harbor ranging from 0 to 7 ft in thickness. It is composed of fine-grained, black to dark brown, odorous surface sediment including abundant wood fragments, chips and sawdust. The contaminants include oil and grease, heavy metals, polyaromatic hydrocarbons (PAH), and polychlorinated biphenyls (PCB).

Although the disposal of dredged materials does not fall within the purview of the Resource Conservation and Recovery Act (RCRA), the act's definition of hazardous waste is useful as a basis of comparison. Extensive laboratory testing (FEIS, DSEIS) has shown that contamination levels in these sediments are well below the concentration associated with hazardous waste designation under RCRA or related Washington State Dangerous Waste Regulations (Chapter 173-303, Washington Administrative Code). However, certain contaminant levels do exceed background levels in Port Gardner and, to a lesser extent, biological effects thresholds observed elsewhere in Puget Sound; therefore, capping of the contaminated sediments is proposed as a means to isolate them from surrounding waters.

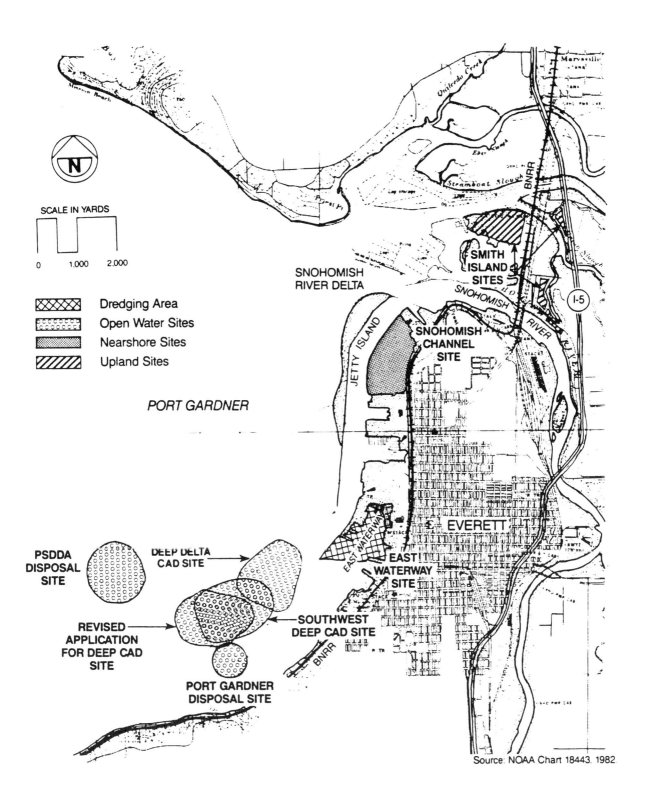

FIGURE 3 Dredging area and alternative disposal sites.

Uncontaminated Soil

The uncontaminated soils are exposed south of the south mole and underlie the contaminated sediments in the harbor area north of the south mole and in limited areas east of the carrier pier. They range in thickness to greater than 50 ft and are composed mainly of native materials in the form of gray and brown sandy silt. Some organic material and wood fragments/chips are present in small amounts in certain areas. Chemical and biological analyses have shown that these sediments meet requirements for unconfined open-water disposal in Port Gardner.

Sediment Chemical Characterization

In June 1984, a contaminated sediments assessment program for the East Waterway of Everett Harbor was developed by the Seattle District, Corps of Engineers (COE) in coordination with key federal and state agencies. Nineteen stations in East Waterway were sampled in July 1984 by the COE using a vibracore sampler. Sediment cores were recovered for depths up to 15 ft. Sediment horizons were visually characterized and subsamples taken for chemical analysis. By comparison to the only Puget Sound sediment criteria in existence at that time, the surface layer of harbor sediment was judged to be unacceptable for unconfined open-water disposal. All chemical values of the native sediment layer were below the reference criteria, and met the chemical guidelines for open-water disposal.

In June 1985, contaminated sediment samples were collected from 16 stations inside the East Waterway and combined to form 8 yd^3 of composited sample, which was provided to the COE Waterways Experiment Station (WES) for physiochemical testing: 1 yd^3 of native sediment was also collected for testing. Subsamples of the composite and native sediments were provided to the Battelle Pacific Northwest Laboratory (PNL) for separate chemical and biological testing.

Priority pollutant analysis of the composite sediment sample collected in the East Waterway by the COE (Table 1) indicated the presence of 33 sediment contaminants of concern. These compounds included chromium (Cr), nickel (Ni), copper (Cu), zinc (Zn), arsenic (As), lead (Pb), cadmium (Cd), mercury (Hg), polychlorinated biphenols (PCBs), polynuclear aromatic hydrocarbons (PAHs), and 1- and 2-methylnapthalene.

Chemically Related Dredge Disposal Considerations

CAD Standard

Elutriate tests were conducted by the COE on the previously noted composite sample of sediments collected from the East Waterway (DSEIS). This information was then used to estimate the potential for dissolved contaminant release to the water column during open-water placement of dredged materials (CAD alternative). Elutriate testing indicated that only 7 of 33 contaminants of concern were detected in the elutriate water: copper (Cu), mercury (Hg), cadmium (Cd), lead (Pb), chromium (Cr), nickel (Ni), and PCB-1254. Of these, only the latter five exceeded Port Gardner background levels. Dissolved concentrations of nickel, lead, and PCB-1254 exceeded EPA water quality criteria.

The standard elutriate procedure was modified to obtain estimates of total contaminant concentrations associated with mass release to the water column during dredging and open-water disposal (CAD) of East Waterway sediments. Results of these tests revealed that concentrations of total Ni and Pb slightly exceeded the measured dissolved concentrations of these metals (i.e., 15 μg/liter dissolved versus 17 μg/liter total for Ni, and 28 μg/liter dissolved versus 30 μg/liter total for Pb). Thus undiluted, the effluent concentration of these two metals would exceed EPA water quality criteria. The total concentration of PCB 1254 was observed to be less than the dissolved concentration (i.e., 0.3 versus 0.4 μg/liter, respectively).

Based on these tests, potential water quality impacts during openwater placement of contaminated sediments (CAD site) appear to be limited to these three pollutants. In this regard, the concentration of Ni in the elutriate was shown to exceed chronic criteria, but was well below the acute exposure value. Because Port Gardner water samples collected by the COE and identified as background or reference waters equal the chronic criteria value for nickel, dilution of the elutriate with this water would not reduce the elutriate concentration below the chronic level. In the case of Pb, a dilution factor of 1 would result in water concentrations below the chronic criteria concentration for the elutriate: for PCB 1254 a dilution factor of approximately 13 would be necessary.

Upland or Intertidal Disposal

Modified elutriate tests were conducted by the COE to estimate contaminant concentrations in effluent discharged from dredge disposal sites located in intertidal or upland areas (DSEIS). These tests were designed to estimate dissolved and particulate-associated contaminant concentrations in the effluent generated during the placement of hydraulically dredged sediments. Results of the modified elutriate test procedure indicates that undiluted discharge of effluent would significantly degrade the local water quality.

The concentration of chromium (Cr) and PCB 1254 were observed to exceed the dissolved concentrations. Total concentrations of PCB 1254 would exceed the EPA's saltwater quality criterion and would require a dilution factor of > 20 to meet the criterion, assuming such effluent was discharged to salt water. If discharged to fresh water (i.e., the Snohomish River) the criteria for Cs, Cr, and PCB 1254 would be exceeded. Dilution factors of 17, 28, and 43 respectively would be necessary to reduce the total concentration of these contaminants to below the acceptable water quality criteria. Obtaining such a dilution could require a diffuser system for the discharge line.

TABLE 1 Bulk Sediment Analyses of Composite Sediment Sample

Parameter	Concentration mg/kg	Parameter	Concentration mg/kg	Parameter	Concentration mg/kg
Arsenic	5.73	4-Nitrophenol	<1.	1 4-Dichlorobenzene	<1.
Copper	73.4	Chloromethane	<0.025	1 2 4-Trichlorobenzene	<1.
Nickel	21.4	Chlordethane	<0.025	Hexachlorocyclopentadiene	<1.
Cadmium	3.30	2-Mthyl-4 6-Dinitrophenol	<10.	1 2-Dichlorobenzene	<1.
Lead	48.1	Bromomethane	<0.025	Naphthalene	8.2
Zinc	148.5	Methylene Chloride	<0.025	2-Chloronaphthalene	<1.
Chromium	39.7	Pentachlorophenol	<1.	Hexachloroethane	<1.
Mercury	0.201	Vinyl Chloride	<0.025	Hexachlorobutadine	<1.
Iron	7864.	1 1-Dichloroethene	<0.025	Acenaphthylene	<1.
Aldrin	<0.0002	1 1-Dichloroethane	<0.025	Dimethyl Phthalate	<1.
Manganese	237.	1 2-Dichloroethane	<0.025	Diethyl Phthalate	<1.
G-BHC	<0.0002	Bromodichloromethane	<0.025	4-Bromophenyl Ether	<1.
Total Phosphorus	789.	Trans-1 2-Dichloroethene	<0.025	Acenaphthene	2.0
A-BHC	<0.0002	1 1 1-Trichloroethane	<0.025	4-Chlorophenyl Phenyl Ether	<1.
D-BHC	<0.0002	1 2 Dichloropropane	<0.025	Hexachlorobenzene	<1.
Ammonia Nitrogen	167.	Chloroform	<0.025	Fluorene	2.2
B-BHC	<0.0002	Carbon Tetrachloride	<0.025	N-Nitrosodiphenyl Amine	<1.
Chlordane	<0.0002	Trans-1 3-Dichloropropene	<0.025	Phenanthrene	5.7
PPDDD	<0.0002	Trichloroethene	<0.025	Anthracene	1.5
Dieldrin	<0.0002	1 1 2-Trichloroethane	<0.025	Pyrene	4.0
Endosulfan Sulfate	<0.0002	Bromoform	<0.025	Benzo(A)Anthracene	2.1
A-Endosulfan	<0.0002	Dibromochloromethane	<0.025	Dibutylphthalate	<1.
Endrin	<0.0002	Benzene	<0.025	Butylbenzylphthalate	<1.
PPDDT	<0.0002	1 1 2 2-Tetrachloroethane	<0.025	Bis(2OEthylhexyl)Phthalate	<1.
B-Endosulfan	<0.0002	Cis-1 3-Dichloropropene	<0.025	Fluoranthene	4.5
Endrin Aldehyde	<0.0002	2-Chloroethylvinylether	<0.025	Chrysene	1.8
Heptachlor	<0.0002	Tetrachloroethene	<0.025	Di-N-Octylphthalate	<1.
Neptachlor Epoxide	<0.0002	Toluene	<0.025	Benzo(B)Fluoranthene	2.5
PCB-1221	<0.002	Acrolein	<0.25	Indeno(1 2 3-C D)Pyrene	<1.
PCB-1248	<0.002	Bis(2-Chlorisopropyl)ether	<1.	Total Organic Carbon	71.540
PCB-1232	<0.002	Chlorobenzene	<0.025	Benzo(K)Fluoranthene	2.5
PCB-1254	0.25	Acrylonitrile	<0.25	Dibenzo(A H)Anthracene	<1.
PCB-1016	<0.002	N-Nitroso-di-n-propylamine	1.	Benzo(A)Pyrene	1.4
PCB-1242	<0.002	Ethylbenzene	<0.025	Benzo(G H I)Perylene	<1.
PCB-1260	<0.002	N-Nitrosodimethylamine	<1.	Total Phenol	<17.1
Toxaphene	<0.002	Nitrobenzene	<1.		
2-Nitrophenol	<1.	Isophorone	<1.		
4-Chloro-3-Methylphenol	<1.	2 4-Dinitrotoluene	<1.		
Phenol	<1.	3 3-Dichlorobenzioine	<5.		
2 4-Dimethylphenol	<1.	Bis(2-Chloroethoxy)Methane	<1.		
2 4 6-Trichlorophenol	<1.	2 6-Dinitrotoluene	<1.		
2-Chlorophenol	<1.	Benzidine	<10.		
2 3-Dichlorophenol	<1.	1 3-Dichlorobenzene	<1.		
2 4 Dinitrophenol	10				

Surface Runoff Impact

Tests were conducted by the COE to estimate the potential impacts to receiving water quality as a result of surface water runoff from a confined upland or nearshore dredge disposal site. A rainfall simulatory-lysimeter was utilized to predict the quality of surface-water runoff from such a disposal site.

Test results showed that if dredged sediments placed in upland or nearshore sites are not capped and are allowed to dry, physiochemical changes will occur. Under such conditions, runoff water from rainfall would potentially carry dissolved contaminants from the site. Studies conducted with East Waterway sediments indicate that under these conditions the concentration of dissolved Cd would substantially exceed EPA water quality criteria.

Leachate Testing

The potential for generation of leachate from an upland disposal site was studied using experimental laboratory testing procedures for sediments collected from the East Waterway. Leachate contaminant levels from these sediments were quantified using batch and column testing techniques.

Based on these leachate tests, the geochemical changes associated with aerobic disposal on land would result in mobilization of a large fraction of some of the contaminants. If the material could be placed below the water table at a given site (usually more of an option for nearshore/intertidal disposal), such mobilization would be significantly reduced. The leaching tests indicated that mobility of metals and organic contaminants is low under anaerobic conditions. Under aerobic conditions, some of the metals were mobilized in large quantities. The fraction of metals that was resistant to anaerobic leaching was generally greater than 90 percent of the bulk sediment concentration. Under aerobic conditions, over 85, 65, and 49 percent of the Zn, Ni, and Cd, respectively, was mobilized in the tests. The higher metal release observed in aerobic testing is related to the pH (i.e., the pH in aerobic testing was lower than the pH in anaerobic testing).

DREDGING AND DISPOSAL ALTERNATIVES

Dredging Equipment Consideration

Based on dissolved contaminants and mass releases, both hydraulic and mechanical dredges appear equally viable. Contaminants associated with East Waterway sediments appear to be strongly bonded to those sediments as long as they remain saturated. Mechanical dredging to remove the contaminated layer is preferred for confined aquatic disposal since the lower in situ water content would encourage clumping of the material. Hydraulic dredging of the native material would assist in maximizing spread for cap placement. Either dredging method could be employed for nearshore confined disposal, although hydraulic dredging appears to be most efficient. Mechanical dredging would require double handling of the material to place it in the site. As water quality problems would not be markedly different from conventional operations, use of specialized dredging equipment is not recommended.

General Disposal Alternatives

A number of disposal methods and site alternatives were considered for the removed sediments. The preferred method is CAD, to dispose of the contaminated sediments at a deepwater site and then cap them with the clean native sediments. The preferred disposal alternative, Revised Application Deep (RAD) CAD, was developed in response to public comments concerning potential significant impacts to the Dungeness crab (*Cancer magister*) resource of Port Gardner. The RAD CAD site is deep enough to minimize short-term and avoid long-term impacts to Dungeness crabs.

A second alternative involved placing either all of the removed sediments or just the contaminated sediments in an intertidal site on the Snohomish River. If only the contaminated sediments were placed there, the clean sediments would be taken to the Port Gardner aquatic disposal site. A third disposal method involves placement of the contaminated sediments on an upland site on Smith Island. With this alternative, clean sediments not required for cover could be disposed of at a deep-water site in Port Gardner. These sites together with other alternative sites are shown on Figure 3.

As part of the evaluation methodology for contaminated materials, seven criteria were used to assess each of the different sites. Criteria included contaminant availability, potential contaminant mobility, site environmental conditions, erosion potential, institutional constraints, site capacity, relative cost, and monitoring capability. A criteria evaluation matrix is given in Table 2. For clean dredged material, five criteria, including site environmental considerations, availability for capping, institutional constraints, site capacity, and relative cost were applied.

Open-Water Capped Disposal Site Considerations

A detailed locational analysis was undertaken within Port Gardner to identify potential site alternatives for disposal of dredged material by the CAD method. An initial step in the site identification process was a bathymetric survey conducted of much of Port Gardner, focusing on areas shallower than 400 ft. Subsequently, core samples were taken throughout the area and a map of sediment types was prepared. Areas of potential geotechnical risk, as indicated by recent slumping and other factors, were identified as well. Other significant characteristics, such as the location of outfalls, were also mapped. The key siting criteria, based on engineering and construction reliability, used to select potential sites included the following:

- <u>Potential for subsequent natural deposition</u>. The site should be in a zone of accretion. That is, natural deposition of sediments that could add to the thickness of the capping material was considered to be beneficial. Conversely, areas of potential erosion that could remove cap material were to be avoided.
- <u>Geotechnical stability</u>. The site should be in an area with no evidence of slope movement. Areas where slumping was identified or where there was a high potential for slumping (in particular, steep slopes) were to be avoided.
- <u>Site configuration</u>. The site should be relatively flat so that the deposited dredge materials would stay in place. An upwardly sloping terrain on one side of the site was considered beneficial, because the slope would function as a natural berm. The natural berm would help confine the cap material and allow a thicker cap to be constructed.
- <u>Site size</u>. The overall size of the disposal sites is governed primarily by the total amount of material being deposited, sediment bulking factors, stable side slope characteristics of the sediments, and existing bottom topography and consolidation characteristics of both the bed and the dredged material. The initial area of deposition for both the barge and hydraulic dredge methods can be expected to increase with increasing depth. For the depth range identified in the general CAD disposal vicinity, the increased depths will not increase initial areas of deposition enough to significantly increase overall site size.
- <u>Other factors</u>. Facilities already in place, such as outfalls, were to be avoided so that there would be no interference with their operation. Dredge disposal sites that have experienced permitting difficulties were considered less desirable. Other potential disposal sites were also avoided because other future disposal activities could potentially violate the integrity of the CAD cap.

Much of the study area was considered unsuitable for a CAD site because of steep slopes or evidence of unstable geotechnical conditions. The RAD CAD site was selected as the preferred alternative, meeting the above criteria and minimizing potential impacts on biological resources (Table 2).

DREDGING AND DISPOSAL DESIGN

Performance Goals

Selection of dredging equipment for the contaminated Everett Harbor sediments was based on the following performance goals:

1. Water entrainment during the dredging operation must be minimized.
2. Dredging equipment must be compatible with the confined aquatic

TABLE 2 Rating Matrix of Alternative Disposal Sites for Contaminated Sediment

	Contaminant Availability	Potential Contaminant Mobility	Site Environmental Considerations	Erosion Potential	Institutional Constraints	Site Capacity	Relative Cost[a]	Monitoring Capability
Revised Application Deep CAD	1	1	2	1	2	1	1	2
Snohomish Channel Nearshore All Sediments	1	1	3	1	2	1	2	1
Smith Island Elevated	1	1	2	1	3	3	3	1

1: Minor or no adverse effects
2: Moderate adverse effects
3: Significant adverse effects

[a]Estimated Construction Costs as follows:

Approach	Cost $ x 1000	Cost $/yd^3
RAD-CAD	$14,500	$ 4.38
Snohomish Channel	$24,000	$ 7.26
Smith Island (Elevated)	$55,000	$16.64

disposal alternative under consideration.
3. Dredging equipment must be capable of removing the sediments at a reasonable cost.

Selection of dredging equipment for the placement of the clean cap sediments was based on the following performance goals:

1. Cap sediment placement would avoid displacement of the contaminated sediments.
2. Cap placement could be controlled and monitored.
3. Dredging equipment must be capable of removing the sediments at a reasonable cost.

Proposed Dredging Equipment

Given the equipment performance goals, a multiphase dredging approach for the CAD alternative was identified. Initial dredging of uncontaminated sediments will be accomplished for purposes of constructing a subaqueous confining mound at the downslope limit of the disposal site. This dredging would also serve as a test to demonstrate the acceptability of the final dredge equipment selection and disposal site design. Depending on the results of the mound construction, the remaining dredging and disposal would be continued as proposed or revised as appropriate.

Mound construction will be accomplished by clamshell dredging and bottom dump of the clean surficial sediments to the disposal site. During the second phase, contaminated materials will be dredged by clamshell dredge, with haul and dumping by split hull barges of 3,000-yd^3 capacity or larger. Uncontaminated capping sediments will then be dredged in the third phase, using a hydraulic cutterhead pipeline dredge with discharge below surface through a diffuser unit. The dredging effort will take place over two years, and the second year will include a repeat of the contaminated phase and capping phase accomplished in the first year.

Mound Construction Phase

Mound construction will be the first phase of dredging to be accomplished. It was originally planned for completion by pipeline dredge with controlled placement of sediments by downpipe in the shape of a confining mound. The intent of this original approach was to control the placement of the clean sediments in a slurry form by the downpipe to create a downslope confining structure to prevent loss of contaminated sediments during the subsequent dredging phase. It is questionable that subaqueous confining berms can be constructed with fine-grained soils to the side slopes proposed using a slurried hydraulic pipeline discharged material. It was further established that the contaminated materials would not require a substantial berm for confinement. It was finally concluded that using clamshelled material for the construction of a mound with flatter side slopes--similar to that proposed for the contaminated materials--was advantageous because of the cohesion and clumping associated with clamshelled sediments. Construction of a mound section with cohesive clumps is considered less of an uncertainty, and construction by surface disposal of clamshelled sediments became a viable option.

The downslope mound still provides a limiting structure for contaminated sediment placement and provides a secondary benefit. Because the sediments are similar in situ to contaminated surficial sediments and will be dredged and dumped in the same manner and with the same equipment, a verification of dumping procedures and sediment fate can be completed in the prototype prior to dredging of the contaminated sediments. This has become an important factor in the regulatory agency considerations to approve the disposal permit because it satisfied the opportunity to check the dredge and disposal design before release of the contaminated sediments at the disposal site.

Contaminated Material Placement

Clamshell dredging for the contaminated sediment is considered the most compatible dredging method for the CAD disposal alternative. Modeling results indicate that the material will tend to mound if dredged by clamshell. Placement of the material by double handling through a submerged discharge such as a vertical downpipe was proposed, but was discarded as unnecessary and undesirable. Use of a vertical downpipe would require mechanical or hydraulic dredge rehandling from the haul barge to the pipe. Based on physical modeling results, the downpipe tended to cause side shear and entrain greater amounts of water in the already reduced 5- to 10-yd^3 clumps of sediment rehandled from the haul barge. This resulted in a lesser sediment strength on the bed to support a cap than the surface release of the barge load.

Disposal of clamshelled material at the surface is a viable option for the contaminated material. For surface disposal, the cohesion and clumping normally associated with clamshelled material would be of benefit in reducing material spread and resuspension and would result in a contaminated sediment mass with more strength for support of a cap. Clamshelled material would also entrain less water, thereby resulting in a smaller subaqueous confined volume.

Cap Placement

Hydraulic dredging of uncontaminated material for the cap placement was recommended. The potential for displacement of the soft contaminated material or bearing-type failure of the cap would require that the cap layer thickness be gradually built up. This process could be accomplished by surface disposal or submerged pipe discharge suspended some variable distance above the contaminated material surface. Use of a submerged diffuser (directly connected to the hydraulic pipeline with flexible hose) is an option that will be used to avoid any potential for jetting action causing erosion of the contaminated sediments.

Final Disposal Configuration--Summary

Several final calculated designs of the disposal cross-section and area spread were accomplished. Since the proposed dredging plan extends over two dredging seasons, the sequence of disposal operations was taken into consideration. This sequence includes initial placement of uncontaminated materials for a mound, then placement of a relatively small amount of contaminated materials and immediate capping with a greater amount (relative to the contaminated materials) of uncontaminated materials. After approximately nine months, an additional larger sequence of contaminated and capping materials would be placed (Figures 1 and 2).

Quantities used in the design were as follows:

- year 1 uncontaminated mound materials, 500,000 yd^3;
- year 1 contaminated materials, 97,000 yd^3;
- year 1 uncontaminated cap materials, 239,000 yd^3;
- year 2 contaminated materials, 831,000 yd^3;
- year 2 uncontaminated cap materials, 1,638,000 yd^3.

Sediment consolidation would occur at a geometric rate, with the greatest amount of consolidation resulting during the first few months following disposal activities. The final design assumed a conservative consolidation of 50 percent of the immediate deposition thickness within three months after placement to establish cap thickness. A minimum of 1-m thickness was required for CAD alternative acceptance by regulatory agencies. It is anticipated that final minimum capping thickness over contaminated materials will be 4.5 ft or more. The total impacted area, where 3 cm or more of dredged material will be deposited, is approximately 380 acres.

ENVIRONMENTAL MONITORING

The dredging and disposal monitoring plan, phase I document outlines a detailed plan for meeting the conditions outlined in the State Water Quality Certification, and COE Section 10/404 permit of September 24, 1987.

Monitoring is divided into five phases:

1. baseline,
2. mound construction,
3. contaminated dredging and disposal,
4. capping dredging and disposal, and
5. long-term monitoring.

The monitoring plan is designed to measure physical, chemical, and biological characteristics of the revised application for deep confined aquatic disposal (RAD/CAD) site and to collect data regarding the effect of project construction on those characteristics. The plan was prepared considering details of the permit application, COE Section 10/404 permit, results of the Washington State Department of Ecology's review process, and the project's construction plans and specifications as they exist. Results of the first-year monitoring and construction activities will then be applied toward development of the second-year monitoring plan (phase II).

A listing of the various monitoring activities associated with the dredging and disposal component of the Navy's Everett Homeport project follows.

- Electronic Positioning: precise sea-surface positions will be established to meet the requirement for absolute accuracy of + 3 m. Also, a seabed positioning system will be used to locate the actual seabed position of the sediment and benthic samples and sediment profile camera photos.
- Bathymetry: precision bathymetry ($\pm$ 20 cm accuracy) with lane spacing of 20 m over the 1150-acre survey area.
- Sidescan sonar surveys: information on the surficial characteristics of the seafloor to either side of the survey trackline out to a range of 500 ft.
- Sediment profile camera: during baseline monitoring, 70 SPC stations will be photographed with three replicates per station. Following sediment disposal, additional photos will be taken to define the limits of the disposed material.
- Sediment cores: sediment box cores at 37 stations will be taken during baseline, with the upper 2 cm of each analyzed for physical properties and 79 PSDDA (Puget Sound Dredge Disposal Analysis) "Chemicals of Concern," all run using Puget Sound Estuary Program (PSEP) protocol. Piston cores will be taken after disposal to identify properties of disposed sediments and thickness of capped material.
- Benthic macroinvertebrate assemblages: using a $0.06m^2$ box corer at 37 stations (5 replicate/station), traditional taxonomic analysis will assess the characteristics of the benthic community. Additionally, complementary Benthic Resource Assessment Technique (BRAT) will also be performed.
- Bioaccumulation: one epifaunal species (*Cancer magister*) and one infaunal species (*Molpadia*) to be collected and investigated at six stations for chemical bioaccumulation in tissue.
- Bioturbation: randomly located on the disposal site, 35 box core stations will be collected and analyzed for density of macrobenthic infauna and number and depth of burrows to determine possible effects on cap material integrity.
- Sea-surface microlayer: microlayer (less than 100 microns) water samples will be collected with a ceramic rotary drum system prior to and during contaminated sediment disposal.

- Current measurements: site-specific current conditions will be collected during mound disposal, during the whole of contaminated sediment disposal and during two days of dredging to determine the current influence on sediment transport. Current meter stations at the RAD CAD site will include arrays of meters at discrete depths (i.e., 90, 120, and 220 ft, and near bottom) as well as an acoustic doppler current profiling meter.
- Water column effects (disposal site): three-step procedure using current drogues and acoustical transponders for plume tracing, acoustical transponders and transmissometers for plume characterization and state water quality standards monitoring.
- Water column effects (dredge site): for dissolved oxygen, percent light transmittance, total suspended solids, and nephelometric turbidity units at various depths and distance from dredging activities.
- Chemical analysis of water samples: conventional parameters will be measured as well as 69 PSDDA chemicals of concern, all run using PSEP protocol.
- Sediment traps: positioned around the disposal site (over 20 locations) will be two pairs of traps on each mooring, one located 1 to 2 m above the bottom and the other 20 m above the bottom. Four locations will include turbidity meters on the moorings as well.
- Bioeffects (shellfish/fish): short- and long-term impacts and changes on density, diversity, and population abundance will be measured in a statistically significant manner.
- Mussel watch: over 2,200 noncontaminated mussels will be deployed at each of three depths on nine moorings for a 30-day period during contaminated sediment disposal, after which mussel tissue chemical concentration analysis will be performed.
- Histopathology: will assess any short- or long-term hepatic pathologic abnormalities in English sole before and after disposal activities.

Second-Year Monitoring

Second-year environmental monitoring activities are expected to be similar to the first-year activities just described, although some adjustments to the monitoring effort will most probably result from a review of the data collected and "lessons learned."

Long-Term Monitoring

Monitoring of environmental conditions at the disposal site is required for at least 10 years following completion of the second-year cap. Monitoring will be conducted at specific intervals, currently scheduled for years 1, 2, 4, 7, and 10. Monitoring will be conducted using the same operational procedures described for baseline and/or disposal monitoring and will minimally include

- electronic positioning,
- bathymetry,
- sediment cores, and
- bioeffects--shellfish/fish, benthic macroinvertebrates, BRAT, bioturbation, and histopathology.

CONCLUSIONS

This paper has presented a summary of the sediment evaluation procedures, disposal alternative assessments, design considerations, and monitoring requirements associated with the dredging and disposal required for construction of the Navy's Carrier Battlegroup Homeport Project in Everett, Washington. The preferred disposal alternative includes the dredging of approximately 1 million yd^3 of marine sediment that will be treated as contaminated, together with an additional volume of clean sediments in excess of 2 million yd^3, to construct an engineered CAD site in water depths of 310 to 430 ft. The work will be completed in three separate, regulated phases, and the success of each preceding phase is the prerequisite to continuing with the next phase. These phases include the clean test mound construction, phase I contaminated sediment and clean cap disposal, and phase II contaminated sediment and clean cap disposal.

Since 1984, the project has undergone extensive environmental review (including 20 public hearings), culminating in the issuance of a Washington State conditional use shoreline permit (May 1988), a COE Section 10/404 permit (September 1987) and attendant Washington State water quality certification (March 1987). The environmental review process resulted in substantive project design modifications to mitigate environmental concerns. Nonetheless, the critical component of the project has always been and continues to be controversy surrounding the technical feasibility and environmental impacts of successfully constructing a CAD site for contaminated marine sediments in a deep-water environment. The quest for the answers to these questions has resulted in administrative permit appeals, legal challenges in federal and state court, repeated schedule revisions and extensions, project funding constraints attributed to environmental concerns, and unprecedented environmental monitoring requirements. As a result, dredging and disposal activities are currently scheduled to start in the summer of 1988, more than a year after the original programmed date.

The preferred disposal alternative (CAD) is a direct extension of existing technology that has been used successfully in other areas of the United States. The CAD procedure has been the subject of extensive research both within the United States as well as abroad, and is a technology sanctioned by the London Dumping Convention scientific group. The levels of contamination encountered in Everett Harbor sediments are handled routinely in other areas of the country. Even with this background to draw upon, regional environmental concern for Puget Sound necessitated the expenditure of substantial time and money, far beyond previous experience. Currently, for this project, predisposal nonconstruction costs are approximately $3.50/yd^3 of contaminated sediment and anticipated environmental monitoring costs exceed $8.00/yd^3 of contaminated sediment. These costs can be compared to anticipated construction costs of $4 to $7/yd^3 of dredged material (Table 3).

The level of monitoring required for this project is particularly extensive, and it should be recognized that such costs would likely not be supportable by smaller projects, privately funded projects, or local governmental entities. A major benefit of the monitoring for this project should be the demonstration of the viability of the CAD technique in deep water. As such, those monitoring activities intended to verify the CAD concept would not be appropriate for future projects. A like conclusion is applicable to other components of this monitoring program that do not demonstrate usefulness in documenting compliance with required levels of environmental performance. Review of the project literature reveals that several of the required sediment evaluation procedures produced inconclusive results. Some evaluations demonstrated poor repeatability and others a potential for significant false positive results when contrasted with controls. Often these results served to confuse rather than clarify the evaluation of possible impacts. Specific aspects of this project which have produced or may produce questionable results include

- mass loss determinations and impacts,
- microtox bioassay,
- standard bioassays,
- sea-surface microlayer,
- chronic toxicity and bioaccumulation,
- plume tracing for water quality impacts, and
- bioturbation.

TABLE 3 Summary of Dredging Costs[a]

Item	Contaminated[b]	Total[c]
Sediment evaluation	$ 1.26	$ 0.35
Disposal alternative evaluation	2.21	0.62
Environmental Monitoring	8.27	2.32
Total nonconstruction costs	11.74	3.29
Total dredging and disposal design costs		0.21
Total Construction Costs		4.00-7.00

NOTES:
[a] Nonconstruction costs, expressed as dollars/yd^3.
[b] 928,000 yd^3
[c] 3,305,000 yd^3

The data summarized by this preliminary case study supports the continued need for a strong national research program relating to marine sediments. The program should be directed at defining appropriate repeatable procedures for sediment evaluation and monitoring efforts. In addition, the process should actively eliminate requirements that demonstrate a significant tendency for inconclusive results. Funding and prioritization for such research should be established by an independent national group and implemented on projects where significant new information could be developed during construction through coordination with the project proponent. It is generally not in the national interest to allow uncertainty with regard to environmental issues to burden a project having significant social or economic benefits. The responsibility of resolving research questions related to issues of environmental impacts should be jointly shared by the research community and project components. Also, a national guideline or standard on the disposal of contaminated dredged sediments should be promulgated. The guideline should draw from the wealth of regional experience, such as the homeport project, to minimize the unilateral research burden on future projects. While the guideline needs to be sensitive to regional differences and concerns, it must at the same time maintain reasonable levels of national uniformity.

APPENDIX A
BIOGRAPHIES OF COMMITTEE MEMBERS

KENNETH S. KAMLET is a senior program manager with A.T. Kearney, Inc., responsible for assessments of hazardous waste problems and practices. A lawyer and biologist, and former director of the Pollution and Toxic Substances Division of the National Wildlife Federation, Mr. Kamlet for a number of years concentrated on the application of scientific and technical knowledge to public policies addressing ocean and land disposal of wastes and other materials. He has published numerous papers on these and related topics and has served on many policy review and planning, including several NRC, committees that have addressed marine environmental issues. Mr. Kamlet has a national reputation for addressing sensitive environmental issues in a fair and open-minded way. Recently, for example, he chaired a successful national meeting on beneficial uses of dredged material. He has twice served on the U.S. delegation to the London Dumping Convention. Mr. Kamlet holds a B.S. degree in biology from the City College of New York, an M. Phil. degree in biology from Yale University, and a J.D. degree from the University of Pennsylvania. He is a member of the District of Columbia bar.

WILLIAM J. ADAMS is an associate fellow in Monsanto's corporate Environmental Sciences Center. Dr. Adams has been at Monsanto for the past 10 years and has worked in the areas of aquatic toxicology, environmental fate, sediment assessment and hazard assessment. He is currently conducting research in the area of specialized uses of microbes for waste site cleanup and treatment of industrial wastes. Dr. Adams received a B.S. degree from Lake Superior State College; he obtained an M.S. degree in wildlife toxicology in 1971 from Michigan State University and his Ph.D. from the same institution in aquatic toxicology.

A. KARIM AHMED is with the Environ Corporation in Princeton, New Jersey. He recently came to Environ from serving as research director and senior staff scientist at the Natural Resources Defense Council, Inc. in New York. Dr. Karim also serves as adjunct professor at the State University of New York, College at Purchase and as Raymond G. Brown Adjunct Professor at the Cooper Union for the Advancement of Science and Art. He received his B.Sc. degree in physics and chemistry, and his M.Sc. degree in chemistry from the University of Karachi, Pakista. Dr. Ahmed received his M.S. degree in organic chemistry and his Ph.D. in biochemistry from the University of Minnesota.

HENRY J. BOKUNIEWICZ is an associate professor in the Marine Sciences Research Center at the State University of New York at Stony Brook. Dr. Bokuniewicz has authored and co-authored numerous papers on sediment transport and deposition, sediment mass balance, and the effects of storm and tidal energy. His current research involves effects of resuspension on containment availability for dredged material, benthic studies associated with containment, prediction of tidal circulation and hydrodynamics, and criteria for selection of dredged material disposal sites. He received his B.A. degree from the University of Illinois and his M. Phil. and Ph.D. degrees from Yale University.

THOMAS A. GRIGALUNAS is a professor in, and is the former chairman of, the Department of Resource Economics at the University of Rhode Island. Dr. Grigalunas has done extensive research in the area of OCS oil and gas economics, including studies of bidding behavior, regional impacts, and environmental effects, including multiple use conflicts. He also has done considerable work on a variety of marine pollution issues, including the development of concepts for measuring the risks and economic damages from oil spills and from spills of hazardous substances. Among other activities, he was principal investigator for the international economic study of the AMOCO CADIZ oil spill, and he served as co-principal investigator for the development of the simplified Natural Resource Damage Assessment Model for Coastal and Marine Environments under CERCLA (Superfund). He also served as co-principal investigator for the analysis of the environmental costs in the most recent Five-year OCS oil and gas leasing program. Dr. Grigalunas received his Ph.D. in economics from the University of Maryland.

JOHN B. HERBICH is professor in the Ocean and Civil Engineering Department and a Graduate Faculty Member at Texas A&M University. Dr. Herbich is an expert in coastal and ocean engineering with a specialty in dredging engineering and technology development and also has consulted extensively for U.S. and international industries and governments on coastal developments and uses. He has served on several committees of the National Research Council, including chairing the Technical Panel of the Committee on National Dredging Issues. A Fellow of the American Society of Civil Engineers, Dr. Herbich received his B.Sc. degree in civil engineering at the University of Edinburgh, his M.S. in hydromechanics at the University of Minnesota, and his Ph.D. in civil Engineering at Pennsylvania State University.

ROBERT J. HUGGETT is professor in the Marine Science Department, assistant director and head of the Division of Chemistry and Toxicology, Virginia Institute of Marine Science, School of Marine Science, College of William and Mary. Dr. Huggett's research interests are directed to the fates and effects of toxic chemicals in aqueous systems. He is a member of the EPA's Science Advisory Board and is assistant editor of the Journal of Environmental Toxicology and Chemistry. Dr. Huggett received a B.S. degree in chemistry from the College of William and Mary, an M.S. degree in marine chemistry from Scripps Institution of Oceanography, and a Ph.D. in marine science from the College of William and Mary.

HOWARD L. SANDERS, a benthic ecologist, has been associated with the Woods Hole Oceanographic Institution for more than 30 years. Dr. Sanders has conducted pioneering research in the biology of marine pollution, the response of benthic communities to environmental insult, and their recovery. He is a member of the American Society of Limnology and Oceanography and a fellow of the American Association for the Advancement of Science, and a member of the National Academy of Sciences. Dr. Sanders received a B.A. degree from the University of British Columbia, an M.S. degree from the University of Rhode Island, and a Ph.D. in zoology from Yale University.

JAMES M. THORNTON is a senior environmentalist for the Washington State Department of Ecology as advisor to the Department Director on policy and technical issues for contaminated sediments. He currently serves as the department's technical expert on development of marine and fresh water sediment evaluation procedures, and also serves as the state's technical and policy liaison with principal planning agencies from other states and appropriate federal agencies concerning dredging and dredge disposal. This includes coordination with the states of Oregon and Idaho for maintenance of the Columbia River Channel and with the Corps of Engineers and EPA in evaluating and designating ocean disposal sites.

APPENDIX B
COASTAL STATE SURVEY OF CONTAMINATED MARINE SEDIMENTS

Introduction

A variety of classification techniques (see attachment) are employed or under development at the federal level for determining when marine sediments are sufficiently contaminated to justify or require action. These techniques include:

1) The hazard ranking system for specifying inactive hazardous waste sites for inclusion on the Superfund National Priorities List;

2) The "Sediment Triad" and "Apparent Effects Threshold" (AET) Methods;

3) The "Equilibrium Partitioning" Approach;

4) Solid Phase Bioassay Procedures; and

5) Screening Level Concentration Approach.

Purpose

This survey is being conducted to determine 1) how states classify and characterize contaminated sediments and 2) the nature and extent of contaminated marine sediments in each state.

Survey Questions

1. Please LIST the estuarine or ocean locations or sites, if any, in your state which are believed to contain contaminated marine sediments, based on any of the following:

   a. Proposed final listing on the Superfund National Priorities List;

   b. Proposed or final listing on a state registry of inactive hazardous waste sites;

   c. Identification as a toxic pollutant "hot spot" under Section 308 of the Water Quality Act of 1987;

   d. Any other classification technique employed by the state to identify or categorize contaminated marine sediments.

2. For each site or location listed in response to question #1, please INDICATE if a classification method(s) was employed, and PROVIDE a DESCRIPTION of each method referenced (feel free to enclose separate documents describing these methods.

3. How comprehensively do you believe the listing provided in response to question #1 covers areas in your state containing contaminated marine sediments? Please SELECT one of the following:

_____ (a) very comprehensively;

_____ (b) coverage is very incomplete;

_____ (c) coverage is partial (intermediate between (a) and (b); and

_____ (d) don't know or unable to evaluate.

4. Please STATE your view of the major strengths and weaknesses of the classification techniques referenced in your response to question #2.

5. Are there other classification techniques known to you--whether or not in use in your state--for identifying or categorizing contaminated marine sediments? If so, please IDENTIFY and/or DESCRIBE such technique(s). Please feel free to enclose documents that describe these methods.

6. Please IDENTIFY coastal areas which you believe may be significantly contaminated, but have not been adequately studied.

7. Please SUMMARIZE or ENCLOSE documents describing methods or criteria to evaluate sediments that are proposed for dredging.

Please attach or enclose additional sheets or documents as required. Thank you very much for your cooperation.

RESPONSES SHOULD BE SENT, ON OR BEFORE APRIL 20, 1988, IF POSSIBLE, TO THE FOLLOWING ADDRESS:

                        Ms. Celia Chen
                        Marine Board, Room GF 250
                        National Research Council
                        2101 Constitution Avenue
                        Washington, DC 20418

APPENDIX C
WORKSHOP PARTICIPANTS

Dr. William J. Adams
Monsanto Company

Mr. James Ahl
Great Lakes United

Dr. A. Karim Ahmed
Natural Resources Defense Council

Mr. Douglas C. Allen
E.C. Jordan

Dr. Herbert E. Allen
Drexel University

Dr. Jack W. Anderson/
Southern California Coastal
  Water Research Project

Dr. Robert G. Arnold
The University of Arizona

Mr. Robert Barrick
PTI

Mr. Brett Betts
Washington State Department of Ecology

Dr. J. Henry Bokuniewicz
State University of New York

Dr. Richard Bopp
Columbia University

Dr. John F. Brown, Jr.
General Electric Corp. R&D

Dr. Mark Brown
New York Department of
  Environmental Conservation

Mr. Fred Calder
Florida Department of
  Environmental Conservation

Dr. Peter M. Chapman
EVS Consultants

Dr. Sharon Christopherson
NOAA Ocean Assessments Division

Dr. William Conner
NOAA National Marine Pollution
  Office

Mr. John Cullinane
U.S. Army Waterways Experiment Station

Ms. Glenda Daniel
Lake Michigan Federation

Dr. Kim Devonald
EPA Office of Marine and Estuarine
  Protection

Dr. Dominic Di Toro
Manhattan College

Dr. Robert Engler
U.S. Army Waterways Experiment Station

Mr. Richard Faris
Naval Facilities Engineering Command

Mr. Jerry Ficklin
Simpson Tacoma Kraft

Mr. Jerry Fitchko
Beak Consultants Ltd.

Mr. Norman R. Francingues
U.S. Army Waterways Experiment Station

Mr. Paul Fuglevand
Hart, Crower, and Associates

Dr. Thomas Ginn
PTI

Dr. Thomas A. Grigalunas
University of Rhode Island

Mr. Daniel E. Haunert
South Florida Water Management
  District

Mr. David J. Hansen
EPA Narragansett Bay Lab/ERL

Dr. Don Hart
Beak Consultants Ltd.

Mr. Gregory L. Hartman
Ogden Beeman & Associates

Dr. John B. Herbich
Texas A&M University

Dr. Robert J. Huggett
Virginia Institute of Marine Sciences

Mr. Alan Ikelainen
E. C. Jordan

Dr. Lucinda Jacobs
PTI

Dr. Gerhard Jirka
Cornell University

Mr. Kenneth S. Kamlet
A. T. Kearney, Inc.

Mr. Anthony G. Kizlauskas
EPA Great Lakes National Program
  Office

Mr. Walter J. Kovalick, Jr.
U.S. Environmental Protection Agency

Mr. Mike Kravitz
Battelle New England Marine Research
  Laboratory

Ms. Jan Kurtz
EPA Science Advisory Board

Dr. Charles R. Lee
U.S. Army Waterways Experiment Station

Mr. Edward Long
NOAA Ocean Assessments Division

Mr. Raymond Luce
TetraTech

Mr. Edward Lukjanowicz
Naval Facilities Engineering Command

Dr. Donald Malins
Pacific Northwest Research Foundation

Mr. David B. Mathis
U.S. Army Corps of Engineers

Mr. Neil McLellan
U.S. Army Waterways Experiment Station

Dr. Robert W. Morton
Science Applications International
  Corp.

Ms. Marcia Kelly Nelson
U.S. Fish & Wildlife Service

Dr. James Opaluch
University of Rhode Island

Mr. Ian Orchard
Environment Canada

Mr. Mark J. Otis
U.S. Army Superfund

Dr. Michael Palermo
U.S. Army Waterways Experiment Station

Dr. Richard K. Peddicord
Battelle Ocean Sciences

Dr. Charles A. Pittenger
Procter and Gamble

Dr. Clifford P. Rice
Patuxent Wildlife Research Center

Mr. Andrew Robertson
NOAA Ocean Assessments Division

Mr. Reginald Rogers
U.S. Environmental Protection Agency

Mr. Steve Schropp
Florida Department of Environmental
  Regulation

Dr. Peter Y. Sheng
University of Florida

Mr. David Smith
Washington State Department of Ecology

Mr. Michael Stoner
U.S. Environmental Protection Agency

Dr. Richard C. Swartz
U.S. Environmental Protection Agency

Mr. John F. Tavalaro
U.S. Army Corps of Engineers

Dr. Louis J. Thibodeaux
Louisiana State University

Mr. James M. Thornton
Washington State Department of Ecology

Dr. Clifford Truitt
U.S. Army Corps of Engineers

Dr. Thomas D. Wright
U.S. Army Corps of Engineers

Mr. Joseph G. Yeasted
IT Corporation

Mr. Christopher H. Zarba
U.S. Environmental Protection Agency

APPENDIX D

SYMPOSIUM/WORKSHOP ON CONTAMINATED MARINE SEDIMENTS

May 31-June 3, 1988
The Harbour Island Hotel
Tampa, Florida
813/229-5000

AGENDA

Tuesday, May 31, 1988 (Ballroom II)

0800 Registration (Ballroom II Foyer)

0830 Welcome Address   Kenneth Kamlet, Chairman

SYMPOSIUM

0845 Session 1 Extent of Contamination

      1.1 Superfund and Contaminated   Christopher Zarba, EPA
          Marine Sites

      1.2 Status and Trends and   Andrew Robertson, NOAA
          HAZMAT Programs

0945 Session 2 Classification of Contaminated Sediments

      2.1 AET and Triad Approaches   Robert Barrick, PTI
                                                Edward Long, NOAA

1030 COFFEE BREAK (Ballroom II Foyer)

1045 2.2 Equilibrium Partitioning   Dominic Di Toro,
                                                    Manhattan College

      2.3 Sediment Bioassays   Richard Swartz, EPA

1145 LUNCH (Fletcher Terrace)

1245 Session 3 Significance of Contamination

      3.1 Benthic Biota Contamination   John Scott, SAIC

      3.2 Risks to Human Health   Donald Malins, Pacific
          and Ecosystem   Research Foundation

1345 <u>Session 4 Mobilization and Resuspension</u>

      4.1   State-of-the-Art Research   Peter Sheng,
                                                                                University of Florida

      4.2   The Long-Term Fate of   Robert Arnold,
              Contaminants   University of Arizona

1445 COFFEE BREAK (Ballroom II Foyer)

1500 <u>Session 5 Assessment and Selection of Remedial Technologies</u>

      5.1   Handling of Contaminated Sediments   John Cullinane, ACOE-WES

      5.2   Management Strategies   Michael Palermo,   "

      5.3   Recent Developments in Equipment   John Herbich, Texas A&M

      5.4   Capping Technology   Robert Morton, SAIC

      5.5   Remedial Technologies Used at IJC   Ian Orchard,
              Areas of Concern   Environment Canada

      5.6   Economic Evaluation of Management   Thomas Grigalunas and
              Strategies   James Opaluch,
                                                                  University of
                                                                  Rhode Island

1745 ADJOURN

1800-1900 Reception (Fletcher Terrace)

<u>Wednesday, June 1, 1988</u> (Ballroom II)

0830 <u>Session 6 Case Studies</u>

      6.1 New Bedford Harbor, Massachusetts

          • Overview   Alan Ikelainen,
                                                       E.C. Jordan

          • Physical Transport Model   Alan Teeter, ACOE-WES

0930   6.2 Hudson River, New York

          • Overview   Clifford Rice, Patuxent
                                                        Wildlife Research Ctr.
          • Sediment Assessment   Richard Bopp,
                                                        Columbia University

1030 COFFEE BREAK (Ballroom II Foyer)

1045   6.3 James River, Virginia   Robert Huggett
                                                               Virginia Institute of
                                                               Marine Sciences

1130   6.4 Commencement Bay, Washington

- Policy Issues — David Bradley, State of Washington
- Defining and Assessing Sediments — Thomas Ginn, PTI
- Remedial Actions — Jerry Ficklin, Simpson Tacoma Kraft

1230 LUNCH (Fletcher Terrace)

Speaker: Walter Kovalick, Jr.
Deputy Director
Office of Emergency and
Remedial Response, USEPA

WORKSHOP
(Ballroom II)

1330 7.0 Workgroup Discussion

7.1 Workgroup I--Extent, Classification and Significance of Contamination — William Adams, Leader; Charles Staples, Rapporteur

1730 ADJOURN

1800 Boat Tour to leave from Harbor Island Dock (behind the hotel), return 2000 hours.

Thursday, June 2, 1988

0830 7.0 Workgroup Discussion (continued)

7.2 Workgroup II--Assessment and Selection of Remedial Technologies — John Herbich, Leader; Michael Palermo, Rapporteur

1200 ADJOURN

COMMITTEE MEETING
(Jackson Room)
(Committee Members and Liaisons Only)

1215 LUNCH (Garrison's)

1300 8.0 Committee Discussion — Kenneth Kamlet

1730 ADJOURN

Friday, June 3, 1988

0830 8.0 Committee Discussion (continued) — Kenneth Kamlet

1200 ADJOURN